FUNCTIONAL ANALYSIS, APPROXIMATION THEORY AND NUMERICAL ANALYSIS

FUNCTIONAL ANALYSIS, APPROXIMATION THEORY AND NUMERICAL ANALYSIS

Editor

John M. Rassias

The National University of Athens

World Scientific
Singapore • New Jersey • London • Hong Kong

Published by

World Scientific Publishing Co. Pte. Ltd.

P O Box 128, Farrer Road, Singapore 9128

USA office: Suite 1B, 1060 Main Street, River Edge, NJ 07661

UK office: 73 Lynton Mead, Totteridge, London N20 8DH

FUNCTIONAL ANALYSIS, APPROXIMATION THEORY AND NUMERICAL ANALYSIS

ISBN 981-02-0737-9

Printed in Singapore by Utopia Press.

Dedicated to

Stefan Banach on his 100th birthday,
Alexander Markowiç Ostrowski on his 99th birthday,
Stanislaw Marcin Ulam on his 83rd birthday

PREFACE

This volume (F.A.N.) contains various parts of *Functional Analysis, Approximation Theory and Numerical Analysis*, namely: A conditional Cauchy equation on rhombuses, spectral properties of matrices with products of binomial coefficients as entries, optimization of functionals and application to differential equations, an alternative Cauchy equation, generalization of the Golab–Schinzel functional equation, and functional equations and exact discrete solutions of ordinary differential equations. Besides it contains part on: Shape from shading problem, approximations to analytic functions, error estimate in non-equi-mesh spline finite strip method for thin plate bending problem, the Hyers-Ulam stability of a functional equation containing partial difference operators, dynamical systems and processes on Banach infinite dimensional spaces, Banach spaces in Bergman operator theory, characterization problems in Hilbert space, fixed point procedure in Banach spaces, Banach algebras of pseudodifferential operators, and Banach spaces. Finally the reader of this volume can find parts on: Ostrowski constant, stability problem of Ulam, Gegenbauer polynomials, quasi-tridiagonal system of linear equations, characterization of Q-algebras, and Landau's type inequalities.

This collection of research works is dedicated to the mathematicians: Stefan BANACH, Alexander Markowiç OSTROWSKI, and Stanislaw Marcin ULAM for their great contributions in Mathematics, Physics, Chemistry, Biology, and many other branches of Science.

Deep gratitude is due to all those friends and scientists who have encouraged me to complete this book in less than four years of continuous work. My very special thanks and appreciation to my family: Katia, Matina, and Vassiliki. Finally I have to thank the consultant editor Professor J. G. Xu and the Scientific editor Dr. Anju Goel of WORLD SCIENTIFIC for their patience and overall cooperation to carry out this project.

John Michael Rassias, Ph.D.
Professor of Mathematics

CONTENTS

Functional Analysis, Approximation Theory
and Numerical Analysis
Ed. John M. Rassias

STEFAN BANACH
ALEXANDER MARKOWIÇ OSTROWSKI
STANISLAW MARCIN ULAM

John Michael Rassias
(Athens, Greece)

BANACH was born in Krakow on March 30, 1892 and died in Lwów on August 31, 1945. He studied in Lwów, Poland. Subsequently, he worked at the University of Lwów and the Polish Academy of Sciences. Banach gave the general definition of normed spaces (1920–22). His motivation was the generalization of integral equations. The essential feature of his work was to set up a space with a norm but one which is no longer defined in terms of an inner product. Whereas in L^2, $\|x\| = \sqrt{(x,x)}$ it is not possible to define the norm of a Banach space in this way because an inner product is no longer available. The axioms for Banach's space B are divided into three groups. The first group contains thirteen axioms which specify that space B is a commutative group under addition; closed under multiplication by a real scalar; and that the familiar associative and distributive relations hold. The second group characterizes a norm on the elements (vectors) of B. The third group contains just a completeness axiom. An important class of operators introduced by Banach is the set of continuous additive ones. An operator f is additive if for all x and y, $f(x+y) = f(x) + f(y)$. Banach (1929) introduced in Functional Analysis, the notion of the dual (or adjoint) space of Banach space. One of his many important theorems is the Hahn-Banach theorem on functionals. Banach's work on functionals leads to the concept of adjoint operator. Banach applied his theory of adjoint operators to Riesz operators (introduced by Riesz in 1918). Banach worked with Stanislaw Marcin Ulam, Stanislaw Mazur, Kazimir Kuratowski, Hugo Steinhaus, W. Orlicz, and many other polish and foreign mathematicians. Besides the afore-mentioned theorem he obtained the following famous theorems: Banach–Alaoglu theorem, Banach fixed point theorem; Banach–Saks theorem. Finally he introduced the notions: Banach indicatrix, Banach limits, Banach algebra, and achieved many other fundamental results in mathematical analysis (for instance, Banach–Steinhaus theorem).

OSTROWSKI died on November 20, 1986 at the age of 93. His collected works amount to some 4,000 pages in six volumes, published by Birkhäuser-Verlag (1986). More than thirty of his students, who are now famous (for instance, Walter Gautschi), maintained close contact with Ostrowski until his final years. Ostrowski's career started in Kiev (1913). He worked with the great mathematicians: Felix Klein, Edmund Landau, and David Hilbert, and with the famous mathematicians: Peter Lancaster, Reich, Rita Jeltsch-Fricker, and many others. Born in Kiev on September 25, 1893, Ostrowski received his initial mathematical education under Ukrainian teachers. When he was working as an assistant to Felix Klein at Göttingen, Ostrowski earned his reputation as one of the world's leading mathematicians. In 1927 he accepted a call to work at Basel University where he remained until 1958. His papers are of interest to algebraists, geometers, analysts, topologists, and computer scientists. Numerical analysts are also indebted to him for the investigations he carried out on the iterative solution of equations. Of particular interest to computer scientists is complexity theory and foundations of symbolic integration due to Ostrowski's results. In the years after the second world war he began a sequence of visits to North America initiated by the U.S. National Bureau of Standards. He received honorary degrees from: University of Waterloo (1967), University of Besançon, and from ETH Zurich. Ostrowski was Honorary Editor-in-Chief of "Aequationes Mathematicae" from its foundation. Theorems: Ostrowski-Reich ("on the iterative techniques in matrix algebra"), and Ostrowski ("on analytic continuation"), are classical. His book "Solution of equations in Euclidean and Banach spaces" (Acad. Press, 1973) is excellent. The writer of this contribution made contact with Ostrowski on August 14, 1983 (11:40 a.m.–4:20 p.m.) at Ostrowski's home at Certenago di Montagnola, Via Sott'Cà 11, Switzerland. My wife and I were very happy to meet one of the last great mathematicians. Ostrowski said to us that his favorite teachers were: A high school teacher and David Hilbert. According to Ostrowski, Hilbert was that time the best mathematician in the world. Quoting Landau's words, Ostrowski said that a mathematical problem must be treated rigorously as it is stated. Finally he quoted Landau's statement, that applied mathematicians duplicate results of pure ones.

ULAM was born in Lwów, Poland on April 3, 1909 and died in Santa Fe, U.S.A. on May 13, 1984. He graduated with a doctorate in pure mathematics from the Polytechnic Institute at Lwów in 1933. Ulam worked at: The Institute for Advanced Study, Princeton (1936), Harvard University (1939–40), University of Wisconsin (1941–43), Los Alamos Scientific Laboratory (1943–65), University of Colorado (1965–76), and University of Florida (1974–). He was a member of the American Academy of Arts and Sciences and the National Academy of Sciences. He made fundamental contributions in mathematics, physics, biology, computer science, and the design of nuclear weapons. His early mathematical work was in set theory, topology, group theory, and measure theory. While still a schoolboy in Lwów, Ulam signed his notebook "S. Ulam, astronomer, physicist and mathematician". As Ulam notes, "the aesthetic appeal of pure mathematics lies not merely in the rigorous logic of the proofs and theorems, but also in the poetic elegance and economy in articulating each step in a mathematical presentation." Ulam worked with Stefan Banach, Kazimir Kuratowski, Karol Borsuk, Stanislaw Mazur, Hugo Steinhaus, John von Neumann, Garrett Birkhoff, Cornelius Everett, Dan Mauldin, D. H. Hyers, Mark Kac, P. R. Stein, Enrico Fermi, John Pasta, Richard Feynman, Ernest Lawrence, J. Robert Oppenheimer, Teller, and many other people of applied and exact sciences. Ulam was invited to Los Alamos by his friend John von Neumann, one of the most influential mathematicians of the twentieth century. Ulam's most remarkable achievement at Los Alamos was his contribution to the postwar development of the thermonuclear or hydrogen (H-) bomb in which nuclear energy is released when two hydrogen or deuterium nuclei fuse together. One of Ulam's early insights was to use the fast computers at Los Alamos to solve a wide variety of problems in a statistical manner using random numbers. This method has become appropriately known as the Monte Carlo method. One example that may have biological relevance is the subfield of cellular automata founded by Ulam and von Neumann. Finally Ulam had a unique ability to raise important unsolved problems. One of these problems was solved by the writer of this contribution (J. Approx. Th., Vol. 57, 268–273, 1989, New York, by Academy Press).

References

For STEFAN BANACH

1. S. Banach, *Théorie des Opérations linéaires*, Monografje Matematyczne, Vol. 1, Warsaw, 1932.
2. M. Kline, *Mathematical Thought from Ancient to Modern Times*, Vol. 3, Oxford, 1972.

3. A. Mukherjea and K. Pothoven, *Real and Functional Analysis*, Plenum Press, New York, Vol. 6, 1978.
4. W. Rudin, *Real and Complex Analysis*, McGraw-Hill Book Co., 1974.

For ALEXANDER MARKOVIÇ OSTROWSKI

1. R. L. Burden and J. D. Faires, *Numerical Analysis*, Prindle, Weber & Schmidt, Boston, 1985.
2. W. Gautschi, *SIAM News, Alexander M. Ostrowski*, Jan., 1987.
3. R. Jeltsch-Fricker, *Alexander M. Ostrowski (1893 bis 1986)* Elemente der Mathematik, (2) **43** Basel, März (1988) 33–38.
4. P. Lancaster, *Alexander M. Ostrowski, 1893–1986*, Aequationes Mathematicae **33** (1987) 121–122.

For STANISLAW MARCIN ULAM

1. D. J. Albers and G. L. Alaxanderson, editors: *Mathematical People, Profiles and Interviews*, Birkhäuser, Boston, 1985.
2. S. M. Ulam, *Collection of Mathematical Problems*, Interscience, New York, 1960; *Sets, Numbers and Universes*, The M.I.T. Press, 1974, Cambridge, Mass.
3. S. M. Ulam, *Adventures of a Mathematician*, Univ. Cal. Berkeley Press, 1991.
4. S. M. Ulam, *John von Neumann, 1903–1957*, Bull. Amer. Math. Soc., (3) **64** (1958) 1–49.

Functional Analysis, Approximation Theory
and Numerical Analysis
Ed. John M. Rassias

ON A CONDITIONAL CAUCHY EQUATION ON RHOMBUSES

C. ALSINA and J-L. GARCIA-ROIG
Secció Matemàtiques, ETSAB, Univ. Politècnica Catalunya
Diagonal 649. 08028 Barcelona, Spain.

ABSTRACT

We solve the conditional Cauchy equation $f(x+y) = f(x) + f(y)$ whenever $\|x\| = \|y\|$ for continuous mappings f from a real inner product space into a topological real linear space.

The aim of this paper is to study the conditional Cauchy equation on rhombuses, i.e.,

$$f(x+y) = f(x) + f(y) \quad \text{whenever} \quad \|x\| = \|y\|,$$

where $f : E \to F$ is a continuous mapping from an inner product space E into a topological real linear space F.

THEOREM 1. Let $f : E \to F$ be a continuous mapping from a real inner product space $(E, \cdot)$ of dimension greater than 1 into a topological real linear space F. Then f satisfies the conditional Cauchy equation

$$f(x+y) = f(x) + f(y) \quad \text{whenever} \quad \|x\| = \|y\|, \tag{1}$$

if and only if f is a continuous linear transformation.

Proof. Obviously any continuous linear mapping satisfies (1). Conversely, if f satisfies (1) then the substitution $x = y = 0$ into (1) yields $f(0) = 0$ and the substitution $y = -x$ yields $f(-x) = -f(x)$. Next we claim that for any real t and any x in E we have the homogeneity of f:

$$f(tx) = tf(x). \tag{2}$$

Since $f(-x) = -f(x)$ and $f(0) = 0$ we need to prove (2) just for $t > 0$ and $x \neq 0$. In effect, when we take in (1) $x = y$ we get $f(2x) = 2f(x)$ and from this we immediately obtain inductively that

$$f(2^n x) = 2^n f(x), \quad \text{for all integers } n. \tag{3}$$

For our x, let x' and x'' be elements in E satisfying the conditions

$$\|x\| = \|x'\| = \|x''\|, \quad x \cdot x' = x \cdot x'' = \frac{1}{2}\|x\|^2 \quad \text{and} \quad x = x' + x'' \tag{4}$$

Such elements exist because dim $E > 1$ and in any plane containing x, by previously considering an orthonormal basis, the problem is easily solved (geometrically speaking) by taking x' and x'' as x rotated 60° and -60° respectively.

By virtue of (1) and (4) we have $\|x + x'\| = \|x + x''\|$ and

$$\begin{aligned} f(3x) &= f((x + x') + (x + x'')) = f(x + x') + f(x + x'') \\ &= f(x) + f(x') + f(x) + f(x'') = 2f(x) + f(x' + x'') = 3f(x). \end{aligned}$$

Again by induction we have

$$f(3^m x) = 3^m f(x)), \qquad \text{for all integers } m. \tag{5}$$

Combining (3) and (5) we deduce $f(2^n 3^m x) = 2^n 3^m f(x)$, for all integers n, m, and since f is continuous and the set $\{2^n 3^m \;|n, m \in \mathbb{Z}\}$ is dense in $\mathbb{R}^+$ we can conclude the validity of (2). Our next step is to show (using the vectors introduced above) that

$$f(ax + bx') = af(x) + bf(x'), \qquad \text{for all } a, b \text{ in } \mathbb{R}. \tag{6}$$

In effect, due to the fact that $\|x\| = \|x'\|$ and $x \cdot x' = \frac{1}{2}\|x\|^2$ and the derivability of the norm in E from an inner product, for all a, b in $\mathbb{R}$ we have:

$$\|ax + bx'\| = \|ax' + bx\| = \|(a + b)x - bx'\| = \|(a + b)x' - ax\|. \tag{8}$$

Therefore by virtue of (1), (2) and (8) we obtain the equalities:

$$f(ax + bx') + f((a + b)x - bx') = f((2a + b)x) = (2a + b)f(x), \tag{9}$$

$$f(ax + bx') + f((a + b)x' - ax) = f((a + 2b)x') = (a + 2b)f(x'), \tag{10}$$

and

$$f(ax + bx') + f(ax' + bx) = f((a + b)(x + x')) = (a + b)(f(x) + f(x')). \tag{11}$$

Bearing in mind (1), (4) and (8), if we add (9) and (10), and then subtract (11), we obtain the desired property (6).

As the preceding procedure can clearly be carried out in any two-dimensional subspace F of E (i.e., we can take x and x' in F) we conclude by (2) and (6) that f is linear.

Let us note that in the proof just given the argument used in order to deduce the full additivity of f from the conditional additivity assumed in (1) went throught the homogeneity of f expressed in (2) where the continuity of f was essential. We will supply now another alternative proof by using a quite strong result which can be found in **2** (problem 25, chapter 11).

LEMMA. A mapping g from a real inner product space $(E, \cdot)$ of dimension greater than 1 into $\mathbb{R}$ is orthogonally additive in the sense that

$$g(x + y) = g(x) + g(y) \text{ whenever } x \cdot y = 0, \tag{13}$$

if and only if there exist additive functions $a : \mathbf{R} \to \mathbf{R}$ and $h : E \to \mathbf{R}$ such that

$$g(x) = a(\|x\|^2) + h(x). \tag{14}$$

Now we can show the following

THEOREM 2. A mapping f from a real inner product space $(E, \cdot)$ of dimension greater than 1 into $\mathbf{R}^n$ satisfies the conditional Cauchy equation

$$f(x+y) = f(x) + f(y) \text{ whenever } \|x\| = \|y\|, \tag{15}$$

if and only if is additive, i.e., $f(x+y) = f(x) + f(y)$, for all x, y in E.

Proof. Assume that $f : E \to \mathbf{R}^n$ satisfies (15). Then, for any couple of orthogonal vectors x and y, we have that $\|x+y\| = \|x-y\| = \|y-x\|$ and, consequently

$$f(x+y) + f(x-y) = f(2x), \tag{16}$$
$$f(x+y) + f(y-x) = f(2y), \tag{17}$$

and since by (15) we know $f(0) = 0$, $f(-x) = -f(x)$ and $f(2z) = 2f(z)$, adding (16) and (17) we obtain at once $f(x+y) = f(x) + f(y)$, i.e., from the conditional Cauchy equation (1) we deduce the orthogonal additivity of f. If for any $i = 1, 2, \ldots, n$, f_i denotes the ith component function of f, then $f_i : E \to \mathbf{R}$ will be orthogonal additive and by the Lemma just quoted there will exist additive functions $a_i : \mathbf{R} \to \mathbf{R}$ and $h_i : E \to \mathbf{R}$ such that

$$f_i(x) = a_i(\|x\|^2) + h_i(x). \tag{18}$$

When we move back from (18) to (15) we see that we would need to require

$$a_i(\|x+y\|^2) = a_i(\|x\|^2) + a_i(\|y\|^2) \text{ whenever } \|x\| = \|y\|, \tag{19}$$

condition which yields (by the additivity of a_i and the derivability of the norm from an inner product) that $a_i(x \cdot y) = 0$ whenever $\|x\| = \|y\|$. Thus with $x = y$ we would have $a_i(\|x\|^2) = 0$ and therefore a_i must be identically zero. Thus by (18) each f_i is additive and so is f.

COROLLARY 1. A mapping f from a real inner product space $(E, \cdot)$ of dimension greater than 1 into $\mathbf{R}^n$ which is continuous at one point, satisfies (15) if and only if f is a linear mapping.

References.

1. J. Aczél., *Lectures on Functional Equations and Their Applications*, (Academic Press, New Yor, 1966).
2. J. Aczél and J. Dhombres, *Functional Equations in several variables*, (Cambridge Univ. Press, 1989).

Functional Analysis, Approximation Theory
and Numerical Analysis
Ed. John M. Rassias

SPECTRAL PROPERTIES OF MATRICES WITH PRODUCTS OF BINOMIAL COEFFICIENTS AS ENTRIES

LOTHAR BERG

Fachbereich Mathematik, Universität Rostock, D-O-2500 Rostock, FR Germany

and

KONRAD ENGEL

Fachbereich Mathematik, Universität Rostock, D-O-2500 Rostock FR Germany

ABSTRACT

For a class of square matrices with two parameters and products of binomial coefficients as entries it is shown that the eigenvalues are also binomial coefficients and that the entries of the corresponding eigenvectors can be expressed as sums over products of binomial coefficients. In the proof the Cauchy integral formula for the coefficients of a Taylor series is used.

1. Preliminaries

We consider the square matrices $A = (a_{ij})$ of order $n+1$ with

$$a_{ij} = \binom{i+j}{j}\binom{m-i-j}{n-j}, \tag{1}$$

$n \geq 1$ and $i, j = 0, 1, \ldots, n$, where m is an additional real parameter. Our main result is that the binomial coefficients

$$\binom{m+1}{0}, \binom{m+1}{1}, \binom{m+1}{2}, \ldots, \binom{m+1}{n} \tag{2}$$

are eigenvalues of A. For the proof of (2) we construct the corresponding eigenvectors explicitly and obtain a series of identities for sums over products of binomial coefficients. Similar identities are well known in the literature, cf. H. Schmidt[6], and in particular in the framework of combinatorics, cf. E. Bannai and T. Ito[1] and J. Riordan[5]. For natural numbers $m \geq 2n$ the matrices A can also be interpreted combinatorially, since the entries (1) are the numbers of monotone functions from $N_{m-n} \to N_n = \{0, 1, 2, \ldots, n\}$ which map i onto j. An example with $n = 2$ and $m = 7$ in lexicographical order of the functions in the columns is

0	0	0	0	0	0	0	0	0	0	0	0	0	0	0	0	0	0	0	0	0	0	1	1	1	1	1	1	2
1	0	0	0	0	0	0	0	0	0	0	0	0	0	0	0	1	1	1	1	1	2	1	1	1	1	1	2	2
2	0	0	0	0	0	0	0	0	0	0	1	1	1	1	2	1	1	1	1	2	2	1	1	1	1	2	2	2
3	0	0	0	0	0	0	1	1	1	2	1	1	1	2	2	1	1	1	2	2	2	1	1	1	2	2	2	2
4	0	0	0	1	1	2	1	1	2	2	1	1	2	2	2	1	1	2	2	2	2	1	1	2	2	2	2	2
5	0	1	2	1	2	2	1	2	2	2	1	2	2	2	2	1	2	2	2	2	2	1	2	2	2	2	2	2

where

$$a_{00} = 21 = \binom{0}{0}\binom{7}{2}, \quad a_{01} = 6 = \binom{1}{1}\binom{6}{1}, \quad a_{02} = 1 = \binom{2}{2}\binom{5}{0},$$

$$a_{10} = 15 = \binom{1}{0}\binom{6}{2}, \quad a_{11} = 10 = \binom{2}{1}\binom{5}{1}, \quad a_{12} = 3 = \binom{3}{2}\binom{4}{0},$$

$$a_{20} = 10 = \binom{2}{0}\binom{5}{2}, \quad a_{21} = 12 = \binom{3}{1}\binom{4}{1}, \quad a_{22} = 6 = \binom{4}{2}\binom{3}{0}.$$

The case $m = 2n-1$ can be treated analogously by means of monotone functions from $N_n \to N_n$ under the additional assumption that the functions have the number n as fixed point. Such functions appear in the framework of linear involutory semigroups, cf. L. Berg and W. Peters[4]. The matrices A can also be used in numerical tests, cf. L. Berg[2,3] and G. Zielke[7].

2. The eigenvectors of A

In order to prove our statement concerning the eigenvalues (2) of A we construct the corresponding eigenvectors.

Theorem 1. *For $j = 0, 1, \ldots, n$ the matrix A has the eigenvalues $\binom{m+1}{j}$ with the eigenvectors $y_j = (y_{ij})$, where*

$$y_{ij} = \sum_{k=0}^{n} (-1)^k \binom{i}{k} \binom{n-k}{j} \binom{m+k-n-j}{k} \quad . \tag{3}$$

Proof. First let us mention that the first component of y_j is $y_{0j} = \binom{n}{j}$, so that y_j cannot be the zero vector. We introduce the matrices $Y = (y_0, y_1, \ldots, y_n)$, $B = (b_{ij})$, $D = (d_{ij})$ and $M = (m_{ij})$ with

$$b_{ij} = \binom{i}{j}, \; d_{ij} = (-1)^i \binom{n-i}{j} \binom{m+i-n-j}{i}, \; m_{ij} = \binom{j}{i} \binom{m+1}{n-j} \tag{4}$$

as well as

$$S = \mathrm{diag}\left(\binom{m+1}{0}, \binom{m+1}{1}, \ldots, \binom{m+1}{n} \right) \quad . \tag{5}$$

Then we have $Y = BD$, and we shall show the equations

$$AB = BM, \; MD = DS \tag{6}$$

which immediately imply our assertion in the form $AY = YS$.

In the following we often use the well known identities

$$\binom{a}{k} = (-1)^k \binom{k-a-1}{k} \tag{7}$$

and

$$\sum_{k=0}^{n} \binom{a}{p-k} \binom{b}{k-q} = \binom{a+b}{p-q} \tag{8}$$

with integers p, q from the interval $[0, n]$, real a, b and $\binom{a}{k} = 0$ in case of $k < 0$.

In order to prove the first equation of (6), we introduce the notations

$$l_{ij} = \sum_{k=0}^{n} \binom{i+k}{k} \binom{m-i-k}{n-k} \binom{k}{j}, \quad r_{ij} = \binom{m+1}{n-j} \sum_{k=0}^{n} \binom{i}{k} \binom{j}{k}$$

for the elements of the left-hand side AB and the right-hand side BM, respectively. In view of $\binom{i+k}{k}\binom{k}{j} = \binom{i+j}{j}\binom{i+k}{k-j}$ we have according to (7) and (8)

$$\begin{aligned} \sum_{k=0}^{n} \binom{m-i-k}{n-k} \binom{i+k}{k-j} &= \sum_{k=0}^{n} (-1)^{n-j} \binom{n-m+i-1}{n-k} \binom{-j-i-1}{k-j} \\ &= (-1)^{n-j} \binom{n-m-j-2}{n-j} = \binom{m+1}{n-j} \end{aligned}$$

and therefore

$$l_{ij} = \binom{i+j}{j}\binom{m+1}{n-j} .$$

Hence the symmetry formula $\binom{i}{k} = \binom{i}{i-k}$ and (8) immediately imply $l_{ij} = r_{ij}$.

In order to prove the second equation of (6) we use analogously the notations $MD = (l_{ij})$, $DS = (r_{ij})$ with

$$l_{ij} = \sum_{k=0}^{n}(-1)^k\binom{k}{i}\binom{m+1}{n-k}\binom{n-k}{j}\binom{m+k-n-j}{k} ,$$

$$r_{ij} = (-1)^i\binom{m+1}{j}\binom{n-i}{j}\binom{m+i-n-j}{i} .$$

According to $\binom{m+1}{n-k}\binom{n-k}{j} = \binom{m+1}{j}\binom{m+1-j}{n-k-j}$, (7) and $\binom{k}{i}\binom{n+j-m-1}{k} = \binom{n+j-m-1}{i}\binom{n+j-m-i-1}{k-i}$ we have

$$l_{ij} = \binom{m+1}{j}\binom{n+j-m-1}{i}\sum_{k=0}^{n}\binom{m+1-j}{n-k-j}\binom{n+j-m-i-1}{k-i} .$$

The sum is equal to $\binom{n-i}{n-i-j} = \binom{n-i}{j}$, so that according to (7) we again obtain $l_{ij} = r_{ij}$, and (6) is proved.

Remark. The matrix B with (4) is a regular triangular matrix, the inverse of which has the entries $(-1)^{i-j}\binom{i}{j}$. Hence $Y = BD$ is equivalent with $D = B^{-1}Y$, i. e. the lines of the matrix D are the differences of the lines of the matrix Y, and the representation (3) considered as a polynomial in m is nothing else than the Newton interpolation representation.

3. The eigenvectors of A^T

It is also possible to construct the eigenvectors of A^T belonging to the eigenvalues (2).

Theorem 2. *For $j = 0, 1, \ldots, n$ the matrix A^T has the eigenvectors $x_j = (x_{ij})$ belonging to the eigenvalues $\binom{m+1}{j}$, where*

$$x_{ij} = \sum_{k=0}^{n}(-1)^k\binom{i}{k}\binom{n+k-j}{k}\binom{m-n-k}{j} . \qquad (9)$$

Proof. First we introduce the notation $N = (n_{ij})$ with

$$n_{ij} = (-1)^{n-j}\binom{i}{n-j}\binom{m+1-j}{n-i} \quad , \tag{10}$$

and show the validity of the equation

$$M^T N = NS \quad . \tag{11}$$

The elements on the left–hand side and right–hand side, respectively, are

$$\begin{aligned} l_{ij} &= (-1)^{n-j}\binom{m+1}{n-i}\sum_{k=0}^{n}\binom{i}{k}\binom{k}{n-j}\binom{m+1-j}{n-k} \quad , \\ r_{ij} &= (-1)^{n-j}\binom{i}{n-j}\binom{m+1-j}{n-i}\binom{m+1}{j} \quad . \end{aligned}$$

In view of $\binom{i}{k}\binom{k}{n-j} = \binom{i}{n-j}\binom{i+j-n}{i-k}$ and (8) we obtain

$$\begin{aligned} l_{ij} &= (-1)^{n-j}\binom{m+1}{n-i}\binom{i}{n-j}\sum_{k=0}^{n}\binom{m+1-j}{n-k}\binom{i+j-n}{k+j-n} \\ &= (-1)^{n-j}\binom{m+1}{n-i}\binom{i}{n-j}\binom{m+i+1-n}{j} \quad , \end{aligned}$$

and it is easy to see that the equation $l_{ij} = r_{ij}$ is satisfied, which proves (11).

The first equation of (6) implies $A^T B^{-T} = B^{-T} M^T$ with $B^{-T} = (B^{-1})^T$, so that according to (11) we have $A^T B^{-T} N = B^{-T} NS$. Introducing the matrix $X = (x_0, x_1, \ldots, x_n)$ with the entries (9) and using (5) the assertion of the theorem can be expressed as $A^T X = XS$. Hence it is proved, if we show that

$$X = B^{-T} N \quad . \tag{12}$$

This representation implies $x_{nj} = (-1)^{n-j}\binom{n}{j}$, so that x_j cannot be the zero vector.

On the other hand, if we introduce the notation $C = (c_{ij})$ with

$$c_{ij} = (-1)^i \binom{n+i-j}{n-j}\binom{m-n-i}{j} \quad , \tag{13}$$

the matrix X with the entries (9) has the representation $X = BC$ with (4). Consequently we have to prove

$$B^T BC = N \quad . \tag{14}$$

For the entries of the left–hand side of (14) we use the notation

$$f_{ij} = \sum_{k,l=0}^{n} (-1)^l \binom{k}{i}\binom{k}{l}\binom{n+l-j}{n-j}\binom{m-n-l}{j} \quad .$$

In view of the Cauchy integral formula for the coefficients of a Taylor series we obtain

$$f_{ij} = -\frac{1}{4\pi^2}\int\int \sum_{k,l=0}^{n} (-1)^l \binom{k}{i}\binom{k}{l} \frac{(1+z)^{n+l-j}(1+w)^{m-n-l}}{z^{n-j+1}w^{j+1}} dzdw \quad ,$$

if we integrate over small circles around the point zero in the positive direction. The sum over l can be calculated explicitly, so that

$$f_{ij} = -\frac{1}{4\pi^2}\int\int \sum_{k=0}^{n} \binom{k}{i} \frac{(1+z)^{n-j}(1+w)^{m-n-k}}{z^{n-j+1}w^{j+1}} (w-z)^k dzdw \quad . \tag{15}$$

According to $(w-z)^k = \sum_{l=0}^{k} \binom{k}{l} w^{k-l}(-z)^l$ and the fact that either $l > n-j$ or $k-l > j$ for $k > n$, the integrals in (15) are zero for these k. Hence we can sum up to infinity and find in view of $\binom{k}{i} = \binom{k}{k-i} = (-1)^{k-i}\binom{-i-1}{k-i}$ that

$$\begin{aligned} f_{ij} &= -\frac{1}{4\pi^2}\int\int \frac{(1+z)^{n-j}(1+w)^{m-n-i}}{z^{n-j+1}w^{j+1}} (w-z)^i \left(1 - \frac{w-z}{1+w}\right)^{-i-1} dzdw \\ &= -\frac{1}{4\pi^2}\int\int \frac{(1+z)^{n-i-j-1}(1+w)^{m+1-n}}{z^{n-j+1}w^{j+1}} (w-z)^i dzdw \quad . \end{aligned}$$

Now, using

$$(w-z)^i = \sum_{l=0}^{i} \binom{i}{l} (1+w)^l (-1)^{i-l} (1+z)^{i-l}$$

and applying once more the Cauchy coefficient formula, we obtain

$$f_{ij} = \sum_{l=0}^{i} (-1)^{i-l} \binom{i}{l}\binom{n-j-l-1}{n-j}\binom{m+l+1-n}{j} \quad ,$$

and in view of

$$\binom{n-j-l-1}{n-j} = (-1)^{n-j}\binom{l}{n-j} , \quad \binom{i}{l}\binom{l}{n-j} = \binom{i+j-n}{i-l}\binom{i}{n-j}$$

and the substitution $k = i - l$ moreover

$$f_{ij} = (-1)^{n-j}\binom{i}{n-j} \sum_{k=0}^{i} (-1)^k \binom{i+j-n}{k}\binom{m+i+1-k-n}{j} \quad .$$

Because of $j \leq n$ it suffices to sum up to $i+j-n$. Applying the symmetry formula to the last binomial coefficient, (8) and two times (7), the sum can be expressed by $\binom{m+1-j}{n-i}$, so that according to (10) we have proved the desired result $f_{ij} = n_{ij}$.

Remark. Analogously as before the consequence $C = B^{-1}X$ of $X = BC$ shows that the lines of C are the differences of the lines of X, and (9) has the form of a Newton interpolation polynomial with respect to the variable m.

4. Determinants

The matrices B and N are triangular matrices, so that (4), (10) and (12) imply $\det B = 1$ and

$$\det X = \det N = \prod_{j=1}^{n} \binom{m+1-j}{j} \quad . \tag{16}$$

This means that X is a regular matrix, if and only if

$$m \notin \{0, 1, \ldots, 2n-2\} \quad . \tag{17}$$

In this case the set of eigenvalues (2) is complete and

$$\det A = \prod_{j=1}^{n} \binom{m+1}{j} \quad .$$

But according to continuity this formula is also valid for the exceptional values of m. The formulas

$$AY = YS \ , \quad A^T X = XS$$

and therefore $AX^{-T} = X^{-T}S$ show that in case of (17) there must exist a diagonal matrix T with $Y = X^{-T}T$, i.e.

$$X^T Y = T \quad . \tag{18}$$

In view of $Y = BD$ and (12) this means $N^T D = T$, and from (4) and (10) we find for the diagonal elements t_j of T that

$$t_j = \sum_{k=0}^{n} (-1)^{n+k-j} \binom{k}{n-j} \binom{m+1-j}{n-k} \binom{n-k}{j} \binom{m+k-n-j}{k} \quad .$$

Since

$$\binom{k}{n-j} \binom{n-k}{j} = \delta_{n-j,k} \quad ,$$

where δ_{ij} is the Kronecker symbol, we obtain moreover

$$t_j = \binom{m+1-j}{j}\binom{m-2j}{n-j} = \binom{n}{j}\binom{m-j}{n}\frac{m-j+1}{m-2j+1} , \tag{19}$$

and as before the first equation is also valid for the exceptional values of m. Finally, (16), (18) and (19) imply

$$\det Y = \prod_{j=1}^{n}\binom{m+2j-2n}{j} , \tag{20}$$

so that Y is also regular for (17).

5. Examples

To illustrate the results we consider the case $n = 2$ in detail, where we have

$$A = \begin{pmatrix} \frac{m(m-1)}{2} & m-1 & 1 \\ \frac{(m-1)(m-2)}{2} & 2m-4 & 3 \\ \frac{(m-2)(m-3)}{2} & 3m-9 & 6 \end{pmatrix}, B = \begin{pmatrix} 1 & 0 & 0 \\ 1 & 1 & 0 \\ 1 & 2 & 1 \end{pmatrix}, B^{-1} = \begin{pmatrix} 1 & 0 & 0 \\ -1 & 1 & 0 \\ 1 & -2 & 1 \end{pmatrix},$$

$$Y = \begin{pmatrix} 1 & 2 & 1 \\ 2-m & 4-m & 1 \\ \frac{(m-2)(m-3)}{2} & 6-2m & 1 \end{pmatrix}, X = \begin{pmatrix} 1 & m-2 & \frac{(m-2)(m-3)}{2} \\ -2 & 4-m & m-3 \\ 1 & -2 & 1 \end{pmatrix},$$

$$D = \begin{pmatrix} 1 & 2 & 1 \\ 1-m & 2-m & 0 \\ \frac{m(m-1)}{2} & 0 & 0 \end{pmatrix}, C = \begin{pmatrix} 1 & m-2 & \frac{(m-2)(m-3)}{2} \\ -3 & 6-2m & -\frac{(m-3)(m-4)}{2} \\ 6 & 3m-12 & \frac{(m-4)(m-5)}{2} \end{pmatrix},$$

$$M = \begin{pmatrix} \frac{(m+1)m}{2} & m+1 & 1 \\ 0 & m+1 & 2 \\ 0 & 0 & 1 \end{pmatrix}, N = \begin{pmatrix} 0 & 0 & \frac{(m-1)(m-2)}{2} \\ 0 & -m & m-1 \\ 1 & -2 & 1 \end{pmatrix}.$$

In the case $m = 1$ the first and the last of the three eigenvalues $1, m+1$ and $\frac{(m+1)m}{2}$ coincide, and the matrix A is not similar to a diagonal matrix as it is seen from

$$P^{-1}AP = \begin{pmatrix} 1 & -3 & 0 \\ 0 & 1 & 0 \\ 0 & 0 & 2 \end{pmatrix}, P = \begin{pmatrix} 1 & 0 & 2 \\ 1 & -2 & 3 \\ 1 & -3 & 4 \end{pmatrix}, P^{-1} = \begin{pmatrix} -1 & 6 & -4 \\ 1 & -2 & 1 \\ 1 & -3 & 2 \end{pmatrix}.$$

This representation shows additionally that there are no further eigenvalues and eigenvectors. In the cases $m = -1$ and $m = -2$ also two of the three eigenvalues (2) coincide, but in view of (17) nevertheless Y is regular and A diagonalizable by $Y^{-1}AY = S$ with (5).

References

1. E. Bannai and T. Ito, *Algebraic Combinatorics*, (Benjamin/Cummings, Menlo Park, 1984).

2. L. Berg, Die Invertierung von Matrizen aus Binomialkoeffizienten, *Z. Angew. Math. Mech.* **63** (1983) 639 - 642.

3. L. Berg, Über eine Folge von Hessenbergmatrizen aus Binomialkoeffizienten, *Rostock. Math. Kolloq.* **23** (1983) 63 - 68.

4. L. Berg und W. Peters, Die linearen involutorischen Halbgruppen achter Ordnung, *Rostock. Math. Kolloq.* (to appear).

5. J. Riordan, *Combinatorial Identities*, (John Wiley, New York, 1968).

6. H. Schmidt, *Analysis der elementaren Funktionen*, (Verlag Technik, Berlin, 1953).

7. G. Zielke, Testmatrizen mit freien Parametern, *Computing* **15** (1975) 87 - 103.

Functional Analysis, Approximation Theory
and Numerical Analysis
Ed. John M. Rassias

OPTIMIZATION OF FUNCTIONALS AND APPLICATION TO DIFFERENTIAL EQUATIONS

P. C. Bhakta
Department of Mathematics, Jadavpur University, Calcutta 700 032

1. Introduction

Let X be a real Banach space and J: $X \to R$. The functional J is said to be convex if for u, v in X and for $\lambda, \mu > 0$ with $\lambda + \mu = 1$, $J(\lambda u + \mu v) \leq \lambda J(u) + \mu J(v)$. If for u, v $(u \neq v)$ in X and λ, $\mu > 0$ with $\lambda + \mu = 1$, the strict inequality $J(\lambda u + \mu v) < \lambda J(u) + \mu J(v)$ holds, then J is called strictly convex. J is said to be coercive if $J(v) \to +\infty$ as $\|v\| \to +\infty$.

Let $u \in X$. If there is a bounded linear functional P_u: $X \to R$ such that for every v in X and α in R we can express

$$J(u + \alpha v) - J(u) = \alpha[P_u(v) + \phi(\alpha, v)] ,$$

where $\phi(\alpha, v) \to 0$ (uniformly in v on every bounded subset of X) as $\alpha \to 0$, then we say that J is differentiable at u and P_u is the Frechet derivative of J at u and write $J'(u)$ for P_u.

In this lecture I have considered the minimization of the functional J: $X \to R$ over a closed convex subset of X when it is coercive and weakly lower semi-continuous. For the uniqueness of the minimum point strict convexity of J is necessary. Then the results are applied to find the solution of some standard Boundary Value Problems.

2. Minimization of Functionals

Throughout this section we assume that X is a real reflexive Banach space, K is a closed convex subset of X and $J : X \to R$. In this direction we have the following important theorem.

Theorem 2.1. *If the functional $J : X \to R$ is coercive, weakly lower semi-continuous and strictly convex, there exists a unique point u in K such that*

$$J(u) = \inf\{J(v) : \ v \in K\} . \tag{1}$$

Furthermore if J is differentiable at u, then

$$J'(u)(v-u) \geq 0 \quad for\ all \quad v \in K . \tag{2}$$

Also if K is a closed subspace of X, then

$$J'(u)(v) = 0 \quad for\ all \quad v \in K . \tag{3}$$

Proof. The proof of the theorem may be found in [1] and [3]. But for completeness, I give the proof here.

Let $m = \inf\{J(v)\colon v \in K\}$. Clearly

$$m < +\infty . \tag{4}$$

There is a sequence $\{v_n\}$ in K such that

$$m = \lim_{n\to\infty} J(v_n) .$$

It will be shown that $\{v_n\}$ is bounded. On the contrary, let us assume that it is not bounded. Then $\{v_n\}$ has a subsequence $\{v_{n_i}\}$ such that $\|v_{n_i}\| \to +\infty$ as $i \to \infty$. Since J is coercive $J(v_{n_i}) \to +\infty$ as $i \to \infty$ which contradicts (4). Hence $\{v_n\}$ is bounded.

Again, since X is reflexive, $\{v_n\}$ has a weakly convergent subsequence. Without loss of generality we may take $\{v_n\}$ to be convergent. Let u be the weak limit of $\{v_n\}$. Since K is closed and convex, it is weakly closed. This gives that $u \in K$.

Furthermore since J is weakly lower semi-continuous we have

$$J(u) \leq \lim_{n\to\infty} J(v_n) = m \leq J(u) .$$

So we obtain

$$J(u) = m = \inf\{J(v) : v \in K\} . \tag{5}$$

Lastly assume that u_1, u_2 ($u_1 \neq u_2$) are two points in K such that $J(u_1) = m = J(u_2)$. Since $\frac{1}{2}u_1 + \frac{1}{2}u_2 \in K$ and J is strictly convex

$$J\left(\frac{1}{2}u_1 + \frac{1}{2}u_2\right) < \frac{1}{2}J(u_1) + \frac{1}{2}J(u_2) = m .$$

This contradicts the definition of m. Hence u is the unique point in K satisfying (5).

Now we deduce (2) and (3).

Let $v \in K$. Take any α in R with $0 < \alpha < 1$. Write $w = \alpha v + (1-\alpha)u$. Then $w \in K$. So by (1)

$$J(u) \leq J(w) = J(\alpha v + (1-\alpha)u) = J(u + \alpha(v-u)) ,$$

or,

$$J(u + \alpha(v-u)) - J(u) \geq 0 . \tag{6}$$

Since J is differentiable at u, we can express

$$J(u + \alpha(v-u)) - J(u) = \alpha[J'(u)(v-u) + \phi(\alpha, v)]$$

where $\phi(\alpha, v) \to 0$ (uniformly in v on every bounded subset of X) as $\alpha \to 0$. Using (6) we get, $J'(u)(v-u) + \phi(\alpha, v) \geq 0$. Letting $\alpha \to 0$,

$$J'(u)(v-u) \geq 0 \tag{7}$$

for every $v \in K$.

Next suppose that K is a closed subspace of X. Then K is automatically convex. Take any v in K, then $v + u \in K$. So from (7)

$$J'(u)(v) \geq 0 .$$

Since $-v \in K$ we also get

$$J'(u)(-v) \geq 0 \quad \text{or} \quad J'(u)(v) \leq 0 \quad [\because J'(u) \text{ is linear}]$$

Hence we obtain $J'(u)(v) = 0$ for every $v \in K$.

Now we consider a special type of functional $J\colon K \to R$ which is suitable for application to linear boundary value problems. We take J as follows.

$$J(v) = \frac{1}{2}\pi(v, v) - L(v) \qquad \text{for} \quad v \in X , \tag{8}$$

where $\pi\colon X \times X \to R$ is bilinear and $L\colon X \to R$ is linear.

Definition 2.1. The bilinear form π on X is said to be continuous if there is a positive number M such that

$$|\pi(u, v)| \leq M\|u\| \cdot \|v\| \quad \text{for all} \quad u, v \in X .$$

The bilinear form π on X is said to be X-elliptic or simply elliptic if there is a positive number α such that

$$\pi(v,v) \geq \alpha\|v\|^2 \quad \text{for all} \quad v \in X\ .$$

If π is elliptic, from definition it follows that $\pi(v,v) \geq 0$ for every $v \in X$ and $\pi(v,v) > 0$ if $v \neq 0$.

Theorem 2.2. *Let π be a continuous symmetric bilinear form on X and L be a bounded linear functional on X, and let $J : X \to R$ be defined by (8).*

(i) *Then J is continuous, differentiable and convex.*

(ii) *If π is elliptic, then J is strictly convex and coercive.*

Proof. (i) Continuity of J is obvious.

Let $u \in X$. Take any $v \in X$ and $\alpha \in R$. We have

$$\begin{aligned} J(u+\alpha v) &= \frac{1}{2}\pi(u+\alpha v, u+\alpha v) - L(u+\alpha v) \\ &= \left\{\frac{1}{2}\pi(u,u) - L(u)\right\} + \alpha\{\pi(u,v) - L(v)\} + \frac{1}{2}\alpha^2\pi(v,v) \\ &= J(u) + \alpha[P_u(v) + \phi(\alpha,v)]\ , \end{aligned}$$

where $P_u(v) = \pi(u,v) - L(v)$ and $\phi(\alpha,v) = \frac{1}{2}\alpha\pi(v,v)$. Clearly P_u is linear.

Since π and L are continuous there are positive numbers M and β such that

$$\left.\begin{aligned} |\pi(v,w)| &\leq M\|v\|\cdot\|w\| \\ |L(v)| &\leq \beta\|v\| \end{aligned}\right\} \quad \text{for } v,w \in X\ . \tag{8a}$$

We have

$$|P_u(v)| \leq |\pi(u,v)| + |L(v)| \leq M\|u\|\cdot\|v\| + \beta\|v\| = (M\|u\| + \beta)\|v\|$$

This gives that P_u is bounded.

Again, $|\phi(\alpha,v)| = \frac{1}{2}|\alpha|\,|\pi(v,v)| \leq \frac{1}{2}|\alpha|\cdot M\|v\|^2$. So $\phi(\alpha,v) \to 0$ (uniformly on every bounded subset of X) as $\alpha \to 0$. Hence J is differentiable at u and $J'(u) = \pi(u,\cdot) - L(\cdot)$.

Now let u, v be any two elements in X and $\lambda, \mu > 0$ with $\lambda + \mu = 1$. Then

$$\begin{aligned} J(\lambda u + \mu v) &= \frac{1}{2}\pi(\lambda u + \mu v, \lambda u + \mu v) - L(\lambda u + \mu v) \\ &= \frac{1}{2}\{\lambda^2\pi(u,u) + \lambda\mu\pi(u,v) + \lambda\mu\pi(v,u) + \mu^2\pi(v,v)\} \\ &\quad - \lambda L(u) - \mu L(v) \end{aligned}$$

and

$$\lambda J(u) + \mu J(v) = \frac{1}{2}\{\lambda\pi(u,u) + \mu\pi(v,v)\} - \lambda L(u) - \mu L(v) .$$

So

$$\begin{aligned}\lambda J(u) + \mu J(v) &- J(\lambda u + \mu v)\\ &= \frac{1}{2}\lambda\mu\{\pi(u,u) - \pi(u,v) - \pi(v,u) + \pi(v,v)\}\\ &= \frac{1}{2}\lambda\mu\pi(u-v,u-v) \geq 0\end{aligned}$$

which gives that J is convex.

(ii) Next suppose that π is elliptic. There is a positive number α such that

$$\pi(v,v) \geq \alpha\|v\|^2 \quad \text{for} \quad v \in X .$$

If $u \neq v$, then from the above,

$$\lambda J(u) + \mu J(v) > J(\lambda u + \mu v) .$$

Hence J is strictly convex.

Take any $v \in X$. Then we have using (8a)

$$\begin{aligned}J(v) &= \frac{1}{2}\pi(v,v) - L(v) \geq \frac{1}{2}\alpha\|v\|^2 - \beta\|v\| = \|v\|^2\left(\frac{1}{2}\alpha - \frac{\beta}{\|v\|}\right)\\ &> \frac{1}{4}\alpha\|v\|^2 \quad \text{if} \quad \|v\| > \frac{4\beta}{\alpha} .\end{aligned}$$

This gives that $J(v) \to +\infty$ as $\|v\| \to +\infty$. Hence J is coercive.

Theorem 2.3. *Let π be a continuous symmetric and elliptic bilinear form on X and L be a bounded linear functional on X. Then there exists a unique element u in K such that*

(i) $\pi(u, v-u) \geq L(v-u)$ for every $v \in K$. If K is a closed subspace of X, then

(ii) $\pi(u,v) = L(v)$ for every $v \in K$.

Note: The result (ii) is known as Lax-Milgram Theorem.

Proof. Let us take $J(v) = \frac{1}{2}\pi(v,v) - L(v)$ for $v \in X$. By Theorem 2.2, the functional J is continuous, differentiable, strictly convex and coercive.

Clearly J is weakly lower semi-continuous. This theorem now follows from Theorem 2.1.

3. Application to Differential Equations

In this section we show that well-known boundary value problems can be transformed to a functional equation of the form

$$\pi(u,v) = L(v) \tag{9}$$

with suitable Banach space X and closed subspace K. It will be shown that every classical solution of the given BVP is a solution of the functional equation (9), but the converse, in general, is not true in ordinary sense. But it is true in distribution sense. So we require some knowledge of distribution and Sobolev spaces which we introduce first.

Space of test functions and distributions

Let Ω be an open subset of R^n and let $f\colon \Omega \to R$ be continuous. We denote by Supp(f) the support of the function f and define

$$\text{Supp}(f) = cl\{x : x \in \Omega \text{ and } f(x) \neq 0\} .$$

If the set Supp(f) is compact we say that f is a function with compact support. We denote by $\mathcal{D}(\Omega)$ the set of all infinitely differentiable functions $f\colon \Omega \to R$ with compact support contained in Ω. It is easy to see that $\mathcal{D}(\Omega)$ is a linear space over R. Certain topology is introduced on $\mathcal{D}(\Omega)$ to make it a linear topological space. A sequence $\{\phi_m\}$ in $\mathcal{D}(\Omega)$ converges to zero iff the following two conditions hold.

(i) There is a compact set $K \subset \Omega$ such that Supp$(\phi_m) \subset K$ for every positive integer m.

(ii) The sequences $\{\phi_m\}$ and $\{D^\alpha \phi_m\}$ converge uniformly to zero on K for every multi-index $\alpha = (\alpha_1, \alpha_2, \ldots, \alpha_n)$ where

$$D^\alpha \phi_m = \frac{\partial^{|\alpha|} \phi_m}{\partial x_1^{\alpha_1} \partial x_2^{\alpha_2} \cdots \partial x_n^{\alpha_n}}, \quad |\alpha| = \alpha_1 + \alpha_2 + \cdots + \alpha_n .$$

A bounded linear functional on $\mathcal{D}(\Omega)$ is called a distribution on Ω. We denote by $\mathcal{D}'(\Omega)$ the set of all distributions on Ω. A distribution on Ω is also called a generalized function on Ω.

Partial derivative of a distribution

Let T be a distribution on Ω. The partial derivative $\frac{\partial T}{\partial x_i}$ of T is defined by

$$\frac{\partial T}{\partial x_i}(\phi) = -T\left(\frac{\partial \phi}{\partial x_i}\right) \quad \text{for all} \quad \phi \in \mathcal{D}(\Omega) .$$

It is easy to see that $\frac{\partial T}{\partial x_i}$ is also a distribution on Ω. For any multi-index $\alpha = (\alpha_1, \alpha_2, \ldots, \alpha_n)$ we define $D^\alpha T$ by

$$D^\alpha T(\phi) = (-1)^{|\alpha|} T(D^\alpha \phi) \quad \text{for all} \quad \phi \in \mathcal{D}(\Omega) .$$

$D^\alpha T$ is also a distribution on Ω.

Locally integrable function

A function f: $\Omega \to R$ is said to be locally integrable if for every compact subset K of Ω

$$\int_K |f| < +\infty .$$

It is easy to see that every continuous function is locally integrable; also every function $f \in L(\Omega)$ is locally integrable. If Ω is bounded, then $f \in L^2(\Omega)$ is locally integrable.

Example. Let f: $\Omega \to R$ be locally integrable. Define the mapping T_f: $\mathcal{D}(\Omega) \to R$ by

$$T_f(\phi) = \int_\Omega f\phi \quad \text{for every} \quad \phi \in \mathcal{D}(\Omega) .$$

Clearly T_f is linear. Take any sequence $\{\phi_m\}$ in $\mathcal{D}(\Omega)$ converging to zero. Then there is a compact set $K \subset \Omega$ such that

(i) $\text{Supp}(\phi_m) \subset K$ for every m.

(ii) $\{\phi_m\}$ and $\{D^\alpha \phi_m\}$ converge uniformly to zero on K. We have

$$T_f(\phi_m) = \int_K f\phi_m \quad (m = 1, 2, 3, \ldots) .$$

Since $\{\phi_m\}$ converges uniformly to zero on K $\lim_{m\to\infty} T_f(\phi_m) = 0$.

So T_f is continuous at zero and hence it is bounded. Therefore T_f is a distribution on Ω.

Let $\mathcal{L}(\Omega)$ denote the set of all locally integrable functions on Ω. If $f_1, f_2 \in \mathcal{L}(\Omega)$ and $\lambda \in R$, we can verify that

$$T_{f_1+f_2} = T_{f_1} + T_{f_2}$$
$$T_{\lambda f_1} = \lambda T_{f_1} .$$

Let $\mathcal{E} = \{T_f\colon f \in \mathcal{L}(\Omega)\}$. Then $\mathcal{E} \subset \mathcal{D}'(\Omega)$ and $\mathcal{E}$ is a linear space over R. Also $\mathcal{E}$ is isomorphic to $\mathcal{L}(\Omega)$. So we can identify T_f with f. With this agreement we have $\mathcal{E} = \mathcal{L}(\Omega) \subset \mathcal{D}'(\Omega)$ and therefore a distribution on Ω is called a generalized function.

Sobolev spaces

Let m be a positive integer. We define the Sobolev space $H^m(\Omega)$ as follows.

$$H^m(\Omega) = \{u : \ u \in L^2(\Omega), \ \ D^\alpha u \in L^2(\Omega) \ \text{ for } \ |\alpha| \le m\} .$$

For u, v in $H^m(\Omega)$, the inner product is defined by $(u, v) = \{\sum_{|\alpha|\le m} \int_\Omega D^\alpha u \cdot D^\alpha v\}^{\frac{1}{2}}$. It can be verified that $H^m(\Omega)$ is a Hilbert space. Clearly $\mathcal{D}(\Omega) \subset H^m(\Omega)$. We denote by $H_0^m(\Omega)$ the closure of $\mathcal{D}(\Omega)$ in the space $H^m(\Omega)$.

Solution of a partial differential equation

Let Ω be a bounded open subset of R^n and let m be a positive integer; and let

$$L = \sum_{|\alpha|\le m} a_\alpha D^\alpha$$

where $a_\alpha \in C^m(\bar{\Omega})$. Now consider the partial differential equation

$$L(u) = f, \ \ \text{where} \ \ f \in C^m(\bar{\Omega}) . \tag{10}$$

If there is a distribution T on Ω such that $L(T) = T_f$, then T is called a distribution solution of the partial differential equation (10). If there is a locally integrable function g on Ω such that $L(T_g) = T_f$, then g is called a weak solution of the partial differential equation (10).

If there is a function $w \in C^m(\bar{\Omega})$ such that $L(w) = f$ pointwise on Ω, then w is called a classical solution of the partial differential equation (10).

Now we consider the important Boundary Value Problems such as Dirichlet Problem, Neumann Problem, etc. and transform them to the form (9).

(A) *Dirichlet Problem*

Let Ω be a bounded open subset of R^3. Consider the following Boundary Value Problem.

$$\begin{aligned} -\nabla^2 u + qu &= f \quad \text{in } \ \Omega \cdots (a) \\ u &= 0 \quad \text{on } \ \Gamma \cdots (b) \end{aligned} \tag{11}$$

where Γ denotes the boundary of Ω and $q, f \in C(\bar{\Omega})$ and $q(x) \ge \alpha > 0$ for all $x \in \bar{\Omega}$.

The boundary value problem (11) is known as Dirichlet Problem.

Let u be a classical solution of the BVP (11). Then $u \in C^2(\bar{\Omega})$ and satisfies 11(a) and 11(b).

Let $v \in H_0^1(\Omega)$. Multiplying 11(a) by v and integrating we get

$$-\int_\Omega (\nabla^2 u)v + \int_\Omega quv = \int_\Omega fv \, .$$

Using Green's Theorem we have

$$\int_\Omega \nabla u \nabla v + \int_\Gamma v\frac{\partial u}{\partial \nu} + \int_\Omega quv = \int_\Omega fv$$

where ν denotes the outward unit normal to Γ at the point $x \in \Gamma$. Since $v(x) = 0$ on Γ, we obtain

$$\int_\Omega (\nabla u \nabla v + quv) = \int_\Omega fv$$

or,

$$\pi(u, v) = L(v) \, , \tag{12}$$

where

$$\pi(u, v) = \int_\Omega (\nabla u \cdot \nabla v + quv), \quad L(v) = \int_\Omega fv \, .$$

It is easy to verify that π is a continuous, symmetric and H_0^1-elliptic bilinear form on $H_0^1(\Omega)$. From above we see that every classical solution of the BVP (11) is a solution of (12).

Next suppose that $u \in H_0^1(\Omega)$ is a solution of (12). For any $\phi \in \mathcal{D}(\Omega) \subset H_0^1(\Omega)$ we have

$$\begin{aligned} &-(\nabla^2 u)(\phi) + (qu)(\phi) - f(\phi) \\ &= (\nabla u)(\nabla \phi) + (qu)(\phi) - f(\phi) = \pi(u, \phi) - L(\phi) = 0 \, , \end{aligned}$$

or,

$$(-\nabla^2 u + qu - f)(\phi) = 0 \, .$$

This gives that

$$-\nabla^2 u + qu - f = 0 \qquad \text{[in distribution sense]}.$$

Hence u is a weak solution of the BVP (11). Since $H_0^1(\Omega)$ is a closed subspace of $H^1(\Omega)$ by Lax-Milgram Theorem the BVP (11) possesses unique weak solution.

(B) *Neumann Problem*

Let Ω be an open bounded subset of R^3. Consider the following Boundary Value Problem.

$$\begin{aligned} -\nabla^2 u + qu = f \quad & \text{in } \Omega \cdots (a) \\ \frac{\partial u}{\partial \nu} = 0 \quad & \text{on } \Gamma \cdots (b) \end{aligned} \tag{13}$$

where $f, q \in C(\bar{\Omega})$ and $q(x) \geq \alpha > 0$ for $x \in \bar{\Omega}$; Γ being the boundary of Ω and ν the outward unit normal at $x \in \Gamma$. The BVP (13) is known as Neumann Problem.

Let u be a classical solution of the BVP (13). Then $u \in C^2(\bar{\Omega})$ and satisfies 13(a) and 13(b). Let $v \in H_0^1(\Omega)$. Multiplying 13(a) by v and integrating we get

$$-\int_\Omega \nabla^2 uv + \int_\Omega quv = \int_\Omega fv\,.$$

Using Green's Theorem we obtain

$$\int_\Omega \nabla u \cdot \nabla v + \int_\Gamma v \frac{\partial u}{\partial \nu} + \int_\Omega quv = \int_\Omega fv$$

or

$$\int_\Omega (\nabla u \cdot \nabla v + quv) = \int_\Omega fv \quad \left[\because \frac{\partial u}{\partial \nu} = 0 \text{ on } \Gamma\right]$$

or

$$\pi(u, v) = L(v) \tag{14}$$

where π and L are same as in previous case.

(C) *The elasticity system*

Let $\Omega \subset R^3$ be a bounded open set representing the volume occupied by an elastic body and Γ be its boundary. Let Γ be partitioned into two parts Γ_0 and Γ_1 with surface measure of Γ_0 being strictly positive.

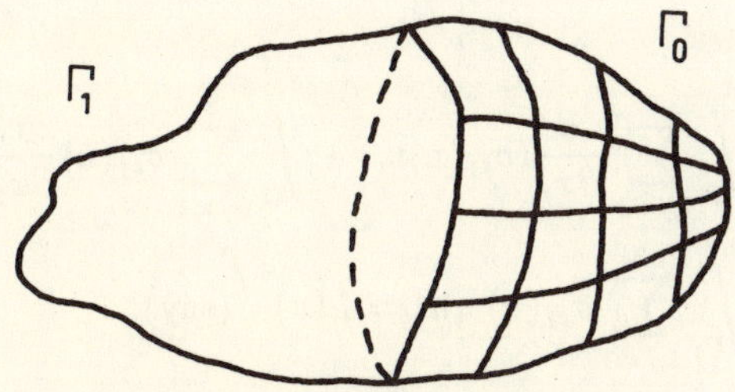

Fig. 1.

Assume that a body force $f = (f_1, f_2, f_3)$ acts on the body and a surface force $g = (g_1, g_2, g_3)$ acts on Γ_1. Let $u = (u_1, u_2, u_3)$ be the displacement vector. Then the strain tensor (ε_{ij}) is defined by

$$\varepsilon_{ij}(u) = \frac{1}{2}\left(\frac{\partial u_i}{\partial x_j} + \frac{\partial u_j}{\partial x_i}\right) \quad (i, j = 1, 2, 3) . \tag{15}$$

Let σ_{ij} denote the stress tensor. The constitutive law relating strain and stress is given by

$$\sigma_{ij}(u) = \lambda\left(\sum_{k=1}^{3} \varepsilon_{kk}(u)\right)\delta_{ij} + 2\mu\varepsilon_{ij}(u) , \tag{16}$$

λ and μ are Lame's coefficients where $\lambda \geq 0$ and $\mu > 0$. The elastic system consists of the following Boundary Value Problem.

$$\begin{cases} \sum_{j=1}^{3} \dfrac{\partial}{\partial x_j}(\sigma_{ij}(u)) = f_i & \text{in } \Omega \cdots (a) \\ u = 0 & \text{on } \Gamma_0 \cdots (b) \\ \sum_{j=1}^{3} \sigma_{ij}(u) n_j = g_i & \text{on } \Gamma_1 \cdots (c) \end{cases} \tag{17}$$

where $n = (n_1, n_2, n_3)$ is the outward unit normal to the point x on Γ_1.

Let $V = \{v\colon\ v = (v_1, v_2, v_3),\ v_i \in H^1(\Omega) \text{ and } v = 0 \text{ on } \Gamma_0\}$. Then V is a closed subspace of $H^1(\Omega)$.

Suppose that u is a classical solution of the BVP (17). Then $u \in C^2(\bar{\Omega})$. Take any $v \in V$. Multiplying 17(a) and 17(b) by v_i, adding the two together and then integrating we get

$$-\int_\Omega \sum_{i,j=1}^{3} \frac{\partial}{\partial x_j}(\sigma_{ij}(u))v_i + \int_{\Gamma_1} \sum_{i,j=1}^{3} \sigma_{ij}(u)v_i n_j = \int_\Omega fv + \int_{\Gamma_1} gv$$

or

$$-\int_\Omega \sum_{i,j=1}^{3} \frac{\partial}{\partial x_j}(\sigma_{ij}(u)v_i) + \int_\Omega \sum_{i,j=1}^{3} \sigma_{ij}(u)\frac{\partial v_i}{\partial x_j}$$
$$+ \int_{\Gamma_1} \sum_{i,j=1}^{3} \sigma_{ij}(u)v_i n_j = L(v) \quad \text{(say)}.$$

Using Green's Theorem we have

$$-\int_{\Gamma_1} \sum_{i,j=1}^{3} \sigma_{ij}(u)v_i n_j + \int_\Omega \sum_{i,j=1}^{3} \sigma_{ij}(u)\frac{\partial v_i}{\partial x_j}$$
$$+ \int_{\Gamma_1} \sum_{i,j=1}^{3} \sigma_{ij}(u)v_i n_j = L(v)$$

or

$$\frac{1}{2}\int_\Omega \sum_{i,j=1}^{3} \sigma_{ij}(u)\left(\frac{\partial v_i}{\partial x_j} + \frac{\partial v_j}{\partial x_i}\right) = L(v) \quad [\because \sigma_{ij} \text{ is symmetric}]$$

or $\int_\Omega \sigma_{ij}(u)\varepsilon_{ij}(v) = L(v)$.

Now substituting the value of $\sigma_{ij}(u)$ obtained from (16) we get

$$\lambda \int_\Omega \operatorname{div}(u) \sum_{i,j=1}^{3} \delta_{ij}\varepsilon_{ij}(v) + 2\mu \int_\Omega \sum_{i,j=1}^{3} \varepsilon_{ij}(u)\varepsilon_{ij}(v) = L(v)$$

or

$$\lambda \int_\Omega \operatorname{div}(u)\operatorname{div}(v) + 2\mu \int_\Omega \sum_{i,j=1}^{3} \varepsilon_{ij}(u)\varepsilon_{ij}(v) = L(v)$$

or

$$\pi(u,v) = L(v)\,, \tag{18}$$

where

$$\pi(u,v) = \lambda \int_\Omega \operatorname{div}(u)\operatorname{div}(v) + 2\mu \int_\Omega \sum_{i,j=1}^{3} \varepsilon_{ij}(u)\varepsilon_{ij}(v)\,.$$

Suppose that $f \in (L^2(\Omega))^3$ and $g \in (L^2(\Gamma))^3$. Then π is continuous, symmetric and V-elliptic bilinear form on V. So the equation (18) has a unique solution in V which is the weak solution of the elastic system (17).

Note. The equation of the form

$$\pi(u,v) = L(v)$$

may be regarded as the abstract formulation of each of the above BVP's.

Method of finding the solution of the equation $\pi(u,v) = L(v)$

Let H be a real Hilbert space and let π: $H \times H \to R$ be symmetric, continuous and H-elliptic bilinear form and let L: $H \to R$ be a bounded linear functional. Then by Lax-Milgram Theorem there exists a unique element $u \in H$ such that

$$\pi(u,v) = L(v) \quad \text{for every } \ v \in H \ . \tag{19}$$

We assume that H is a separable space. Then H has a countable complete orthonormal basis $\{w_i\}_{i=1}^{\infty}$ (say).

Write

$$\left.\begin{aligned} c_n &= (u, w_n) \\ s_n &= c_1 w_1 + c_2 w_2 + \cdots + c_n w_n \end{aligned}\right\} \quad n = 1, 2, 3, \ldots$$

Then

$$\|u - s_n\| \to 0 \quad \text{as } \ n \to \infty \ . \tag{20}$$

Take any positive integer n and consider the subspace V_n spanned by $w_1, w_2, \ldots, w_n$. Then V_n is a closed subspace of H. So by Lax-Milgram Theorem there exists a unique element $u_n \in V_n$ such that

$$\pi(u_n, v) = L(v) \quad \text{for } \ v \in V_n \ . \tag{21}$$

We show that lim $u_n = u$. Take any $v \in V_n$. From (19) and (20) we get

$$\pi(u,v) = L(v) = \pi(u_n, v)$$

or

$$\pi(u - u_n, v) = 0 \quad \text{for } \ v \in V_n \ . \tag{22}$$

Since π is continuous and H-elliptic there are positive numbers α and M such that

$$|\pi(w,v)| \le M\|w\| \cdot \|v\|, \quad \pi(v,v) \ge \alpha\|v\|^2, \quad \text{for } \ w, v \in H \ .$$

We have

$$\begin{aligned} \alpha\|u - u_n\|^2 &\le \pi(u - u_n, u - u_n) = \pi(u - u_n, u - s_n + s_n - u_n) \\ &= \pi(u - u_n, u - s_n) + \pi(u - u_n, s_n - u_n) \\ &= \pi(u - u_n, u - s_n) \quad \text{[By (22)]} \\ &\le M\|u - u_n\| \cdot \|u - s_n\| \ . \end{aligned}$$

or $\|u - u_n\| \le \frac{M}{\alpha}\|u - s_n\|$.

Using (20) we obtain $\lim_{n\to\infty} u_n = u$.

Method of finding u_n.

We have $u_n = \lambda_1 w_1 + \lambda_2 w_2 + \cdots + \lambda_n w_n$, where $\lambda_1, \lambda_2, \ldots, \lambda_n$ are re numbers to be determined. Substituting in (21) and taking $v = w_1, w_2, \ldots, w_n$ respectively we obtain the following system of linear equations

$$\sum_{i=1}^{n} \lambda_i \pi(w_i, w_j) = L(w_j) \quad (j = 1, 2, \ldots, n)\,. \tag{23}$$

Write

$$\left.\begin{aligned} a_{ij} &= \pi(w_1, w_j) \\ b_j &= L(w_j) \end{aligned}\right\} \quad i, j = 1, 2, \ldots, n\,.$$

Then the system (23) can be written as

$$\begin{aligned} a_{11}\lambda_1 + a_{12}\lambda_2 + a_{13}\lambda_3 + \cdots + a_{1n}\lambda_n &= b_1 \\ a_{21}\lambda_1 + a_{22}\lambda_2 + a_{23}\lambda_3 + \cdots + a_{2n}\lambda_n &= b_2 \\ \vdots \qquad \vdots \qquad \vdots \qquad\qquad \vdots \quad &\quad \vdots \\ a_{n1}\lambda_1 + a_{n2}\lambda_2 + a_{n3}\lambda_3 + \cdots + a_{nn}\lambda_n &= b_n \end{aligned}$$

We now show that the coefficient matrix $A = (a_{ij})$ of the system (24) is positive definite and hence non-singular.

Let $\xi = (\xi_1, \xi_2, \ldots, \xi_n) \in R^n$ and $v = \xi_1 w_1 + \xi_2 w_2 + \cdots + \xi_n w_n$. Then $v \in V_n$. We have

$$\pi(v, v) = \sum_{i=j}^{n} \sum_{j=1}^{n} \pi(w_i, w_j)\xi_i\xi_j = \sum_{i=1}^{n} \sum_{j=1}^{n} a_{ij}\xi_i\xi_j = \xi A \xi'\,.$$

By the ellipticity we have

$$\xi A \xi' = \pi(v, v) \ge \alpha\|v\|^2 > 0 \quad \text{for } \xi \ne 0\,.$$

So A is positive definite and hence non-singular. Therefore the system (24) has a unique solution for $(\lambda_1, \lambda_2, \ldots, \lambda_n)$ which determines u_n and hence the solution u.

References

1. Chow and Hale, *Methods of Bifurcation Theory*. Springer-Verlag, 1982.
2. S. Kesavan, *Topics in Functional Analysis and Applications*. Wiley Eastern Limited, 1989.
3. J. L. Lions, *Optimal Controls of Systems Governed by Partial Differential Equations*. Springer-Verlag, 1971.

Functional Analysis, Approximation Theory
and Numerical Analysis
Ed. John M. Rassias

ON AN ALTERNATIVE FUNCTIONAL EQUATION IN $\mathbb{R}^n$.

Costanza Borelli-Forti
and
Gian Luigi Forti
*Universita' di Milano, Dipartimento di Matematica,
via Saldini 50, I-20133 Milano, ITALY.*

ABSTRACT

In this paper we study some properties of a special class of solutions of an alternative Cauchy equation for functions from a stable group G in R^n. In the case n=3 the class of solutions assuming the value zero in the unity of G is completely described.

1.- During the last years the alternative Cauchy equation

$$f(xy)-f(x)-f(y) \in \mathbf{V}$$

has been extensively studied in different situations concerning the domain of the function and the given set **V**.
The problem was posed by the first time by R. Ger[7] in the case f real valued function defined on a group and $\mathbf{V}=\{0,1\}$ and, in more general form, by M. Kuczma[8].

Many results about the equation above have been published (see references) for different finite sets **V** and in most of them the main tool used to obtain the description of the solutions has been the stability in Hyers-Ulam sense.

In a former paper[5] the case where **V** is a set of independent vectors in a Banach space has been considered, so it is natural to ask for the solutions of the previous equation when the vectors in **V** are not independent.

In the present paper we study the special case where **V** is given by all the vertices of the unit cube in $\mathbb{R}^n$ but one and the function f, defined on a stable group G, takes the value 0 on the unity of G.

2.- In this short section we recall some definitions and results we use in the following. They can be found in[5].

DEFINITION 1.- Let G be a group and B a Banach space. The group G is called *stable* (in the sense of Hyers-Ulam) if for every function $f:G\longrightarrow B$ such that $\|f(xy)-f(x)-f(y)\|\le\delta$ for all $x,y\in G$ and some $\delta>0$, there exists a (unique) $\phi_f\in Hom\,(G,B)$ such that $\|\phi_f(x)-f(x)\|\le\delta$ for all $x\in G$.

It is well known[5] that every amenable group (and so every commutative group) is stable.

THEOREM 1.- *Let G be a stable group, B a Banach space and M a bounded subset of B. Assume the function $f:G\longrightarrow B$ is such that $f(xy)-f(x)-f(y)\in M$ for all $x,y\in G$. Then the range of the function $h:=f-\phi_f$ is contained in $\overline{C(-\overline{M})}$ (the closure of the convex hull of the set $-M$).*

Moreover if ε is the identity of G and $h(\varepsilon)=-\mu\in-M$ then $h(x)\in\{-(\mu+M)+\overline{C(\overline{M})}\}\cap\overline{C(-\overline{M})}$.

3.- We consider the functional equation

$$f(xy)-f(x)-f(y)\in \mathbf{V}_n \qquad (1_n)$$

where $f:G\longrightarrow\mathbb{R}^n$, G is a stable group, $\mathbf{V}_n$ is the subset of $\mathbb{R}^n$ given by $\mathbf{V}_n=\{0,e_1,\dots,e_n,e_1+e_2,\dots,e_{n-1}+e_n,\dots,e_2+\dots+e_n\}$, where $\{e_1,\dots,e_n\}$ is the canonical bases of $\mathbb{R}^n$, that is $\mathbf{V}_n$ is the set of all vertices of the unit cube of $\mathbb{R}^n$ but $e_1+e_2+\dots+e_n$.

Note that if $n=1$ we have the equation

$$f(xy)-f(x)-f(y)\in \mathbf{V}_1=\{0,1\} \qquad (1_1)$$

which has been completely solved by G.L. Forti[4]; we shall describe a special class of solutions of Eq. 1_n in term of the solutions of Eq. 1_1.

Since G is stable, by Theorem 1, in order to get all solutions of Eq. 1_n it is enough to determine all solutions f of Eq. 1_n with the additional condition that their range is a subset of $C(-\mathbf{V}_n)=\overline{C(-\mathbf{V}_n)}$ and then add to each of them an arbitrary homomorphism of G in $\mathbb{R}^n$.

If $f=(f_1,f_2,\dots,f_n)$ is a solution of Eq. 1_n, then obviously each component f_j is a solution of Eq. 1_1 and if the range of f is contained in $C(-\mathbf{V}_n)$ then the range of every f_j is a subset of the interval $[-1,0]$.

We denote with $p_j:\mathbb{R}^n \longrightarrow \mathbb{R}$ the projection on the j-th coordinate axis; for every $j=1,\dots,n$ and $\alpha=0,1$ we define $S_{j,\alpha}:=\{x\in G:\ f_j(x)=-\alpha\}$.

We have the following.

Lemma 1.- *Let* $f:G\longrightarrow C(-\mathbf{V}_n)$ *be a solution of Eq.*1_n*.Then*
i) the sets $S_{j,\alpha}$ *are either empty or subsemigroups of* G*;*
ii) $H_j:=S_{j,0}\cup S_{j,1}$ *is a non empty normal subgroup of* G.
iii) if $y\in H_j$ *and* $x\in G\backslash H_j$*, then* $f_j(xy)=f_j(x)=f_j(yx)$.

Proof- We prove the lemma for $j=1$.Let $x,y\in S_{1,0}$, that is $f_1(x)=f_1(y)=0$.By Eq. 1_n, $f(xy)=f(x)+f(y)+v$, for some $v\in\mathbf{V}_n$; if $p_1(v)=0$, then $f_1(xy)=f_1(x)+f_1(y)=0$, so $xy\in S_{1,0}$.
The case $p_1(v)=1$ is impossible, otherwise we get $f_1(xy)=f_1(x)+f_1(y)+1=1$, contrary to our assumption that the range of f_1 is in $[-1,0]$.

Analogously we prove that $S_{1,1}$, if not empty, is a semigroup.

Let now $x\in S_{1,0}$ and $y\in S_{1,1}$ (or vice-versa); then $f_1(xy)=f_1(x)+f_1(y)+p_1(v)=-1+p_1(v)$, since $p_1(v)$ equals 0 or 1, we have $xy\in H_1$.If ε denote the identity of G, then by Eq.1_n we have $f(\varepsilon)\in-\mathbf{V}_n$, so $f_1(\varepsilon)\in\{0,-1\}$, i.e. $\varepsilon\in H_1$.
Now take $x\in H_1$; we have $f(x^{-1})=-f(x)+f(\varepsilon)-v$, for some $v\in\mathbf{V}_n$.
If $f_1(\varepsilon)=0$, then

$$f_1(x^{-1})=-f_1(x))-p_1(v)\in\{-1,0,1\}\cap C(-\mathbf{V}_n)=\{-1,0\};$$

if $f_1(\varepsilon)=-1$, then

$$f_1(x^{-1})=-f_1(x)-1-p_1(v)\in\{-2,-1,0\}\cap C(-\mathbf{V}_n)=\{-1,0\};$$

thus $x^{-1}\in H_1$, i.e. H_1 is a subgroup of G.
Let now $x\in H_1$ and $y\in G\backslash H_1$; we have

$$f(y^{-1}xy)-f(y^{-1})-f(xy)=v_1 \quad , \quad f(xy)=f(x)+f(y)+v_2,$$

$$f(\varepsilon)-f(y)-f(y^{-1})=v_3,$$

for some $v_1,v_2,v_3\in\mathbf{V}_n$; hence

$$f(y^{-1}xy)=f(y^{-1})+f(xy)+v_1=f(x)+f(\varepsilon)-v_3+v_1+v_2,$$

and so $f_1(y^{-1}xy)=f_1(x)+f_1(\varepsilon)-p_1(v_3)+p_1(v_1)+p_1(v_2)$.

From $f_1(x)+f_1(\varepsilon)-p_1(v_3)+p_1(v_1)+p_1(v_2)\in\{-3,-2,-1,0,1,2\}$ and $-1\le f_1(y^{-1}xy)\le 0$, we get $f_1(y^{-1}xy)\in\{-1,0\}$, i.e. $y^{-1}xy\in H_1$; thus H_1 is normal. Let now $y\in H_1$ and $x\in G\setminus H_1$:

$f_1(xy)=f_1(x)+f_1(y)+p_1(v)=f_1(x)+\{-1,0\}+\{0,1\}\in[-1,0]$ and this implies $f_1(y)+p_1(v)=0$, i.e. iii). ∎

We now look for the solutions f of Eq. 1_n such that $f(\varepsilon)=0$. From now on we intend that the function f is such a solution.

By Theorem 1 we have the following sharper condition on the range of f:

$$f(x)\in\{-\mathbf{V}_n+C(\mathbf{V}_n)\}\cap C(-\mathbf{V}_n).$$

If $u\in\{-\mathbf{V}_n+C(\mathbf{V}_n)\}\cap C(-\mathbf{V}_n)$, then $u=-v+t=-z$, for some $v\in\mathbf{V}_n$, $t,z\in C(\mathbf{V}_n)$; since $p_i(v)=0$ for some i, from $p_i(t),p_i(z)\in[0,1]$ we get $p_i(u)=0$, that is the point u belongs to the coordinate hyperplane

$\mathcal{P}_i:=\{t\in\mathbb{R}^n:\ p_i(t)=0\}$.

We can summarize as follows.

Lemma 2.– *Let $f:G\longrightarrow C(-\mathbf{V}_n)$ be a solution of Eq. 1_n with $f(\varepsilon)=0$. Then the range of f is contained in $C(-\mathbf{V}_n)\cap\left(\bigcup_{i=1}^{n}\mathcal{P}_i\right)$.*

Among the solutions of Eq. 1_n with $f(\varepsilon)=0$ there are those with range contained in a single hyperplane $\mathcal{P}_i$. The description of this class is very easy and is given by the following.

Theorem 2.– *A function $f:G\longrightarrow C(-\mathbf{V}_n)$ is a solution of Eq. 1_n with $f(\varepsilon)=0$ and $f(G)\subset\mathcal{P}_i$ if and only if for every $j\neq i$ the function $f_j:G\longrightarrow[-1,0]$ is a solution of Eq. 1_1 with $f_j(\varepsilon)=0$.*

We now look for the solutions of Eq. 1_n whose range is contaned in more than one coordinate hyperplane.

The first step is given by the following lemma.

Lemma 3.- *Let* $f:G \longrightarrow C(-\mathbf{V}_n)$ *be a solution of Eq.* 1_n *with* $f(\varepsilon)=0$. *If for some* $x \in G$ *we have* $f_i(x)=0$ *and* $f_j(x)\in(-1,0)$, *then for every* $y\in H_i$ *it is* $f_i(y)=0$.

Proof- Assume, on the contrary, the existence of $y\in H_i$ with $f_i(y)=-1$. We have $f(y)-f(x)-f(yx^{-1})\in\mathbf{V}_n$, i.e. $-f(yx^{-1})=f(x)-f(y)+v$, $v\in\mathbf{V}_n$ and so $-f_i(yx^{-1})=1+p_i(v)$. This implies $p_i(v)=0$ and $-f_i(yx^{-1})=1$. If $j\neq i$, then

$$-f_j(yx^{-1})=f_j(x)-f_j(y)+p_j(v)\in[0,1],$$

hence we get

$$-f_j(yx^{-1})=f_j(x)-\begin{cases} f_j(y) & \text{, if} \quad f_j(y)\leq f_j(x) \\ f_j(y)+1 & \text{, if} \quad f_j(y)\geq f_j(x). \end{cases}$$

Again by Eq. 1_n we have $f(x)-f(xy^{-1})-f(y)\in\mathbf{V}_n$, i.e. $-f(xy^{-1})=f(y)-f(x)+u$, $u\in\mathbf{V}_n$ and so $-f_i(xy^{-1})=-1-f_i(x)+p_i(u)=$ $=-1+p_i(u)\in[0,1]$; this implies $p_i(u)=1$ and $-f_i(xy^{-1})=0$.

If $j\neq i$, then $-f_j(xy^{-1})=f_j(y)-f_j(x)+p_j(u)\in[0,1]$, hence we get

$$-f_j(xy^{-1})=-f_j(x)+\begin{cases} f_j(y)+1 & \text{, if} \quad f_j(y)\leq f_j(x) \\ f_j(y) & \text{, if} \quad f_j(y)\geq f_j(x). \end{cases}$$

Consider now the difference

$$f(\varepsilon)-f(xy^{-1})-f(yx^{-1})=-f(xy^{-1})-f(yx^{-1})=w\in\mathbf{V}_n.$$

We have $p_i(w)=-f_i(xy^{-1})-f_i(yx^{-1})=0+1=1$ and for $j\neq i$,

$$p_j(w)=-f_j(xy^{-1})-f_j(yx^{-1})=-f_j(x)+f_j(x)+1=1,$$

thus $w=e_1+\ldots+e_n\notin\mathbf{V}_n$; a contradiction. ■

Assume now there exist two elements $x,y\in G$ such that

$$\left.\begin{array}{l} \text{there is } i \text{ such that } f_i(x)=0 \text{ and } f_k(x)\in(-1,0) \text{ for } k\neq i \\ \text{there is } j\neq i \text{ such that } f_j(y)=0 \text{ and } f_k(y)\in(-1,0) \text{ for } k\neq j \end{array}\right\} \quad (2)$$

that is we assume that the solution f takes values in at least two different coordinate hyperplanes.

We consider now the value $f(yx^{-1})$; by the equation we have

$$f(y)-f(x)-f(yx^{-1})=v\in\mathbf{V}_n, \text{ i.e. } f(yx^{-1})=f(y)-f(x)-v,$$

and so

$$f_i(yx^{-1})=f_i(y)-f_i(x)-p_i(v)=f_i(y)-p_i(v)\neq 0 \tag{3}$$

$$f_j(yx^{-1})=f_j(y)-f_j(x)-p_j(v)=-f_j(x)-p_j(v)\neq 0. \tag{4}$$

Hence by Lemma 2 there exists k, different from i and j, such that $f(yx^{-1})\in\mathcal{P}_k$.

Therefore we have the following

THEOREM 3.- *Doesn't exist any solution f of Eq. 1_n such that $f(\varepsilon)=0$, with range contained in two (but not one) coordinate hyperplanes.*

By still assuming Eq.2 we obtain the following properties.

By Lemma 3 we can conclude that

for every $y\in H_i\cap H_j\cap H_k=:H_{ijk}$ we have $f(y)\in\mathcal{P}_i\cap\mathcal{P}_j\cap\mathcal{P}_k$.

Eq.3 and 4 imply

$$p_i(v)=0 \text{ and } f_i(yx^{-1})=f_i(y)$$

$$p_j(v)=1 \text{ and } f_j(yx^{-1})=-f_j(x)-1.$$

The condition $f_k(yx^{-1})=f_k(y)-f_k(x)-p_k(v)=0$ implies

$$p_k(v)=0 \text{ and } f_k(y)=f_k(x).$$

Consider now the value of f in xy: $f(xy)=f(x)+f(y)+w$, for some $w\in\mathbf{V}_n$.The relations

$$f_i(xy)=f_i(y)+p_i(w)\in[-1,0], \tag{5}$$

$$f_j(xy)=f_j(x)+p_j(w)\in[-1,0] \tag{6}$$

imply $p_i(w)=p_j(w)=0$ and

$$f_i(xy)=f_i(y)\neq 0,\ f_j(xy)=f_j(x)\neq 0;$$

moreover it is

$$f_k(xy))=2f_k(x)+p_k(w). \tag{7}$$

We now look for the solutions of Eq. 1_n with the range contained in three hyperplanes $\mathcal{P}_i$, $\mathcal{P}_j$ and $\mathcal{P}_k$.
We prove the following.

THEOREM 4.- *Let* $f:G\longrightarrow C(-V_n)$ *be a solution of Eq.* 1_n *such that* $f(\varepsilon)=0$ *and with range contained in* $\mathcal{P}_i\cup\mathcal{P}_j\cup\mathcal{P}_k$ *(but not in a single hyperplane).Then for every* $x\notin H_{ijk}$ *we have:*

$$\text{if } f_i(x)=0 \text{ then } f_j(x)=f_k(x)=-1/2;$$

$$\text{if } f_j(x)=0 \text{ then } f_i(x)=f_k(x)=-1/2;$$

$$\text{if } f_k(x)=0 \text{ then } f_i(x)=f_j(x)=-1/2.$$

PROOF-Assume $f(x)\in\mathcal{P}_i$, $f(x)\notin\mathcal{P}_j\cup\mathcal{P}_k$, $f(y)\in\mathcal{P}_j$, $f(y)\notin\mathcal{P}_i\cup\mathcal{P}_k$.From Eq.5, Eq.6 and the hypothesis we get immediately $f_k(xy)=0$; therefore Eq.7 yields $p_k(w)=1$and $f_k(x)=f_k(y)=-1/2$.From $f(x^2y)=f(xy)+f(x)+v$, $v\in V_n$ and Eq.6 we get

$$f_k(x^2y)=f_k(xy)+f_k(x)+p_k(v)=-1/2\ +p_k(v)\neq0,$$

$$f_i(x^2y)=f_i(xy)+f_i(x)+p_i(v)=f_i(y)+p_i(v)\neq0$$

(note that $f_i(y)\in(-1,0)$).Thus we must have $f_j(x^2y)=0$:

$$0=f_j(x^2y)=f_j(xy)+f_j(x)+p_j(v)=2f_j(x)+p_j(v),$$

and so, since $f_j(x)\neq0$, $p_j(v)=1$ and $f_j(x)=-1/2$.
Proceeding in the same way we get the theorem. ∎

LEMMA 4.- *Assume the hypotheses of Theorem 4.If* $x\notin H_{ijk}$ *then* $f(x)$ *belongs to only one of the three hyperplanes* $\mathcal{P}_i$, $\mathcal{P}_j$ *and* $\mathcal{P}_k$.
PROOF- Let $f_i(x)=f_j(x)=0$ and take $y\in G$ such that $f_k(y)=0$ and $f_i(y)=f_j(y)=-1/2$; then $f(yx^{-1})=f(y)-f(x)-v$ and so
$f_i(yx^{-1})=f_i(y)-p_i(v)=-1/2-p_i(v)$

$$f_j(yx^{-1})=f_j(y)-p_j(v)=-1/2-p_j(v)$$

$$f_k(yx^{-1})=-f_k(x)-p_k(v).$$

Thus we get $p_i(v)=p_j(v)=0$ and $f_i(yx^{-1})=f_j(yx^{-1})=-1/2$; this implies $f_k(yx^{-1})=-f_k(x)-p_k(v)=0$, i.e. $x\in H_{ijk}$; a contradiction. ∎

Lemma 5.- *Assume the hypotheses of Theorem 4.If $x \notin H_{ijk}$ then $x^2 \in H_{ijk}$.*

Proof- Let $f_i(x)=f_j(x)=-1/2$ and $f_k(x)=0$; then

$$f_i(x^2)=-1+p_i(v)\in\langle-1,0\rangle,\ f_j(x^2)=-1+p_j(v)\in\langle-1,0\rangle$$

and $f_k(x^2)=p_k(v)=0$, that is $x^2\in H_{ijk}$. ∎

We can now conclude with the following.

Theorem 5.- *A function $f:G\longrightarrow C(-\mathbf{V}_n)$ is a solution of Eq. 1_n with $f(\varepsilon)=0$ and range contained in $\mathcal{P}_i\cup\mathcal{P}_j\cup\mathcal{P}_k$ (but not in a single hyperplane) if and only if:*

i) *for every $t=1,\dots,n$ the function $f_t:G\longrightarrow[-1,0]$ is a solution of Eq. 1_1;*

ii) *H_{ijk} is a normal subgroup of G of index 4 such that $(G\setminus H_{ijk})^2\subseteq H_{ijk}$;*

iii) *if $H^{(i)}$, $H^{(j)}$ and $H^{(k)}$ are the cosets of H_{ijk}, then:*

$$\forall\ x\in H_{ijk}\quad f_i(x)=f_j(x)=f_k(x)=0,$$

$$\forall\ x\in H^{(i)}\quad f_j(x)=f_k(x)=-1/2,$$

$$\forall\ x\in H^{(j)}\quad f_i(x)=f_k(x)=-1/2,$$

$$\forall\ x\in H^{(k)}\quad f_i(x)=f_j(x)=-1/2.$$

Proof-If f is a solution with the properties in the statement, then by iii) of Lemma 1 and by Theorem 4 it has the form given above.Conversely a simple check shows that any function of that form is solution of Eq. 1_n. ∎

4.- In this section we confine ourselves to the case $n=3$.In this case Theorems 2 and 5 give the complete description of the solutions of Eq. 1_3 such that $f(\varepsilon)=0$.

It is natural to ask for the solutions of the equation analogous to Eq. 1_3 obtained by replacing the set $\mathbf{V}_3$ with a subset $\mathbf{U}$ of it with the following properties: $\mathbf{U}$ spans $\mathbb{R}^3$, $0\in\mathbf{U}$, $\mathbf{U}\setminus\{0\}$ is not a set of independent vectors.We have to consider the following four cases:

$$U_1=\{0,e_1,e_2,e_3,e_1+e_2,e_1+e_3\}$$

$$U_2=\{0,e_1,e_2,e_1+e_2,e_1+e_3,e_2+e_3\}$$

$$U_3=\{0,e_1,e_2,e_1+e_2,e_1+e_3\}$$

$$U_4=\{0,e_1,e_2,e_1+e_3,e_2+e_3\}.$$

By Theorem 1, with $\mu=0$, for each of the four cases above we have that the range of the solutions is a subset of

$[-U_1+C(U_1)]\cap C(-U_1)=$

$$=\{\lambda_1e_1+\lambda_2e_2:\ -1\le\lambda_i\le 0,\ i=1,2\}\cup\{\lambda_1e_1+\lambda_3e_3:\ -1\le\lambda_i\le 0,\ i=1,3\},$$

$$[-U_2+C(U_2)]\cap C(-U_2)=\{\lambda_1e_1+\lambda_2e_2:\ -1\le\lambda_i\le 0,\ i=1,2\}\cup$$
$$\cup\{\lambda(e_1+e_3):\ -1\le\lambda\le 0\}\cup\{\lambda(e_2+e_3):\ -1\le\lambda\le 0\},$$

$$[-U_3+C(U_3)]\cap C(-U_3)=\{\lambda_1e_1+\lambda_2e_2:\ -1\le\lambda_i\le 0,\ i=1,2\}\cup$$
$$\cup\{\lambda(e_1+e_3):\ -1\le\lambda\le 0\},$$

$$[-U_4+C(U_4)]\cap C(-U_4)=\{\lambda(e_1+e_3):\ -1\le\lambda\le 0\}\cup\{\lambda(e_2+e_3):\ -1\le\lambda\le 0\}\cup$$
$$\cup\{\lambda e_1:\ -1\le\lambda\le 0\}\cup\{\lambda e_2:\ -1\le\lambda\le 0\}$$

respectively.

By Theorems 2 and 3 we have immediately the following

THEOREM 6.- *A function* $f:G\longrightarrow C(-U_1)$ *is a solution of the equation* $f(xy)-f(x)-f(y)\in U_1$ *with* $f(\varepsilon)=0$ *if and only if it has the form* $f=(f_1,f_2,0)$ *or the form* $f=(f_1,0,f_3)$ *where* $f_i:G\longrightarrow[-1,0]$, $i=1,2,3$, *are solutions of Eq.* 1_1 *with* $f_i(\varepsilon)=0$.

Much more complicate is the situation for the sets U_2 and U_3. In these cases one can ask if there exist any solution whose range has point in both sets $\{\lambda(e_1+e_3):-1\le\lambda\le 0\}$ and $\{\lambda e_1:-1\le\lambda\le 0\}$. If such a solution $f=(f_1,0,f_3)$ exists then define the sets

$$A=\{x\in G\ :\ f_1(x)=f_3(x)=0\}\neq\emptyset,$$
$$B=\{x\in G\ :\ f_1(x)\neq 0,\ f_3(x)=0\},$$
$$C=\{x\in G\ :\ f_1(x)=f_3(x)=-1\},$$
$$D=\{x\in G\ :\ f_1(x)=f_3(x)\notin\{0,-1\}\}.$$

By Lemma 1 C is a semigroup, $A\cup B$ is a semigroup and $A\cup B\cup C$ is a subgroup of G. Moreover it is easy to show that the following additional properties hold:

$$x,y\in A \Rightarrow xy,yx\in A \ ; \ x\in A,\ y\in B \text{ (or vice-versa)} \Rightarrow xy,yx\in A;$$

$$x,y\in D \Rightarrow xy,yx\in A\cup C\cup D \ ; \ x\in A,\ y\in C \text{ (or vice-versa)} \Rightarrow xy,yx\in A\cup C;$$

$$x\in A,\ y\in D \text{ (or vice-versa)} \Rightarrow xy,yx\in D \ ;$$

$$x\in B,\ y\in C \text{ (or vice-versa)} \Rightarrow xy,yx\in B \ ;$$

$$x\in B,\ y\in D \text{ (or vice-versa)} \Rightarrow xy,yx\in D \ ;$$

$$x\in C,\ y\in D \text{ (or vice-versa)} \Rightarrow xy,yx\in D \ .$$

A simple check proves that the conditions above are also sufficient for $f=(f_1,0,f_3)$ to be a solution.

We summarize the previous discussion in the following.

THEOREM 7.- *A function* $f:G\longrightarrow C(-\mathbf{U}_2)$ *is a solution of the equation* $f(xy)-f(x)-f(y)\in\mathbf{U}_2$ *with* $f(\varepsilon)=0$ *if and only if it has one of the following forms:*

i) $f=(f_1,f_2,0)$;

ii) $f=(f_1,0,f_1)$; $f=(0,f_2,f_2)$;

iii) $f=(f_1,0,f_3)$; $f=(0,f_2,f_3)$, *with the sets A, B, C and D (and the analogous where the role of* f_1 *is assumed by* f_2*) satisfying the conditions listed above,*

where $f_i:G\longrightarrow[-1,0]$, $i=1,2,3$, *are solutions of Eq.* 1_1 *with* $f_i(\varepsilon)=0$.

In the case of the set $\mathbf{U}_3$ *we have the same result without the functions of the forms* $f=(0,f_2,f_2)$ *and* $f=(0,f_2,f_3)$.

Whether solutions of the form iii) of Theorem 7 actually exist is not known and it may depend on the structure of the group G.It is easy to show that for $G=\mathbb{Z}$ (the integers) they do not exist.

To finish this section it remains to consider the case of the set $\mathbf{U}_4$.By the former discussion and by Theorem 12 of the paper[5] we obtain:

THEOREM 8.- *A function* $f:G\longrightarrow C(-\mathbf{U}_4)$ *is a solution of the equation* $f(xy)-f(x)-f(y)\in\mathbf{U}_4$ *with* $f(\varepsilon)=0$ *if and only if it has one of the following forms:*

i) $f=(f_1,0,0)$; $f=(0,f_2,0)$;

ii) $f=(f_1,0,f_1)$; $f=(0,f_2,f_2)$;

iii) $f=(f_1,0,f_3)$; $f=(0,f_2,f_3)$, *with the sets A, B, C and D (and the analogous where the role of* f_1 *is assumed by* f_2*) satisfying the conditions listed above,*

where $f_i:G \longrightarrow [-1,0]$, $i=1,2,3$, *are solutions of Eq.* 1_1 *with* $f_i(\varepsilon)=0$.

5.- To finish the paper we present some open problems concerning Eq. 1_n.

If $n\geq 4$ the results contained in Section 3 do not describe the whole class of the solutions with the property $f(\varepsilon)=0$, but only those with range in at most three coordinate hyperplanes.
A simple check shows that if G has a normal subgroup H of index s, $4\leq s\leq n$, such that each element of G/H has order 2, then generalizing in an obvious way the construction of Theorem 5 we get a solution of Eq. 1_n assuming values in s coordinate hyperplanes.It is open the question if these solutions are, as for $s=3$, the only possible.

The condition $f(\varepsilon)=0$ is a very strong restriction; if we look for other classes of solutions, from Theorem 1 we get that the range of these solutions is no more contained in the coordinate hyperplanes: in the case $n=3$, for instance, it contains a "layer" between two planes.In this case there are no results about the solutions of Eq. 1_n.

REFERENCES

1 J. Aczèl and J. Dhombres, *Functional equations containing several variables*, Encyclopedia of Mathematics and its Applications 30, Cambridge Univ. Press, 1988.

2 C. Borelli Forti, *Soluzioni di una equazione funzionale alternativa sui gruppi abeliani*, Rend. Sem. Mat. Univers. Politecn. Torino **48** (1990),(to appear).

3 C. Borelli Forti and G.L. Forti, *On a class of alternative functional equations of Cauchy type*, in "Topics in Mathematical Analysis", T. Rassias (Ed.), World Scientific Publ. Co., Singapore, 1989, p. 273-293.

4 G.L. Forti, *On an alternative functional equation related to the Cauchy equation*, Aequationes Math. **24** (1982), p. 195-206.

5 G.L. Forti, *The stability of homomorphisms and amenability,with applications to functional equations*, Abh. Math. Sem. Univ. Hamburg **57** (1987), p. 215-226.

6 G.L. Forti and L. Paganoni, *A method for solving a conditional Cauchy equation on abelian groups*, Ann. Mat. Pura Appl. (IV) **127** (1981), p. 79-99.

7 R. Ger, *On a method of solving of conditional Cauchy equations*, Univ. Beograd. Publ. Elektrotechn. Fak. Ser. Mat. Fiz. No. **544-576** (1976), p. 159-165.

8 M. Kuczma, *Functional equations on restricted domains*, Aequationes Math. **18** (1978), p. 1-34.

9 L. Paganoni, *Soluzione di una equazione funzionale su dominio ristretto*, Boll. Un. Mat. Ital. (5) **17-B** (1980), p. 979-993.

10 L. Paganoni, *On an alternative Cauchy equation*, Aequationes Math. **29** (1985), p. 214-221.

Functional Analysis, Approximation Theory
and Numerical Analysis
Ed. John M. Rassias

ON A GENERALIZATION OF THE GOLAB-SCHINZEL FUNCTIONAL EQUATION

by Nicole BRILLOUET-BELLUOT
Ecole Centrale de Nantes
Dèpartement de Mathèmatiques
1 rue de la Noé - 44 072 NANTES Cedex 03 - FRANCE

ABSTRACT

Let E be a real Hausdorff topological vector space. We consider on **R**xE the following binary law : $(a,x)*(b,y) = (\lambda ab, b^k x + a^l y)$ for $(a,x),(b,y) \in$ **R**xE where λ is a fixed real number , k and l are fixed nonnegative integers.

We find here all the subgroupoïds of (**R**xE,*) which depend faithfully and continuously on a set of parameters . The two related functional equations are solved when the functions have some regularity property. The greatest part of this paper consists in solving one of them which is a generalization of the Gołąb-Schinzel functional equation.

1. INTRODUCTION

Let E be a real Hausdorff topological vector space.

The following functional equation :

$$f(f(x)y + x) = f(x)f(y) \qquad (x,y \in E) \tag{GS}$$

where f is a mapping from E into **R** , is called ***functional equation of Gołąb-Schinzel***. It has been first considered by J.Aczél in 1957 , and then by S.Gołąb and A.Schinzel in 1959 . The general solution of (GS) has been characterized [1] and all the continuous solutions of (GS) have been explicitely obtained [3] .

We consider now the following binary law on **R**xE :

$$(a,x) * (b,y) = (\lambda ab\ ,\ b^k x + a^l y\ ) \qquad \text{for } (a,x),(b,y) \in \mathbf{R}\text{x}E \qquad \text{(L)}$$

where λ is a fixed real number , k and l are fixed nonnegative integers .

Let us recall the following definition given by J.Dhombres[3] :

DEFINITION 1 *A subset H of **R**xE* ***depends faithfully and continuously*** *on a set F of parameters if F is a topological space and if there exists a mapping g from F onto H :*

$$g(u) = (\alpha(u),\beta(u)) \qquad \text{for } u \in F$$

such that we have either :

(i) $\beta(F) = E$ *and* $\beta(u) = \beta(u')$ *implies* $\alpha(u) = \alpha(u')$ *,*
α *is continuous and* β *admits locally a continuous lifting*

or

(ii) $\alpha(F) = \mathbf{R}$ *and* $\alpha(u) = \alpha(u')$ *implies* $\beta(u) = \beta(u')$ *,*
β *is continuous and* α *admits locally a continuous lifting*

We look for the subgroupoïds of (**R**xE,*) which depend faithfully and continuously on a set F of parameters .

In the case (i) , the relation : $f(\beta(u)) = \alpha(u)$ $(u \in F)$ defines a continuous function f from E into **R** which satisfies the following functional equation :

$$f(f(y)^k x + f(x)^l y) = \lambda f(x) f(y) \qquad (x,y \in E) \qquad (1)$$

In the case (ii) , the relation : $f(\alpha(u)) = \beta(u)$ $(u \in F)$ defines a continuous function f from **R** into E which satisfies the following functional equation :

$$f(\lambda xy) = y^k f(x) + x^l f(y) \qquad (x,y \in \mathbf{R}) \qquad (2)$$

The main part of this paper will consist in solving the functional equation (1) . This functional equation is a generalization of the Gołąb-Schinzel functional equation since (GS) corresponds to the particular case of (1) where $k = 0$, $l = 1$, $\lambda = 1$. It has been studied by many authors in various cases . Among them , let us notice that the author[4] found all the solutions of (1) having some regularity property in the case where λ is a nonnegative real number and W.Benz [2] determined the cardinality of the set of discontinuous solutions $f : \mathbf{R} \to \mathbf{R}$

of (1) for infinitely many real numbers λ. In the case where λ is a nonzero real number, J.Brzdęk [7] obtained all the continuous solutions of (1) when k and l are distinct positive integers, and the author [5] found all the solutions of (1) having some regularity property when $k = l$.

The present paper is a survey of this problem. It gives all the solutions of (1) having some regularity property when λ is an arbitrary real number and k, l are arbitrary nonnegative integers. It is mainly based on the papers written by the author [3,4,5] and J.Brzdęk [8]. Some further references concerning this problem may be found in these papers.

2. INVESTIGATION OF FUNCTIONAL EQUATION (1)

Following A.M.Bruckner and J.C.Ceder [6], we shall denote by DB_1 the set of all functions from **R** into **R** which are in class I of Baire and have the Darboux property.

The need of the set DB_1 is explained by the following result:

LEMMA 2 *Let f be a function in* DB_1. *Let us define the function* $\varphi : \mathbf{R}^2 \to \mathbf{R}$ *by* :

$$\varphi(x,y) = f(y)^k\, x + f(x)^l\, y \qquad (x,y \in \mathbf{R})$$

Then, for every fixed real numbers x and y, the functions $\varphi(.,y)$ *and* $\varphi(x,.)$ *have the Darboux property.*

Proof of Lemma 2

Since f is in DB_1, the function $x\,f(.)^k$ belongs also to DB_1 if x is any nonzero real number. Therefore, the graph of the function $x\,f(.)^k$ is connected [6]. The continuity of the function :

$\mathbf{R}^2 \to \mathbf{R}^2$

$(t,s) \to (t, f(x)^l t + s)$ implies that the graph of the function $\varphi(x,.)$ is also connected. Therefore, the function $\varphi(x,.)$ has the Darboux property.

The proof is the same for the function $\varphi(.,y)$.

Let us first study some particular cases of the functional equation (1).

2.1. Case $\lambda = 0, k \geq 0, l \geq 0$

(1) is just : $f(f(y)^k x + f(x)^l y) = 0 \quad (x,y \in E)$

For $k = 0$ and $l \geq 0$, it is obvious that the unique solution of (1) is $f = 0$.

So we consider now the case where k and l are positive integers. Let us suppose that there exists an element x_0 in E such that $f(x_0) \neq 0$. By taking $x = y = 0$ in (1), we get $f(0) = 0$. Therefore, x_0 is different from 0. Let us suppose also that ***the function*** $g: \mathbf{R} \to \mathbf{R}$ ***defined by*** : $g(t) = f(t\, x_0) \quad (t \in \mathbf{R})$ ***belongs to*** DB_1. By taking $x = x_0$ and $y = t\, x_0$ ($t \in \mathbf{R}$) in (1), we obtain :

$$f(f(t\, x_0)^k x_0 + f(x_0)^l t\, x_0) = 0 \qquad \text{for every } t \text{ in } \mathbf{R}$$

Let us define : $\psi(t) = g(t)^k + t\, f(x_0)^l \quad (t \in \mathbf{R})$.

We have : $f(\psi(t)\, x_0) = 0$ for every t in $\mathbf{R}$.

Since g is in DB_1, we may prove as in Lemma 2 that ψ has the Darboux property. Therefore, $\psi(\mathbf{R})$ is an interval of $\mathbf{R}$ which contains 0, but does not contain 1. So $\psi(\mathbf{R})$ is included in $(-\infty,1)$.

Let us suppose that ψ is bounded below by b. The relation :

$f(t\, x_0)^k = \psi(t) - t\, f(x_0)^l \quad (t \in \mathbf{R})$ shows that : $f(\mathbf{R}\, x_0)^k = \mathbf{R}$.

This implies that k is an odd integer and so $f(\mathbf{R}\, x_0) = \mathbf{R}$. Let c be the unique point of (0,1) which satisfies : $c^k + c^l = 1$. Then, there exists a nonzero real number s such that $f(s\, x_0) = c$. By taking $x = y = s\, x_0$ in (1)), we obtain : $f(s\, x_0) = 0$, which brings a contradiction. Therefore, $\psi(\mathbf{R})$ contains $(-\infty,0]$ and we have $f(t\, x_0) = 0$ for every nonpositive real number t. Since ψ is bounded above by 1, we deduce first from : $\psi(t) = t\, f(x_0)^l \quad (t \leq 0)$ that $f(x_0)^l$ is a positive real number, and then that $g(t)^k = \psi(t) - t\, f(x_0)^l$ tends to $-\infty$ when t goes to $+\infty$. In view of the Darboux property of g, we deduce that $g([0,+\infty))^k$ contains $(-\infty,0]$. By taking now $x = t x_0$, $t < 0$, and $y = r x_0$, $r > 0$, in (1), we get : $f(g(r)^k t x_0) = 0$ and therefore : $f(s x_0) = 0$ for every positive real number s. This contradicts $f(x_0) \neq 0$.

PROPOSITION 3 *In the class of functions $f: E \rightarrow \mathbf{R}$ which have the property that for every x in E the function f_x defined by : $f_x(t) = f(tx)$ $(t \in \mathbf{R})$ belongs to DB_1, the unique solution of (1) in the case $\lambda = 0$ is $f = 0$.*

2.2. Case $\lambda \neq 0$, $k = l = 0$.

In this case, (1) is : $f(x+y) = \lambda f(x) f(y)$ $(x,y \in E)$.
So, λf is a solution of Cauchy's exponential equation. Therefore, all the solutions of (1) are given by :

(i) $f = 0$

(ii) $f(x) = 1/\lambda \cdot e^{g(x)}$ $(x \in E)$ where $g: E \rightarrow \mathbf{R}$ is an arbitrary additive function.

2.3. Case $\lambda \neq 0$, $k = 0$, $l > 0$.

We give first some property of the function φ defined in Lemma 2 when f is a non identically zero solution in DB_1 of functional equation (1) in the general case where k and l are nonnegative integers and λ is a nonzero real number.

LEMMA 4 *Let us suppose that λ is a nonzero real number and k, l are nonnegative integers.*

If f is a non identically zero solution of (1) in DB_1, the function ψ_1 defined by : $\psi_1(x) = \varphi(x,x_0)$ $(x \in \mathbf{R})$ and the function ψ_2 defined by : $\psi_2(x) = \varphi(x_0,x)$ $(x \in \mathbf{R})$ are one-to-one and continuous when x_0 is any real number satisfying $f(x_0) \neq 0$.

Proof of Lemma 4

By Lemma 2, the functions ψ_1 and ψ_2 have the Darboux property.

Let us suppose now for example that ψ_1 is not one-to-one. Then there exist x and y in **R** such that $x \neq y$ and

$$\varphi(x,x_0) = \varphi(y,x_0) \qquad (3)$$

We deduce by (1) : $\lambda f(x) f(x_0) = \lambda f(y) f(x_0)$.

Since $\lambda f(x_0) \neq 0$, this implies : $f(x) = f(y)$. With (3), we obtain : $f(x_0)^k x = f(x_0)^k y$ and therefore $x = y$. This is a contradiction.

So, ψ_1 and ψ_2 are one-to-one and have the Darboux property. Therefore, they are continuous [6].

COROLLARY 5 *Under the same hypotheses as in Lemma 4, if f is a solution of (1) in DB_l, the functions $f(.)^l$ and $f(.)^k$ are continuous.*

Proof of Corollary 5

If f is a non identically zero solution of (1) in DB_l, there exists x_0 in $\mathbf{R}-\{0\}$ such that $f(x_0) \neq 0$. By Lemma 4, the functions $\psi_1(x) = \varphi(x,x_0)$ $(x \in \mathbf{R})$ and $\psi_2(x) = \varphi(x_0,x)$ $(x \in \mathbf{R})$ are continuous. We deduce immediately that the functions $f(.)^l$ and $f(.)^k$ are continuous.

Let us consider now the functional equation (1) when $k = 0$, l is a positive integer and λ is a nonzero real number.

If f is a non identically zero solution of (1) in DB_l, the function $g(x) = f(x)^l$ $(x \in \mathbf{R})$ is continuous by Corollary 5. Moreover, g is a solution of :

$$g(x + g(x)y) = \lambda^l g(x) g(y) \qquad (x,y \in \mathbf{R}) \qquad (4)$$

which is similar to the Gołąb-Schinzel functional equation.

By taking $x = y = 0$ in (4), we obtain either $g(0) = 0$ or $g(0) = \lambda^{-l}$.

When $g(0) = 0$, we get $g = 0$ as we can see by taking $y = 0$ in (4).

So, we consider now the case where $g(0) = \lambda^{-l}$. By taking $x = 0$ in (4), we get :

$$g(y) = g(\lambda^{-l} y) \qquad (y \in \mathbf{R}) \tag{5}$$

and therefore :

$$g(y) = g(\lambda^{-nl} y) \qquad (y \in \mathbf{R}) \tag{6}$$

for every positive integer n

When $|\lambda|$ is different from 1 , (6) implies :

$g = g(0) = \lambda^{-l}$ and therefore $f = 1/\lambda$.

When λ^l is equal to 1 , (4) is just the functional equation of Gołąb-Schinzel for which we know all the continuous solutions [3] .

When λ^l is equal to -1 (i.e $\lambda = -1$ and l odd) , (5) implies by changing y into -y in (4) :

$$g(x - g(x)y) = -g(x)g(y) \qquad (x,y \in \mathbf{R})$$

This means that -g is a continuous solution of the functional equation of Gołąb-Schinzel .

So , we obtain the following result :

PROPOSITION 6 *When λ is a nonzero real number and l is a positive integer , all the solutions in the class of functions DB_l of the following functional equation :*

$$f(x + f(x)^l y) = \lambda f(x) f(y) \qquad (x,y \in \mathbf{R}) \tag{7}$$

are given by :

(i) $f = 0$

and (ii) if $|\lambda| \neq 1$, $f = 1/\lambda$

(iii) if $\lambda = 1$ *and if* l *is odd*

$f(x) = (1 + ax)^{1/l}$ $(x \in \mathbf{R})$ *and* $f(x) = (\mathrm{Sup}(1 + ax, 0))^{1/l}$ $(x \in \mathbf{R})$

(iv) if $\lambda = 1$ *and if* l *is even*

$$f(x) = (\mathrm{Sup}(1 + ax, 0))^{1/l} \qquad (x \in \mathbf{R})$$

(v) if $\lambda = -1$ *and if* l *is odd*

$f(x) = -(1 + ax)^{1/l}$ $(x \in \mathbf{R})$ *and* $f(x) = -(\mathrm{Sup}(1 + ax, 0))^{1/l}$ $(x \in \mathbf{R})$

(vi) if $\lambda = -1$ *and if* l *is even*

$$f(x) = -(\mathrm{Sup}(1 + ax, 0))^{1/l} \qquad (x \in \mathbf{R})$$

where a is an arbitrary real number .

With the same proof we obtain also all the continuous solutions $f : E \to \mathbf{R}$ of (7) when E is a real Hausdorff topological vector space . Namely :

PROPOSITION 7 *When λ is a nonzero real number and l is a positive integer, all the continuous solutions $f : E \to \mathbf{R}$ of the following functional equation :*

$$f(x + f(x)^l y) = \lambda f(x) f(y) \qquad (x, y \in E) \qquad (7)$$

are given by :

(i) $f = 0$

and (ii) if $|\lambda| \neq 1$, $f = 1/\lambda$

(iii) if $\lambda = 1$ and if l is odd

$$f(x) = (1 + \langle x, x^* \rangle)^{1/l} \quad (x \in E) \quad \text{and} \quad f(x) = (\mathrm{Sup}(1 + \langle x, x^* \rangle, 0))^{1/l} \quad (x \in E)$$

(iv) if $\lambda = 1$ and if l is even

$$f(x) = (\mathrm{Sup}(1 + \langle x, x^* \rangle, 0))^{1/l} \qquad (x \in E)$$

(v) if $\lambda = -1$ and if l is odd

$$f(x) = -(1 + \langle x, x^* \rangle)^{1/l} \quad (x \in E) \quad \text{and} \quad f(x) = -(\mathrm{Sup}(1 + \langle x, x^* \rangle, 0))^{1/l} \quad (x \in E)$$

(vi) if $\lambda = -1$ and if l is even

$$f(x) = -(\mathrm{Sup}(1 + \langle x, x^* \rangle, 0))^{1/l} \qquad (x \in E)$$

where x^* is an arbitrary element of the topological dual of E .

Let us finally mention that J.Brzdęk studied in detail the functional equation (7) in the case $\lambda = 1$ in his Doctor Thesis [7] .

2.4. Case $\lambda \neq 0, k > 0, l > 0$.

We start with a preliminary remark concerning the solutions of the functional equation (1) in DB_l .

LEMMA 8 *Let f be a solution of (1) in DB_1 .*

If λ is positive and f is bounded above on ***R***
or if λ is negative and f is bounded below on ***R*** *,*
then f is constant .

Proof of Lemma 8

Let us suppose that λ is a positive real number . The proof is similar when λ is negative .

For an indirect proof , we suppose that f is a solution of (1) in DB_1 bounded above on **R** and that f is not constant .

Let M be an upper bound of f(**R**) . By taking x = y in (1) , we obtain : $\lambda f(x)^2 \le M$ for every x in **R** . Since f is not identically zero , M is a positive real number .

By taking x = y in (1) , we get successively :

$|f(x)| \le (M/\lambda)^{1/2}$ for every x in **R**

$|f(x)| \le (M^{1/4}) / (\lambda^{(1/2) + (1/4)})$ for every x in **R**

....

$|f(x)| \le (M^{1/2^n}) / (\lambda^{(1/2) + (1/4) + ... + (1/2^n)})$ for every x in **R** and every positive integer n

As n goes to +∞ , we obtain :

$$|f(x)| \le 1/\lambda \quad \text{for every } x \text{ in } \mathbf{R} \tag{8}$$

Since f is bounded and non identically zero , we have , by the Darboux property of the function φ(.,t) (Lemma 2) , φ(**R**,t) = **R** for each t ε **R** such that f(t) ≠ 0 . Therefore , for every real number x , there exists a real number s such that φ(s,t) = x . In view of the Darboux property of f , we may choose x and t in **R** such that : 0 < |f(t)| < |f(x)| . By using (1) and (8) , we obtain :

$$0 < |f(t)| < |f(x)| = | f (\varphi(s,t)) | = \lambda\, |f(s)|\, |f(t)| \le |f(t)|$$

which brings a contradiction . Therefore , f is constant .

In order to obtain all the solutions of (1) in DB_l , we shall consider the two cases : $k \neq l$ and $k = l$

2.4.1. Solutions of (1) in DB_l in the case $k \neq l$

We are going to prove that the only solutions of (1) in DB_l in this case are the constant functions . The following results are due to J.Brzdęk [8] :

PROPOSITION 9 *Let k and l be distinct positive integers and let f be a non identically zero solution of (1) in* DB_l . *We define* $F = \{ x \in \mathbf{R} \ / \ f(x) \neq 0 \}$.

Then , $f_{/F}$ *is not one-to-one , where* $f_{/F} : F \to \mathbf{R}$ *is the function defined by :* $f_{/F}(x) = f(x)$ *for all x in F .*

Proof of Proposition 9

For an indirect proof , let us suppose that $f_{/F}$ is one-to-one .

Let us remark that , by (1) , if x and y belong to F , $\varphi(x,y)$ belongs also to F . Therefore , the symmetry in x and y of the second member of (1) implies :

$$f(y)^k x + f(x)^l y = f(x)^k y + f(y)^l x \qquad \text{for all } x,y \text{ in } F$$

We deduce :

$$(f(x)^l - f(x)^k)\, y = (f(y)^l - f(y)^k)\, x \qquad \text{for all } x,y \text{ in } F$$

Since f has the Darboux property and is not constant , there exists y_0 in F-{0} such that $p = (f(y_0)^l - f(y_0)^k) \ / \ y_0 \neq 0$. Therefore , we obtain :

$$f(x)^l - f(x)^k = p\,x \qquad \text{for every } x \text{ in } F \qquad (9)$$

Let us define : $g(x) = (x^l - x^k) \ / \ p$ for every x in $\mathbf{R}$.
We have by (9) :

$$g\,(\,f(F)\,) = F \qquad (10)$$

Let us suppose that $F = \mathbf{R}$. Then , since f has the Darboux property , f(F) is an interval of $\mathbf{R}$ which does not contain 0 . So , we have either $f(F) \subset (-\infty, 0)$ or $f(F) \subset (0, +\infty)$. It is not difficult to see that we have $g((-\infty, 0)) \neq \mathbf{R}$ and $g((0, +\infty)) \neq \mathbf{R}$. This contradicts (10).

Therefore, we have $F \neq \mathbf{R}$. So, there exists x_0 in $\mathbf{R}$ such that $f(x_0) = 0$. By taking $x = y = x_0$ in (1), we obtain :

$$f(0) = 0 \tag{11}$$

So, F does not contain 0. Then, by (9), f(F) does not contain 1. Since f has the Darboux property, $f(\mathbf{R}) = f(F) \cup \{0\}$ is an interval of $\mathbf{R}$ which does not contain 1. So, it is included either in $(-\infty, 1)$ or in $(1, +\infty)$. (11) implies :

$$f(\mathbf{R}) \subset (-\infty, 1) \tag{12}$$

If λ is a positive real number, f is constant by Lemma 8. But, it is not the case. Therefore, λ is a negative real number. Since f is not constant, we have by Lemma 8 :

$$(-\infty, 0] \subset f(\mathbf{R}) \tag{13}$$

Let us suppose that there exists a in (0,1) such that a belongs to $f(\mathbf{R})$. By (13), there exists b in $f(\mathbf{R})$ such that $b < 1/\lambda a$. By (1), $\lambda a b$ belongs to $f(\mathbf{R})$ and we have : $\lambda a b > 1$. This contradicts (12). We deduce : $f(\mathbf{R}) = (-\infty, 0]$ and therefore :

$$f(F) = (-\infty, 0) \tag{14}$$

If k and l are both odd or both even, (9) implies that f(F) does not contain -1, which contradicts (14). Therefore, either k is odd and l is even, or k is even and l is odd. By (10), we have either $F = (0, +\infty)$ or $F = (-\infty, 0)$. We deduce :

$$\mathbf{R} - F = [0, +\infty) \quad \text{or} \quad \mathbf{R} - F = (-\infty, 0] \tag{15}$$

Let us suppose for example that k is odd (and l is even).
Let us define : $A_k = \{ f(x)^k \,;\, x \in F \}$
(14) implies : $A_k = (-\infty, 0)$

By (1), we have :

$$f(f(y)^k x) = 0 \qquad \text{for all } x \text{ in } \mathbf{R}-F \text{ and all } y \text{ in } F$$

We deduce : $A_k . (\mathbf{R}-F) \subset \mathbf{R}-F$

where $A_k . (\mathbf{R}-F) = \{ xy \ ; \ x \in A_k , y \in \mathbf{R}-F \}$

This is impossible by (15).

This completes the proof.

LEMMA 10 *Let k and l be distinct positive integers. If f is a non identically zero solution of (1) in DB_l, there exist a, b in **R**, a < b, such that f is constant and nonzero on the interval [a, b].*

Proof of Lemma 10

By Proposition 9, $f_{/F}$ is not one-to-one. Therefore, there exist a, b in **R**, a < b, such that $f(a) = f(b) \neq 0$.

Let us suppose that f is not constant on [a, b].
So, there exists c in (a, b) such that $f(c) \neq f(a)$. Since f has the Darboux property, we may choose c in (a, b) such that $f(c)^k \neq f(a)^k$.

Without loss of generality, we may suppose ***l < k***.
We consider the following function :

$$V(x,y,z) = (f(x)^l / x)(y-z) + f(y)^k - f(z)^k \qquad (x \in \mathbf{R}-\{0\} ; y,z \in \mathbf{R})$$

We shall prove first that :

for every $r>0$, there exists x_r in $F-\{0\}$ such that $|f(x_r)^l / x_r| < r$ (16)

For an indirect proof of (16), let us suppose that there exists a positive real number r such that :

$$|f(x)^l / x| \geq r \qquad \text{for every } x \text{ in } F-\{0\} \qquad (17)$$

Let us suppose that F is bounded. Since f is a non identically zero solution of (1) in DB_l, the function $f(.)^l$ is continuous and not identically zero by Corollary 5. Therefore, F is a non empty open subset of **R**. So, Sup F and Inf F are not both equal to 0. Let a be an element of {Sup F, Inf F} − {0}. There exists a sequence $\{x_n\}_{n\in\mathbf{N}}$ in $F-\{0\}$ converging to a. The definition of F and the continuity of the function $f(.)^l$ imply that the sequence $\{ f(x_n)^l / x_n \}_{n\in\mathbf{N}}$ converges to 0. This contradicts (17).

Therefore, F is unbounded and there exists a sequence $\{x_n\}_{n\in\mathbf{N}}$ in $F-\{0\}$ such that $\{ |x_n| \}_{n\in\mathbf{N}}$ tends to $+\infty$. By (1) and (17), we have for all x and y in $F-\{0\}$:

$$| f(\varphi(x,y))^l/\varphi(x,y) | = | (\lambda^l f(x)^l f(y)^l) / (f(y)^k x + f(x)^l y)| \geq r \qquad (18)$$

We have also for every x in $F-\{0\}$ and for every n in $\mathbf{N}$:

$$| f(\varphi(x,x_n))^l/\varphi(x,x_n) | = | (\lambda^l f(x)^l) / (f(x_n)^{k-l} x + f(x)^l f(x_n)^{-l} x_n)|$$

(17) implies that :
$\{ | f(x_n)^{k-l} | \}_{n\in\mathbf{N}}$ tends to $+\infty$ and $| x_n f(x_n)^{-l}| \leq 1/r$ for every n in $\mathbf{N}$.
We deduce that the sequence $\{ | f(\varphi(x,x_n))^l/\varphi(x,x_n) | \}_{n\in\mathbf{N}}$ converges to 0. This contradicts (18). Therefore, we have proved (16).

By (16), there exists x_0 in $F-\{0\}$ such that :

$$| (f(x_0)^l / x_0) (c - a)| < | f(c)^k - f(a)^k |$$

and $$| (f(x_0)^l / x_0) (b - c)| < | f(b)^k - f(c)^k |$$

We deduce : $\quad \operatorname{sign} V(x_0,c,a) = \operatorname{sign} (f(c)^k - f(a)^k)$

and $\quad \operatorname{sign} V(x_0,b,c) = \operatorname{sign} (f(b)^k - f(c)^k)$

where $\quad \operatorname{sign} t = \begin{cases} 1 & \text{if } t > 0 \\ -1 & \text{if } t < 0 \end{cases}$

Since $\quad \operatorname{sign} (f(c)^k - f(a)^k) \neq \operatorname{sign} (f(b)^k - f(c)^k)$

we have either $\quad \operatorname{sign} V(x_0,b,a) \neq \operatorname{sign} V(x_0,c,a) \qquad (19)$

or $\quad \operatorname{sign} V(x_0,b,a) \neq \operatorname{sign} V(x_0,b,c) \qquad (20)$

Let us suppose that (19) occurs. The proof would be similar with (20). By Corollary 5, the function : $y \in \mathbf{R} \dashrightarrow V(x_0,y,a)$ is continuous. Therefore, (19) implies that there exists y_0 in (c,b)

such that $V(x_0,y_0,a) = 0$. We deduce : $\varphi(x_0,y_0) = \varphi(x_0,a)$. This is not possible since , by Lemma 4 , the function $\varphi(x_0,.)$ is one-to-one . Therefore , f is constant on the interval $[a, b]$.

THEOREM 11 *If k and l are distinct positive integers , the only solutions of (1) in DB_l are : $f = 0$ and $f = 1/\lambda$.*

Proof of Theorem 11

Let f be a non identically zero solution of (1) in DB_l . By Lemma 10 , there exist a , b in **R** , $a < b$, such that :

$$f(x) = f(a) \neq 0 \qquad \text{for every } x \text{ in } [a,b]$$

Let us suppose that there exists x_0 in **R** such that $f(x_0) \neq f(a)$. By the Darboux property of f , there exists a nontrivial interval **I** of **R** included in f(**R**) which contains f(a) , but does not contain 0 . On the set $f^{-1}(\mathbf{I})$, we consider the following equivalence relation :

$$x \sim y \quad \Longleftrightarrow \quad f(x) = f(y)$$

By the axiom of choice , there exists a bijection g from the quotient set $(f^{-1}(\mathbf{I})/\sim)$ onto a subset Y of $f^{-1}(\mathbf{I})$. The function $f^{\sim} : (f^{-1}(\mathbf{I})/\sim) \to \mathbf{I}$ defined by :

$$f(y) = f^{\sim}(g^{-1}(y)) \qquad \text{for every } y \text{ in } Y$$

is a bijection from $(f^{-1}(\mathbf{I})/\sim)$ onto **I** . Therefore , we have :

$$\text{Card } Y = \text{Card } \mathbf{I} > \text{Card } \mathbf{N} \tag{21}$$

$$f(Y) = f^{\sim}(f^{-1}(\mathbf{I})/\sim) = \mathbf{I} \tag{22}$$

$$f(x) \neq f(y) \qquad \text{for } x, y \in Y, \ x \neq y \tag{23}$$

By (1) , we have :

$$f(\varphi(x,y)) = \lambda f(a) f(y) \qquad \text{for every } x \text{ in } [a,b] \text{ and every } y \text{ in } \mathbf{R} \tag{24}$$

By (22) and Lemma 4 , for each y in Y , $\varphi([a,b], y)$ is a non trivial interval of **R** since **I** does not contain 0 . Moreover , by (23) and (24) , if y and z are distinct elements of Y , the intervals

$\varphi([a,b], y)$ and $\varphi([a,b], z)$ are disjoint. Therefore, $A = \{\varphi([a,b], y) \,;\, y \in Y\}$ is a family of disjoint intervals. By (21), we have : Card A = Card Y > Card **N**. This is impossible.

Therefore, f is a nonzero constant function on **R**. (1) implies : $f = 1/\lambda$.

This completes the proof of Theorem 11.

2.4.2. Solutions of (1) in DB_1 in the case $k = 1$.

The following results are due to the author [3,4,5].

The proofs are based on the following known result [9]:

PROPOSITION 12 *All the continuous solutions $h : \mathbf{R} \to \mathbf{R}$ of the functional equation :*

$$h(h(x)) = (\gamma + 1)\, h(x) - \gamma x \qquad (x \in \mathbf{R}) \qquad (25)$$

where γ is a given nonzero real number, are given by :

a) if $\gamma > 0$, $\gamma \neq 1$

(i)

$$h(x) = \begin{cases} \gamma x + (1-\gamma)\, a & \text{for } x \le a \\ x & \text{for } a \le x \le b \\ \gamma x + (1-\gamma)\, b & \text{for } x \ge b \end{cases} \qquad \text{with } -\infty \le a < b \le +\infty$$

(ii) $h(x) = \gamma x + \delta \qquad (x \in \mathbf{R}) \qquad$ *with* $\delta \in \mathbf{R}$

b) if $\gamma = 1$

$h(x) = x + \delta \qquad (x \in \mathbf{R}) \qquad$ *with* $\delta \in \mathbf{R}$

c) if $\gamma < 0$, $\gamma \neq -1$

(i) $h(x) = \gamma x + \delta \qquad (x \in \mathbf{R}) \qquad$ *with* $\delta \in \mathbf{R}$

(ii) $h(x) = x \qquad (x \in \mathbf{R})$

d) if $\gamma = -1$

(i) $h(x) = x \qquad (x \in \mathbf{R})$

(ii) $h(x) = \begin{cases} \Phi(x) & \text{for } x \in (-\infty, c] \\ \Phi^{-1}(x) & \text{for } x \in [c, +\infty) \end{cases}$

where c *is an arbitrary real number and* Φ *is an arbitrary continuous and strictly decreasing function mapping* $(-\infty, c]$ *onto* $[c, +\infty)$

If $k = l$ and if f is a solution of (1) in DB_l, the function : $g(x) = f(x)^l \quad (x \in \mathbf{R})$ is continuous by Corollary 5 and satisfies the following functional equation :

$$g(g(y)x + g(x)y) = \lambda^l g(x)g(y) \qquad (x, y \in \mathbf{R}) \tag{26}$$

Let us remark that the functional equation (26) corresponds to the particular case of (1) where $k = l = 1$ and λ is replaced by λ^l.

We shall solve (26) and we consider the two cases : $\lambda^l < 0$ and $\lambda^l > 0$.

We start with the first case and we shall prove the following result :

PROPOSITION 13 *All the continuous solutions* $g : \mathbf{R} \to \mathbf{R}$ *of the functional equation :*

$$g(g(y)x + g(x)y) = \mu\, g(x)g(y) \qquad (x, y \in \mathbf{R}) \tag{27}$$

where μ *is a given negative real number are given by :*

(i) $g = 0$ *(ii)* $g = 1/\mu$

Proof of Proposition 13

By taking $x = y = 0$ in (27), we have either $g(0) = 1/\mu$ or $g(0) = 0$.

a) ***Let us first consider the case*** $g(0) = 1/\mu$.

By taking $y = 0$ in (27), we get :

$$g(x) = g(x/\mu) \qquad \text{for every } x \text{ in } \mathbf{R} \tag{28}$$

Therefore, we have for every x in **R** and for every positive integer n:

$$g(x) = g(x/\mu^n) = g(x\mu^n)$$

If $\mu \neq -1$, we see, as n goes to $+\infty$ and by using the continuity of g at 0, that: $g(x) = g(0) = 1/\mu$ for every x in **R**.

If $\mu = -1$, (28) becomes: $g(x) = g(-x)$ for every x in **R**. By taking $y = -x$ in (27), we get:

$$g(g(-x)x - g(x)x) = -g(x)g(-x) \qquad \text{for every } x \text{ in } \mathbf{R}$$

or, with (28): $\quad g(0) = -1 = -g(x)^2 \qquad$ for every x in **R**.

Using the continuity of g, we deduce:

$$g(x) = -1 \qquad \text{for every } x \text{ in } \mathbf{R}$$

b) Let us now consider the case $g(0) = 0$.

Let g be a non identically zero continuous solution of (27). By taking $y = x$ in (27), we see that the set of all real numbers x such that $g(x) < 0$ is a non empty open subset of **R**. The continuity of g implies then that $g(\mathbf{R})$ is an interval of **R** containing an interval of the form $(\alpha,0]$. By Lemma 8, g is not bounded below and therefore ***g(R) is an interval of R which contains $(-\infty, 0]$***.

So, there exists a nonzero real number x_0 such that $g(x_0) = 1/\mu$. Let us denote:

$$h(x) = g(x_0)x + x_0 g(x) = x/\mu + x_0 g(x) \qquad (x \in \mathbf{R})$$

h is continuous and satisfies the following functional equation:

$$h(h(x)) = (1/\mu + 1)\, h(x) - 1/\mu\ x \qquad (x \in \mathbf{R}) \tag{29}$$

Now, all the continuous solutions of (29) are given by Proposition 12.

The solution: $h(x) = x \quad (x \in \mathbf{R})$ of (29) gives: $g(x) = (1 - 1/\mu)\, x/x_0 \quad (x \in \mathbf{R})$ which does not satisfy (27).

The solution : $h(x) = 1/\mu \; x + \delta$ $\quad (x \in \mathbf{R})$ of (29) leads to a constant function g . This is not possible since we have supposed that $g(0) = 0$ and g is not identically zero .

So, we have necessarily $\mu = -1$ and

$$h(x) = \begin{cases} \Phi(x) & \text{for } x \in (-\infty, c] \\ \Phi^{-1}(x) & \text{for } x \in [c, +\infty) \end{cases} \tag{30}$$

where Φ is a continuous and strictly decreasing function mapping $(-\infty, c]$ onto $[c, +\infty)$.

The function : $x \dashrightarrow h(x) - x$ is continuous and strictly decreasing on $\mathbf{R}$. Therefore, it vanishes at most once . From $h(c) = c$ and $h(0) = 0$, we deduce : $c = 0$.

By taking $y = x_0$ in (27) , we get :

$$g(h(x)) = g(x) \qquad \text{for every } x \text{ in } \mathbf{R} \tag{31}$$

Therefore , we may suppose that x_0 is a positive real number .

By taking $y = h(x)$ in (27) and using (31) , we get :

$$g(x\,g(x) + g(x)(x_0\,g(x) - x)) = -g(x)^2 \qquad (x \in \mathbf{R})$$

or
$$g(x_0\,g(x)^2) = -g(x)^2 \qquad (x \in \mathbf{R}) \tag{32}$$

Since $g(\mathbf{R})$ is an interval of $\mathbf{R}$ which contains $(-\infty, 0]$, the set $\{x_0\,g(x)^2 \; ; \; x \in \mathbf{R}\}$ is the interval $[0, +\infty)$. (32) implies :

$$g(x) = -x/x_0 \qquad \text{for every } x \text{ in } [0, +\infty) \tag{33}$$

For x in $(-\infty, 0]$, $h(x)$ belongs to $[0, +\infty)$ by (30) . Therefore, by using (31) and (33) , we obtain :

$$g(x) = g(h(x)) = -h(x)/x_0 = -g(x) + x/x_0 \qquad \text{for every } x \text{ in } (-\infty, 0]$$

or
$$g(x) = x/2x_0 \qquad \text{for every } x \text{ in } (-\infty, 0]$$

It is now easy to check that the function defined by :

$$g(x) = \begin{cases} -x/x_0 & \text{for } x \geq 0 \\ x/2x_0 & \text{for } x \leq 0 \end{cases} \qquad \text{does not satisfy (27) .}$$

Therefore , there does not exist a non identically zero continuous solution of (27) satisfying $g(0) = 0$. This ends the proof of Proposition 13 .

When μ is a positive real number, we have the following result [3]:

PROPOSITION 14 *All the continuous solutions $g : \mathbf{R} \to \mathbf{R}$ of the functional equation :*

$$g(g(y)x + g(x)y) = \mu\, g(x)\, g(y) \qquad (x, y \in \mathbf{R}) \qquad (27)$$

where μ is a given positive real number

are given by :

(i) $g = 0$ *(ii)* $g = 1/\mu$

and, in the case $\mu = 2$ only :

(iii) $g(x) = \alpha x \qquad (x \in \mathbf{R})$

(iv) $g(x) = \mathrm{Sup}(\alpha x, 0) \qquad (x \in \mathbf{R})$

where α is an arbitrary nonzero real number.

Proof of Proposition 14

As in Proposition 13, we have either $g(0) = 1/\mu$ or $g(0) = 0$

a) ***Let us first consider the case $g(0) = 1/\mu$.***

As in Proposition 13, we have the equality (28) which implies $g(x) = 1/\mu$ for every x in $\mathbf{R}$ when μ is different from 1.

So, we consider the case where $\mu = 1 = g(0)$. If f is not identically equal to 1, there exists x_0 in $\mathbf{R}$ such that $g(x_0) = 1 + \varepsilon$ where ε is a nonzero real number.

By taking $x = y = x_0$ in (27), we get with $x_1 = 2 x_0 g(x_0)$:

$$g(x_1) = (1 + \varepsilon)^2$$

By taking $x = y = x_1$ in (27), we get with $x_2 = 2 x_1 g(x_1)$:

$$g(x_2) = (1 + \varepsilon)^4$$

By this way, we build a sequence of real numbers x_n such that : $g(x_n) = (1 + \varepsilon)^{2^n}$ for every positive integer n.

If $g(x_0) > 1$, ε is a positive real number and the sequence

$\{g(x_n)\}_{n\in\mathbf{N}}$ tends to $+\infty$. The continuity of g implies : $[1,+\infty) \subset g(\mathbf{R})$

If $g(x_0) < 1$, ε is a negative real number and we can assume $-1 < \varepsilon < 0$. Therefore, the sequence $\{g(x_n)\}_{n\in\mathbf{N}}$ converges to 0. The continuity of g implies : $(0,1] \subset g(\mathbf{R})$.

We notice that $g(\mathbf{R})$ does not contain 0 since, if there exists x_0 in $\mathbf{R}$ such that $g(x_0) = 0$, we get, by taking $x = y = x_0$ in (1), $g(0) = 0$, which is not the case.

So, by Lemma 8, $g(\mathbf{R})$ satisfies one of the two following conditions :

(i) $g(\mathbf{R}) = [1,+\infty)$

(ii) $g(\mathbf{R}) = (0,+\infty)$

In the case (i), let us choose a nonzero real number t such that : $g(t) > 1$.

Let us denote as in Lemma 2 :

$$\varphi(x,y) = g(y)\,x + g(x)\,y \qquad (x,y \in \mathbf{R})$$

We have : $\varphi(-t,t) = t\,(g(-t) - g(t))$

If $g(-t) = g(t)$, we have by (27) :

$g(\varphi(-t,t)) = 1 = g(-t)\,g(t) \geq g(t) > 1$, which is not possible.

If $g(-t) < g(t)$, $\varphi(-t,t)$ and $\varphi(0,t)$ do not have the same sign. By the continuity of the function $\varphi(.,t)$, there exists a nonzero real number u such that $\varphi(u,t) = 0$. Then, by (27), we have :

$g(\varphi(u,t)) = 1 = g(u)\,g(t) \geq g(t) > 1$, which is impossible.

Therefore, we have : $g(-t) > g(t)$. $\varphi(-t,t)$ and $\varphi(-t,0)$ do not have the same sign. By the continuity of the function $\varphi(-t,.)$, there exists a nonzero real number u such that $\varphi(-t,u) = 0$. Then, by (27), we have :

$g(\varphi(-t,u)) = 1 = g(-t)\,g(u) \geq g(-t) > g(t) > 1$, which is also impossible.

Therefore, the case (i) cannot occur.

Let us consider now the case (ii). There exists a nonzero real number x_0 satisfying $g(x_0) = 1/2$. By taking $x = y = x_0$ in (27), we obtain: $1/2 = g(x_0) = g(x_0)^2 = 1/4$, which is not possible. Therefore, the case (ii) cannot occur either.

In conclusion, when $g(0) = 1/\mu = 1$, g is identically equal to 1.

Therefore, if μ is any positive given real number, the only continuous solution of (27) satisfying $g(0) = 1/\mu$ is: $g = 1/\mu$.

b) Let us now consider the case $g(0) = 0$.

Let g be a non identically zero continuous solution of (27). By taking $y = x$ in (27), we see as in Proposition 13 that the set of all real numbers x such that $g(x) > 0$ is a non empty open subset of $\mathbf{R}$. The continuity of g implies then that $g(\mathbf{R})$ is an interval of $\mathbf{R}$ containing an interval of the form $[0,\alpha)$. By Lemma 8, g is not bounded above on $\mathbf{R}$ and, therefore, ***$g(R)$ is an interval of R which contains $[0, +\infty)$.***

So, there exists a nonzero real number x_0 such that $g(x_0) = 1/\mu$. Let us denote:

$$h(x) = \varphi(x,x_0) = x/\mu + x_0\, g(x) \qquad (x \in \mathbf{R})$$

h is continuous and satisfies the following functional equation:

$$h(h(x)) = (1/\mu + 1)\ h(x) - 1/\mu\ x \qquad (x \in \mathbf{R}) \qquad (29)$$

Now, all the continuous solutions of (29) are given by Proposition 12.

The solution: $h(x) = 1/\mu\ x + \delta \quad (x \in \mathbf{R})$ of (29) leads to a constant function g. This is not possible since we have supposed that $g(0) = 0$ and g is not identically zero.

So, we have necessarily $\mu \neq 1$ and

$$h(x) = \begin{cases} x/\mu + (1 - 1/\mu)\, a & \text{for } x \leq a \\ x & \text{for } a \leq x \leq b \\ x/\mu + (1 - 1/\mu)\, b & \text{for } x \geq b \end{cases} \qquad \text{with } -\infty \leq a < b \leq +\infty$$

We deduce :

$$g(x) = \begin{cases} (1 - 1/\mu)\ a/x_0 & \text{for } x \le a \\ (1 - 1/\mu)\ x/x_0 & \text{for } a \le x \le b \\ (1 - 1/\mu)\ b/x_0 & \text{for } x \ge b \end{cases}$$

$g(0) = 0$ implies : $a \le 0 \le b$.

Since g is not constant , g is not bounded above on **R** by Lemma 8 . Therefore , we have : $b = +\infty$ if $(1 - 1/\mu)/x_0$ is positive and $a = -\infty$ if $(1 - 1/\mu)/x_0$ is negative

Let us suppose $(1 - 1/\mu)/x_0 > 0$ and $b = +\infty$. By taking $x = y > 0$ in (27) , we get :

$$g(g(y)x + g(x)y) = g(2(1 - 1/\mu)\ x^2/x_0) = 2(1 - 1/\mu)^2\ x^2/x_0^2$$
$$= \mu(1 - 1/\mu)^2\ x^2/x_0^2$$

This implies $\mu = 2$ and we have the following expression for g :

$$g(x) = \begin{cases} a/2x_0 & \text{for } x \le a \\ x/2x_0 & \text{for } x \ge a \end{cases} \qquad \text{with } -\infty \le a \le 0 \text{ and } x_0 > 0$$

Let us suppose $a > -\infty$. By taking $x < a$ and $y > x_0$ in (27), we obtain :

$$g(g(y)x + g(x)y) = g(yx/2x_0 + ay/2x_0) = g(y.(a+x)/2x_0) = a/2x_0$$
$$= ay/2x_0^2$$

This implies $a = 0$ and g has the expression (iv) with $\alpha > 0$.

If $a = -\infty$, g has the expression (iii) with $\alpha > 0$.

If $(1 - 1/\mu)/x_0 < 0$ and $a = -\infty$, we obtain similarly $\mu = 2$ and either $b = 0$ or $b = +\infty$. This gives the expressions (iii) and (iv) for g with $\alpha < 0$.

It is easy to verify that the expressions (iii) and (iv) of g are solutions of (27) .

This ends the proof of Proposition 14 .

So, when f is a solution of (1) in DB_l in the case $k = l$, we obtain first from the Propositions 13 and 14 all the possible expressions of $g(x) = f(x)^l$ $(x \in \mathbf{R})$ and we deduce then all the possible expressions of f.

2.4.3. *Regular solutions of (1)*

Using Theorem 11 and Propositions 13 and 14, we deduce first all the solutions of (1) in DB_l.

THEOREM 15 *When λ is a nonzero real number and k, l are positive integers, all the solutions $f : \mathbf{R} \to \mathbf{R}$ in the class of functions DB_l of the functional equation:*

$$f(f(y)^k x + f(x)^l y) = \lambda f(x) f(y) \qquad (x, y \in \mathbf{R}) \qquad (1)$$

are given by:

a) if $k \neq l$ or if $k = l$ and $\lambda^l \neq 2$

(i) $f = 0$ (ii) $f = 1/\lambda$

b) if $k = l$ is an odd integer and $\lambda = 2^{1/l}$

(i) $f = 0$ (ii) $f = 1/\lambda$ (iii) $f(x) = \alpha\, x^{1/l}$ $(x \in \mathbf{R})$

(iv) $f(x) = \mathrm{Sup}(\alpha\, x^{1/l}, 0)$ $(x \in \mathbf{R})$ where α is an arbitrary nonzero real number.

c) if $k = l$ is an even integer and $\lambda = 2^{1/l}$

(i) $f = 0$; (ii) $f = 1/\lambda$; iii) $f(x) = (\mathrm{Sup}(\alpha x, 0))^{1/l}$ $(x \in \mathbf{R})$ where α is an arbitrary nonzero real number.

d) if $k = l$ is an even integer and $\lambda = -2^{1/l}$

(i) $f = 0$; (ii) $f = 1/\lambda$; (iii) $f(x) = -(\mathrm{Sup}(\alpha x, 0))^{1/l}$ $(x \in \mathbf{R})$ where α is an arbitrary nonzero real number.

We obtain now all the continuous solutions $f : E \to \mathbf{R}$ of (1) when E is a real Hausdorff topological vector space.

THEOREM 16 *Let E be a real Hausdorff topological vector space. When λ is a nonzero real number and k, l are positive integers, all the continuous solutions $f : E \to \mathbf{R}$ of the functional equation :*

$$f(f(y)^k x + f(x)^l y) = \lambda f(x) f(y) \qquad (x, y \in E) \qquad (1)$$

are given by :

a) if $k \neq l$ or if $k = l$ and $\lambda^l \neq 2$

(i) $f = 0$ (ii) $f = 1/\lambda$

b) if $k = l$ is an odd integer and $\lambda = 2^{1/l}$

(i) $f = 0$ (ii) $f = 1/\lambda$ iii) $f(x) = (\langle x, x^* \rangle)^{1/l}$ $(x \in E)$

(iv) $f(x) = \mathrm{Sup}((\langle x, x^* \rangle)^{1/l}, 0)$ $(x \in E)$ *where x^* is a nonzero element of the topological dual E^* of E.*

c) if $k = l$ is an even integer and $\lambda = 2^{1/l}$

(i) $f = 0$; (ii) $f = 1/\lambda$; (iii) $f(x) = (\mathrm{Sup}(\langle x, x^* \rangle, 0))^{1/l}$ $(x \in E)$

where x^* is a nonzero element of E^*.

d) if $k = l$ is an even integer and $\lambda = -2^{1/l}$

(i) $f = 0$; (ii) $f = 1/\lambda$; (iii) $f(x) = -(\mathrm{Sup}(\langle x, x^* \rangle, 0))^{1/l}$ $(x \in E)$

where x^* is a nonzero element of E^*.

Proof of Theorem 16

Let $f : E \to \mathbf{R}$ be a continuous solution of (1). Then, $f(0)$ is either 0 or $1/\lambda$. For every $x \neq 0$ in E, we consider the function $f_x : \mathbf{R} \to \mathbf{R}$ defined by : $f_x(t) = f(tx)$ $(t \in \mathbf{R})$. It is easy to see that f_x is a continuous solution of (1).

By Theorem 15, if $k \neq l$ or if $k = l$ and $\lambda^l \neq 2$, f_x is a constant function for every $x \neq 0$ in E. Therefore, we have :

$$f_x = f_x(0) = f(0) = f_x(1) = f(x) \qquad \text{for every } x \neq 0 \text{ in } E$$

So, f is identically equal either to 0 or to $1/\lambda$.

So, we consider now the case where $k = l$ and $\lambda^l = 2$.

The function $g: E \to \mathbf{R}$ defined by : $g(x) = f(x)^{l}$ $(x \in E)$ is continuous and satisfies the following functional equation :

$$g(\,g(y)\,x + g(x)\,y\,) = 2\,g(x)\,g(y) \qquad (x,y \in E) \qquad (34)$$

Now , all the continuous solutions $g: E \to \mathbf{R}$ of (34) are known [3] and are given by :

$g = 0$; $g = 1/2$; $g(x) = \langle x,x^*\rangle$ $(x \in E)$; $g(x) = \mathrm{Sup}(\langle x,x^*\rangle, 0)$ $(x \in E)$

where x^* is a nonzero element of the topological dual E^* of E . (Let us notice that this result is stated in the reference for a real Hausdorff locally convex topological vector space . But , K . Baron observed in a private communication that this result is also true for a general real Hausdorff topological vector space .)

We deduce then all the possible expressions of f given in Theorem 16 .

3. SUBGROUPOIDS OF (RxE , *)

In our problem , we consider the groupoïd $(\mathbf{R}\times E,*)$ where E is a real Hausdorff topological vector space and the binary operation * is defined by :

$$(a,x) * (b,y) = (\lambda ab\,,\, b^k x + a^l y) \qquad \text{for } (a,x),(b,y) \in \mathbf{R}\times E \qquad (L)$$

When we look for the subgroupoïds of $(\mathbf{R}\times E,*)$ which depend faithfully and continuously on a topological space F of parameters , we have to find (Definition 1) :

in the case (i) , all the continuous functions $f: E \to \mathbf{R}$ defined by : $f(\beta(u)) = \alpha(u)$ $(u \in F)$ which satisfy the functional equation (1) .

in the case (ii) , all the continuous functions $f: \mathbf{R} \to E$ defined by : $f(\alpha(u)) = \beta(u)$ $(u \in F)$ which satisfy the functional equation (2) .

The continuous solutions $f: E \to \mathbf{R}$ of (1) are given by Theorem 16 .

For the functional equation (2), we have the following result :

PROPOSITION 17 *All the solutions* $f: \mathbf{R} \dashrightarrow E$ *of the functional equation :*

$$f(\lambda xy) = y^k f(x) + x^l f(y) \qquad (x, y \in \mathbf{R}) \qquad (2)$$

are given by :

a) $f = 0$

b) if $k \neq l$ *and* $\lambda^k = \lambda^l = 1$

$f(x) = (x^l - x^k)\, v \quad (x \in \mathbf{R})$, *where* v *is an arbitrary nonzero element of* E.

c) if $k = l$ *and* $\lambda^l = 1$

(i) if $l = 0$, $f(x) = h(\lambda x) \quad (x \in \mathbf{R})$

where h *is a homomorphism from* $(\mathbf{R}, .)$ *into* $(E, +)$.

(ii) if $l > 0$, $f(x) = \begin{cases} x^l h(x) & \text{if } x \neq 0 \\ 0 & \text{if } x = 0 \end{cases}$

where h *is a homomorphism from* $(\mathbf{R} - \{0\}, .)$ *into* $(E, +)$.

d) if $k = l > 0$ *and* $\lambda^l = 2$

$f(x) = x^l v \quad (x \in \mathbf{R})$, *where* v *is an arbitrary nonzero element of* E.

Proof of Proposition 17

Let $f: \mathbf{R} \dashrightarrow E$ be a non identically zero solution of (2). By inverting x and y in (2), we get :

$$f(\lambda xy) = x^k f(y) + y^l f(x) \qquad (x, y \in \mathbf{R}) \qquad (2\,\text{bis})$$

(2) and (2 bis) imply :

$$(x^l - x^k)\, f(y) = (y^l - y^k)\, f(x) \qquad (x, y \in \mathbf{R})$$

If $k \neq l$, there exists a nonzero real number y_0 such that $y_0^{\,l} \neq y_0^{\,k}$. We deduce :

$f(x) = (x^{l} - x^{k})\ v \qquad (x \in \mathbf{R})$, where v is a nonzero element of E . It is easy to check that this function is a solution of (2) if , and only if , $\lambda^k = \lambda^l = 1$.

Let us suppose now $k = l$. If $\lambda = 0$, it is easy to see that f is identically zero , which is not the case . Therefore , λ is a nonzero real number . By taking $x = y = 1/\lambda$ in (2) , we get : $f(1/\lambda)(1 - 2/\lambda^l) = 0$, which implies either $f(1/\lambda) = 0$ or $\lambda^l = 2$.

Let us suppose that $f(1/\lambda) = 0$. By taking $y = 1/\lambda$ in (2) , we obtain : $f(x)\ (1 - 1/\lambda^l) = 0$ for every x in $\mathbf{R}$. Since f is not identically zero , this implies : $\lambda^l = 1$.

If $l = 0$, (2) becomes :

$f (\lambda xy) = f(x) + f(y) \qquad (x,y \in \mathbf{R})$ where λ is an arbitrary nonzero given real number .

Let us define : $h(x) = f(x/\lambda) \qquad (x \in \mathbf{R})$

Then , by (2) , h is a homomorphism from $(\mathbf{R},.)$ into $(E , +)$. This gives the solution c) (i) of (2) .

Let us suppose now $l > 0$. We define :

$$h(x) = f(x) / x^l \qquad (x \in \mathbf{R} - \{0\}) \tag{35}$$

f is a solution of (2) if , and only if , h satisfies the following functional equation :

$$h (\lambda xy) = h(x) + h(y) \qquad (x,y \in \mathbf{R} - \{0\}) \quad \text{with } \lambda^l = 1 \tag{36}$$

By taking $y = 1$ in (36) , we obtain :

$$h (\lambda x) = h(x) + h(1) \qquad (x \in \mathbf{R} - \{0\}) \tag{37}$$

Since $\lambda^2 = 1$, we get with $x = \lambda$ in (37) :

$h(1) = h(\lambda) + h(1)$ which implies : $h(\lambda) = 0$

Also , by taking $x = 1$ in (37) , we obtain :

$h(\lambda) = 0 = 2\,h(1)$ which implies : $h(1) = 0$

Therefore , (37) becomes : $h (\lambda x) = h(x) \qquad (x \in \mathbf{R} - \{0\})$, and we deduce from (36) that h is a homomorphism from $(\mathbf{R} - \{0\},.)$ into $(E , +)$. This gives the solution c) (ii) of (2) .

Finally, let us suppose $\lambda^l = 2$. Then, l must be a positive integer. f is a solution of (2) in this case if, and only if, the function h defined by (35) on $\mathbf{R} - \{0\}$ satisfies the following functional equation :

$$h(\lambda xy) = 1/2\,(h(x) + h(y)) \qquad (x,y \in \mathbf{R} - \{0\}) \tag{38}$$

Taking $y = 1/\lambda$ in (38), we see that h is a constant function. Therefore, we obtain : $f(x) = x^l v$ $(x \in \mathbf{R})$, where v is a nonzero element of E.

In the sequel, we shall denote by $\mathbf{R}^*$ the set of all nonzero real numbers.

From Theorem 16 and Proposition 17 and by using the expression of the continuous homomorphisms from $(\mathbf{R}^*, .)$ into $(E, +)$ [1], we deduce easily the following results :

COROLLARY 18 *Let λ be a nonzero real number.*
We consider the groupoïd $(\mathbf{R}^* \times E, *)$ where the binary law $*$ is defined by (L) with $k = l = 0$. All the subgroupoïds of $(\mathbf{R}^* \times E, *)$ which depend faithfully and continuously on a set of parameters, are :

the sets $G_{x^*} = \{(1/\lambda\ e^{\langle \beta, x^* \rangle}, \beta)\ ;\ \beta \in E\}$ where x^* is an element of the topological dual of E

and the sets $G_v = \{(\alpha, Log(|\lambda\alpha|).v)\ ;\ \alpha \in \mathbf{R}^*\}$ where v is an element of E.

COROLLARY 19 *Let k be a nonnegative integer, let l be a positive integer and let λ be a nonzero real number such that λ^l is different from 1 and 2.*
We consider the groupoïd $(\mathbf{R}^* \times E, *)$ where the binary law $*$ is defined by (L). All the subgroupoïds of $(\mathbf{R}^* \times E, *)$ which depend faithfully and continuously on a set of parameters, are :

the set $\{(1/\lambda, \beta)\ ;\ \beta \in E\}$

and the set $\{(\alpha, 0)\ ;\ \alpha \in \mathbf{R}^*\}$

COROLLARY 20 *Let k be a nonnegative integer, let l be a positive integer and let λ be a nonzero real number such that* $\lambda^l = 1$.

We consider the groupoïd $(\mathbf{R}^* \times E, *)$ *where the binary law* $*$ *is defined by (L). All the subgroupoïds of* $(\mathbf{R}^* \times E, *)$ *which depend faithfully and continuously on a set of parameters, are:*

the set $\{(1/\lambda, \beta) ; \beta \in E\}$

and, if $k \neq l$ and $\lambda^k = 1$,

the sets $G_v = \{(\alpha, (\alpha^l - \alpha^k) v) ; \alpha \in \mathbf{R}^*\}$

where v is an element of E

if $k = l$,

the sets $G_v = \{(\alpha, \alpha^l \operatorname{Log}(|\alpha|). v) ; \alpha \in \mathbf{R}^*\}$

where v is an element of E.

COROLLARY 21 *Let k be a nonnegative integer, let l be a positive integer and let λ be a nonzero real number such that* $\lambda^l = 2$.

We consider the groupoïd $(\mathbf{R}^* \times E, *)$ *where the binary law* $*$ *is defined by (L). All the subgroupoïds of* $(\mathbf{R}^* \times E, *)$ *which depend faithfully and continuously on a set of parameters, are:*

the set $\{(1/\lambda, \beta) ; \beta \in E\}$

the set $\{(\alpha, 0) ; \alpha \in \mathbf{R}^*\}$

and, if $k = l$, the sets $G_v = \{(\alpha, \alpha^l v) ; \alpha \in \mathbf{R}^*\}$

where v is an element of E

REFERENCES

1. **J. Aczél and J. Dhombres**, *Functional Equations in Several Variables*, Encyclopaedia for Mathematics, Cambridge University Press, Cambridge, 1989.

2. **W. Benz**, *The Cardinality of the set of Discontinuous Solutions of a class of Functional Equations*, Aequationes Mathematicae 32 (1987) p. 58-62.

3. **N. Brillouët and J. Dhombres**, *Equations Fonctionnelles et Recherche de Sous-Groupes*, Aequationes Mathematicae 31 (1986) p. 253-293.

4. **N. Brillouët-Belluot**, *On some Functional Equations of Gołąb-Schinzel type*, Aequationes Mathematicae, 42 (1991) p. 239-270

5. **N. Brillouët-Belluot**, *On the Solutions of the Functional Equation :* $f(f(y)^k x + f(x)^l y) = \lambda f(x) f(y)$, to appear in Publicationes Mathematicae Debrecen.

6. **A. M. Bruckner and J. G. Ceder**, *Darboux Continuity*, Jahresbericht Deutsche Mathematiker-Vereiningung 67 (1965) p. 93-117.

7. **J. Brzdęk**, *O Rozwiazaniach Rownania Funkcyjnego :* $f(x + f(x)^n y) = f(x) f(y)$, Uniwersytet Slaski, Instytut Matematyki, Katowice, 1990.

8. **J. Brzdęk**, *On the Solutions of the Functional Equation :* $f(x f(y)^l + y f(x)^k) = t f(x) f(y)$, Publicationes Mathematicae Debrecen 38/3-4 (1991) p. 175-183.

9. **J. Dhombres**, *Some Aspects of Functional Equations*, Lecture Notes, Chulalongkorn University (Bangkok, Thailand) Department of Mathematics (1979).

Functional Analysis, Approximation Theory
and Numerical Analysis
Ed. John M. Rassias

FUNCTIONAL EQUATIONS AND EXACT DISCRETE SOLUTIONS OF ORDINARY DIFFERENTIAL EQUATIONS

E. CASTILLO
Dept. of Applied Mathematics and Computational Sciences
University of Cantabria
Avd. de los Castros, 39005, Santander, SPAIN

and

R. RUIZ-COBO
Dept. of Applied Mathematics and Computational Sciences
University of Cantabria

ABSTRACT

The paper presents an application of functional equations to the discrete solution of ordinary differential equations. First, we deal with linear differential equations with constant coefficients, and from one special functional equation we design an algorithm to obtain the exact values of the solution at equally spaced points. Then, a method to obtain approximate and exact solutions and an equivalent functional equation of a linear differential equation is given.

1. Introduction

In this paper, two systems of differential and functional equations are said to be equivalent if they share the same set of solutions. The existence of equivalent systems of differential and functional equations, in the above sense, allows us not only to use differential equations for solving functional equations but also functional equations for solving differential equations. In the next two sections, we shall show how a functional equation, that it is equivalent to the whole family of linear differential equations with constant coefficients, can be used to obtain discrete exact solutions of a differential equation problem. In the last section, we shall explain how a sequence of approximate equations to a linear differential equation, in the sense of having approximate solutions on a grid, can be found.

To clarify the abovementioned, as well as the relations between functional and differential equations and their exact and approximate solutions, we include figure 1.

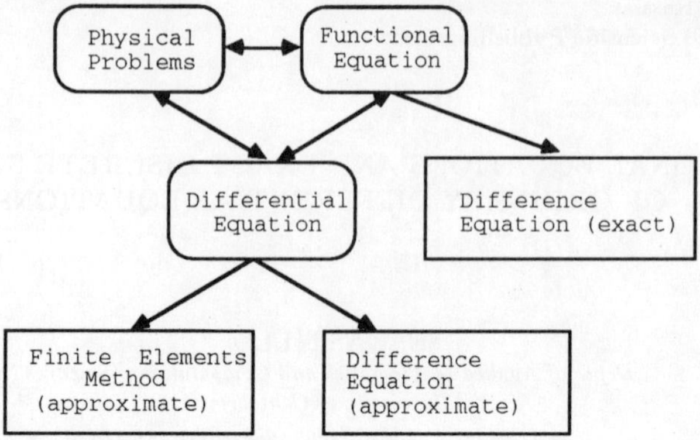

Figure 1: Illustration of the relation between differential equations, functional equations and their associated numerical methods.

The functional equation

$$h(x\Delta y) = \sum_{k=1}^{n} f_k(x)\, g_k(y) = \mathbf{g}^T(y) * \mathbf{f}(x)$$

$$\mathbf{f}(x) = \begin{pmatrix} f_1(x) \\ f_2(x) \\ \dots \\ f_n(x) \end{pmatrix} \; ; \; \mathbf{g}\,(y) = \begin{pmatrix} g_1(y) \\ g_2(y) \\ \dots \\ g_n(y) \end{pmatrix} \tag{1}$$

where Δ is any commutative internal law of composition defined on $\mathbf{R}$ and h, f_k and g_k $(k = 1, 2, ..., n)$ are unknown real functions of real variable such that the set of functions $\{f_1(x), f_2(x), \dots, f_n(x)\}$, on the one hand, and the set of functions $\{g_1(y), g_2(y), \dots, g_n(y)\}$, on the other, are linearly independent has been used by many researches as Stephanos ([6], [7]), Levi-Civita ([4]), Stäkel([5]), Aczél ([1]), etc. We show, without loss of generality, that Eq. 1 can be considerably simplified and we give several equivalent functional equations. Then we demonstrate that when Δ="+" it can be solved by its reduction to an homogeneous differential equation of order n with constant coefficients (see Aczél [1], pp. 197-199). Conversely, every solution $h(x)$ of an homogeneous differential equation of order n with constant coefficients satisfies Eq. 1. Finally, we show how functional equations can be used to identify the differential equation associated with a practical problem and to obtain exact discrete solutions. We also give an algorithm to obtain discrete exact solutions when the value of function $h(x)$ is known at $2n$ points and some method for identifying the coefficients of the associated differential equation.

2. Simplifications

The following theorem demonstrates that Eq. 1 can be simplified if we take into account the commutative property of Δ.

THEOREM 1 *(Symmetry).- Functional Eq. 1 can be written as*

$$h(x\Delta y) = \sum_{i,j=1}^{n} a_{ij}\, f_i(x)\, f_j(y) = \mathbf{f}^T(x)\mathbf{A}\mathbf{f}(y) \;\; ; \;\; a_{ij} = a_{ji} \;\; \forall i,j \tag{2}$$

where $\mathbf{A}$ *is the symmetric matrix of coefficients* a_{ij} $(i,j = 1,2,\ldots,n)$.

Proof: Because of the commutativity of Δ , we have

$$\begin{aligned} h(x\Delta y) &= \sum_{i=1}^{n} f_i(x)g_i(y) = \sum_{i=1}^{n} f_i(y)g_i(x) \Rightarrow \\ &\Rightarrow \sum_{i=1}^{n} [\, f_i(x)g_i(y) - f_i(y)g_i(x)] = 0 \end{aligned} \tag{3}$$

which is an equation of the form

$$\sum_{i=1}^{m} p_i(x)\, q_i(y) = 0 \tag{4}$$

Thus, according to a result of Aczél [1] (see Castillo and Ruiz-Cobo [2]):

$$\begin{pmatrix} \mathbf{f}(x) \\ \mathbf{g}(x) \end{pmatrix} = \begin{pmatrix} \mathbf{I} \\ \mathbf{A} \end{pmatrix} \mathbf{f}(x) \; ; \; \begin{pmatrix} \mathbf{g}(y) \\ -\mathbf{f}(y) \end{pmatrix} = \begin{pmatrix} \mathbf{B} \\ -\mathbf{I} \end{pmatrix} \mathbf{f}(y) \tag{5}$$

where $\mathbf{A}$ and $\mathbf{B}$ are constant matrices such that

$$(\mathbf{I} \;\; \mathbf{A}^T) \begin{pmatrix} \mathbf{B} \\ -\mathbf{I} \end{pmatrix} = \mathbf{0} \quad \Rightarrow \quad \mathbf{B} = \mathbf{A}^T \tag{6}$$

From Eq. 5 and Eq. 6, we get

$$\mathbf{g}(x) = \mathbf{A}\mathbf{f}(x) = \mathbf{B}\mathbf{f}(x) = \mathbf{A}^T\mathbf{f}(x) \;\Rightarrow\; \mathbf{A} = \mathbf{A}^T \tag{7}$$

Finally, substitution of Eq. 7 into Eq. 1 leads to Eq. 2. ∎

We shall now study the uniqueness of representation of Eq. 2, i.e. we try to answer the following question: given $h(x)$, is there a unique set of functions $\{f_1(x), f_2(x), \ldots, f_n(x)\}$ and a unique matrix $\mathbf{A}$, such that Eq. 2 is true?, and if the answer to this question is negative, what is the relation between different solution sets of functions and matrices? The answer to the above questions is given by the following theorem.

THEOREM 2 .- *If there are two sets of linearly independent functions* $\{f_1(x), f_2(x), \ldots, f_n(x)\}$ *and* $\{f_1^*(x), f_2^*(x), \ldots, f_n^*(x)\}$ *and two symmetric matrices* $\mathbf{A}$ *and* $\mathbf{A}^*$ *such that*

$$h(x\Delta y) = \sum_{i,j=1}^{n} a_{ij}\ f_i(x)\ f_j(y) = \sum_{i,j=1}^{n} a_{ij}^*\ f_i^*(x)\ f_j^*(y) \tag{8}$$

then there exists a regular constant matrix $\mathbf{B}$ *of order* n *such that*

$$\mathbf{f}^*(x) = \mathbf{B}\mathbf{f}(x)\ ;\ \mathbf{A} = \mathbf{B}^T\mathbf{A}^*\mathbf{B} \tag{9}$$

where $\mathbf{A}$ *and* $\mathbf{A}^*$ *are the matrices of coefficients* a_{ij} *and* a_{ij}^**, respectively.*

Proof: In effect, equation

$$\sum_{i,j=1}^{n} a_{ij}\ f_i(x)\ f_j(y) = \sum_{i,j=1}^{n} a_{ij}^*\ f_i^*(x)\ f_j^*(y) \tag{10}$$

can be written as

$$\sum_{i=1}^{n}\left[f_i(x)\sum_{j=1}^{n} a_{ij}\ f_j(y)\right] - \sum_{i=1}^{n}\left[f_i^*(x)\sum_{j=1}^{n} a_{ij}^*\ f_j^*(y)\right] = 0 \tag{11}$$

which is of the form Eq. 4. Thus, we have (see Aczél [1])

$$\begin{pmatrix} \mathbf{f}(x) \\ -\mathbf{f}^*(x) \end{pmatrix} = \begin{pmatrix} \mathbf{I} \\ -\mathbf{B} \end{pmatrix} \mathbf{f}(x)\ ;\ \begin{pmatrix} \mathbf{A}\mathbf{f}(y) \\ \mathbf{A}^*\mathbf{f}^*(y) \end{pmatrix} = \begin{pmatrix} \mathbf{A} \\ \mathbf{C} \end{pmatrix} \mathbf{f}(y) \tag{12}$$

where $\mathbf{B}$ and $\mathbf{C}$ are non-singular constant matrices satisfying the equation

$$(\mathbf{I}\quad -\mathbf{B}^T)\begin{pmatrix} \mathbf{A} \\ \mathbf{C} \end{pmatrix} = \mathbf{0} \quad \Rightarrow \quad \mathbf{A} - \mathbf{B}^T\mathbf{C} = \mathbf{0} \tag{13}$$

From Eq. 12 and Eq. 13, we get

$$\mathbf{A}^*\mathbf{B}\mathbf{f}(x) = \mathbf{A}^*\mathbf{f}^*(x) = \mathbf{C}\mathbf{f}(x) = \mathbf{B}^{T^{-1}}\mathbf{A}\mathbf{f}(x) \Rightarrow \mathbf{A} = \mathbf{B}^T\mathbf{A}^*\mathbf{B} \tag{14}$$

■

COROLLARY 1 .- *Functional Eq. 1 can be written as*

$$h(x\Delta y) = \sum_{i=1}^{p} f_i(x)\ f_i(y) - \sum_{i=p+1}^{n} f_i(x)\ f_i(y) = \mathbf{f}^T(x)\begin{pmatrix} \mathbf{I}_p & \mathbf{0} \\ \mathbf{0} & -\mathbf{I}_q \end{pmatrix}\mathbf{f}(y) \tag{15}$$

where $p + q = n$.

Proof: Expression on the right of Eq. 14 shows that matrices **A** and $\mathbf{A}^*$ are congruent, but we know that any non-singular symmetric matrix $\mathbf{A}^*$ of rank n can be transformed by congruence to a matrix of the form

$$\mathbf{D} = \begin{pmatrix} \mathbf{I}_p & \mathbf{0} \\ \mathbf{0} & -\mathbf{I}_q \end{pmatrix} \quad ; \quad p + q = n \tag{16}$$

and then Eq. 15 holds. ■

In the following we assume Δ="+" and we demonstrate that equation

$$h(x+y) = \mathbf{f}^T(y)\,\mathbf{D}\mathbf{f}(x) \tag{17}$$

is equivalent to an homogeneous differential equation.

Taking separated derivatives with respect to x and y in Eq. 17 and equaling we get

$$h'(x+y) = \mathbf{f}^T(y)\,\mathbf{D}\mathbf{f}'(x) = \mathbf{f}'^T(y)\,\mathbf{D}\mathbf{f}(x) \tag{18}$$

Due to the fact that the set of functions $\{f_1(y), f_2(y), \ldots, f_n(y)\}$ is linearly independent, there exist constants y_m $(m = 1, 2, \ldots, n)$ such that det $f_k(y_m) \neq 0$. Consequently, it can be written

$$h'(x+y_m) = \mathbf{f}^T(y_m)\mathbf{D}\mathbf{f}'(x) = \mathbf{f}'^T(y_m)\,\mathbf{D}\mathbf{f}(x) \;\; ; \;\; m = 1, 2, ..., n \tag{19}$$

which in matrix form becomes

$$\mathbf{f}'(x) = \mathbf{G}^{-1}\mathbf{G}'\mathbf{f}(x) = \mathbf{F}\mathbf{f}(x) \quad ; \quad \mathbf{F} = \mathbf{G}^{-1}\mathbf{G}' \tag{20}$$

where **G** and **G**$'$ are matrices with elements depending on **D** and $f_m(y_k)$ and $f'_m(y_k)$, respectively.

Making now $y = 0$ in Eq. 17 we get

$$h(x) = \mathbf{f}^T(0)\mathbf{D}\mathbf{f}(x) = \mathbf{C}\mathbf{f}(x) \quad ; \quad \mathbf{C} = \mathbf{f}^T(0)\mathbf{D} \tag{21}$$

and taking derivatives and using Eq. 20 we get

$$\begin{aligned} h'(x) &= \mathbf{C}\mathbf{f}'(x) = \mathbf{C}\mathbf{F}\mathbf{f}(x) \\ h''(x) &= \mathbf{C}\mathbf{F}\mathbf{f}'(x) = \mathbf{C}\mathbf{F}\mathbf{F}\mathbf{f}(x) = \mathbf{C}\mathbf{F}^2\mathbf{f}(x) \\ &\ldots\ldots\ldots\ldots\ldots\ldots\ldots\ldots\ldots\ldots\ldots \\ h^{(n)}(x) &= \mathbf{C}\mathbf{F}^n\mathbf{f}(x) \end{aligned} \tag{22}$$

which, in matrix form, becomes

$$\mathbf{H}(x) = \begin{pmatrix} h'(x) \\ h''(x) \\ \ldots \\ h^{(n)}(x) \end{pmatrix} = \begin{pmatrix} \mathbf{CF} \\ \mathbf{CF}^2 \\ \ldots \\ \mathbf{CF}^n \end{pmatrix} \mathbf{f}(x) = \mathbf{U}\mathbf{f}(x) \tag{23}$$

Finally, from Eqs. 21 and 23, taking into account that the above functions are linearly independent, we have

$$\mathbf{d}\begin{pmatrix} h(x) \\ h'(x) \\ h''(x) \\ \dots \\ h^{(n)}(x) \end{pmatrix} = 0 \tag{24}$$

where $\mathbf{d}$ is a constant nonzero vector. Thus Eq. 24 is a homogeneous differential equation of order n with constant coefficients.

We now show that every solution $h(x)$ of an homogeneous differential equation of order n with constant coefficients satisfies Eq. 17. In effect, every solution is of the form

$$h(x) = \sum_{k=1}^{m} P_k(x)\ \exp\{w_k x\} \tag{25}$$

where w_k, $(k = 1, 2, ..., m)$ are complex constants (the roots of the characteristic equation) and $P_k(x)$ are polynomials of degree $(n_k - 1)$ where $\sum_{k=1}^{m} n_k = n$.

From Eq. 25, we get

$$\begin{aligned} h(x+y) &= \sum_{k=1}^{m} P_k(x+y)\exp\{w_k(x+y)\} = \\ &= \sum_{k=1}^{n} \lambda_k x^{\alpha_k} y^{\beta_k}\ \exp\{w_k x\}\exp\{w_k y\} = \sum_{k=1}^{n} f_k(x)\ g_k(y) \end{aligned} \tag{26}$$

Thus, Eq. 17 gives a representation of every solution $h(x)$ of an homogeneous differential equation of order n with constant coefficients.

3. Exact Discrete and Numerical Solutions

In the following we call $h_n = h(n\Delta x)$ and we consider that functions $h(x)$ and $f_i(x)$ $(i = 1, 2, \dots, n)$ are defined on a discrete subset of $\mathbf{R}$ $\{n\Delta x, n \in Z\}$. In order to have a unique solution for Eq. 15 we also assume that $h_0, \dots, h_{2n-1}$ are known. Note that we give $2n$ values because Eq. 15 is equivalent not to a single differential equation but to the familly of all differential equations and the constant coefficients should be determined by data.

From Eq. 15, we have

$$h_{m+n} = \mathbf{f}^T(n\Delta x)\begin{pmatrix} \mathbf{I}_p & \mathbf{0} \\ \mathbf{0} & -\mathbf{I}_q \end{pmatrix}\mathbf{f}(m\Delta x) \tag{27}$$

where $y = m\Delta x$ and $x = n\Delta x$ and

$$\begin{pmatrix} h_m \\ h_{m+1} \\ \dots \\ h_{m+n-1} \end{pmatrix} = \mathbf{F}(n\Delta x) \begin{pmatrix} \mathbf{I}_p & \mathbf{0} \\ \mathbf{0} & -\mathbf{I}_q \end{pmatrix} \mathbf{f}(m\Delta x) \tag{28}$$

where

$$\mathbf{F}(n\Delta x) = \begin{pmatrix} f_1(0) & f_2(0) & \dots & f_n(0) \\ f_1(\Delta x) & f_2(\Delta x) & \dots & f_n(\Delta x) \\ \dots & \dots & \dots & \dots \\ f_1[(n-1)\Delta x] & f_2[(n-1)\Delta x] & \dots & f_n[(n-1)\Delta x] \end{pmatrix} \tag{29}$$

Due to the non-singularity of **F**, from Eq. 27 and Eq. 28, we can write

$$h_{m+n} = \mathbf{K}(n\Delta x) \begin{pmatrix} h_m \\ h_{m+1} \\ \dots \\ h_{m+n-1} \end{pmatrix} \quad ; \quad \mathbf{K}(n\Delta x) = \mathbf{f}^T(n\Delta x)\mathbf{F}^{-1}(n\Delta x) \tag{30}$$

and taking into account Eq. 28 for $m = n$, and Eq. 30 we have

$$\mathbf{K}(n\Delta x) = \begin{pmatrix} h_n & h_{n+1} & \dots & h_{2n-1} \end{pmatrix} \left(\mathbf{F}(n\Delta x) \begin{pmatrix} \mathbf{I}_p & \mathbf{0} \\ \mathbf{0} & -\mathbf{I}_q \end{pmatrix} \mathbf{F}^T(n\Delta x) \right)^{-1} \tag{31}$$

but making $m = 0, 1, 2, \dots, n-1$ in Eq. 28, we get

$$\mathbf{F}(n\Delta x) \begin{pmatrix} \mathbf{I}_p & \mathbf{0} \\ \mathbf{0} & -\mathbf{I}_q \end{pmatrix} \mathbf{F}^T(n\Delta x) = \begin{pmatrix} h_0 & h_1 & \dots & h_{n-1} \\ h_1 & h_2 & \dots & h_n \\ \dots & \dots & \dots & \dots \\ h_{n-1} & h_n & \dots & h_{2n-2} \end{pmatrix} \tag{32}$$

and then

$$h_{m+n} = \\ = (h_n h_{n+1} \dots h_{2n-1}) \begin{pmatrix} h_0 & h_1 & \dots & h_{n-1} \\ h_1 & h_2 & \dots & h_n \\ \dots & \dots & \dots & \dots \\ h_{n-1} & h_n & \dots & h_{2n-2} \end{pmatrix}^{-1} \begin{pmatrix} h_m \\ h_{m+1} \\ \dots \\ h_{m+n-1} \end{pmatrix} \tag{33}$$

for $m \geq n$, which is a difference equation of order n.

Hence, exact discrete solutions of Eq. 17 can be obtained by difference Eq. 33. Consequently, *given an homogeneous differential equation of order n, there exists a difference equation of the same order such that their solution coincide at the common* *points.* It is remarkable that *we can obtain the unknown differential equation governing a problem if we know experimental values on a large enough set of equally spaced points.* Note that the difference Eq. 33 depends on Δx and the coefficients of the differential equation.

Eq. 33 can be written as

$$h(x+ny) = \{\, h(ny)\, h[(n+1)y] \ldots h[(2n-1)y]\,\}\bullet$$

$$\bullet \begin{pmatrix} h(0) & h(y) & \ldots & h[(n-1)y] \\ h(y) & h(2y) & \ldots & h(ny) \\ \ldots & \ldots & \ldots & \ldots \\ h[(n-1)y] & h(ny) & \ldots & h[(2n-2)y] \end{pmatrix}^{-1} \begin{pmatrix} h(x) \\ h(x+y) \\ \ldots \\ h[x+(n-1)y] \end{pmatrix} \qquad (34)$$

which is a functional equation equivalent to Eq. 1 with $\Delta =$"+".

For a differential and a difference equations to have the same solution at common points, their characteristic equations must be of the form

$$\prod_{i=1}^{n} (x-\alpha_i) = 0 \quad \text{and} \quad \prod_{i=1}^{n} [x - \exp(\alpha_i \Delta x)] = 0, \qquad (35)$$

respectively. Thus, once one of both equations is known, the other can be immediately obtained from Eq. 35.

EXAMPLE 1 *(Castillo, Ruiz-Dávila and Ruiz-Cobo [3]).- Let us consider the differential equation of a string on an elastic foundation with no load on it:*

$$h''(x) - \frac{K}{T}\, h(x) = 0 \qquad (36)$$

where $h(x)$ is the vertical displacement of the string at the point x, K is the Winkler constant and T is the horizontal tension in the string. In order to simplify we assume that $K/T = 1$.

The exact solution for the case $h(0) = 0; h(1) = 1$ is

$$h(x) = \frac{\exp(x) - \exp(-x)}{\exp(1) - \exp(-1)} \qquad (37)$$

An approximation to Eq. 36, by means of the finite difference method is

$$h_{n+1} - [\,2 + (\Delta x)^2]\, h_n + h_{n-1} = 0 \;;\; h_n = h(n\,\Delta x) \qquad (38)$$

which has as characteristic equation and roots

$$r^2 - (2+\Delta^2 x)\, r + 1 = 0$$

$$r_1 = \frac{2 + \Delta^2 x + \Delta x\sqrt{4+\Delta^2 x)}}{2} \;;\; r_2 = \frac{2 + \Delta^2 x - \Delta x\sqrt{4+\Delta^2 x)}}{2} \qquad (39)$$

Thus, its general differentiable solution is

$$h_n = C_1 \left(\frac{2+\Delta^2 x + \Delta x\sqrt{4+\Delta^2 x}}{2}\right)^n + C_2 \left(\frac{2+\Delta^2 x - \Delta x\sqrt{4+\Delta^2 x}}{2}\right)^n \tag{40}$$

which for $h(0) = 0; h(1) = 1$ *becomes*

$$h_n = \frac{1}{\Delta x\sqrt{4+\Delta^2 x}} \left(\frac{2+\Delta^2 x + \Delta x\sqrt{4+\Delta^2 x}}{2}\right)^n - \\ - \frac{1}{\Delta x\sqrt{4+\Delta^2 x}} \left(\frac{2+\Delta^2 x - \Delta x\sqrt{4+\Delta^2 x}}{2}\right) \tag{41}$$

Assume now that we do not know what the differential equation governing the problem of a string on an elastic foundation with no load on it is, but we run an experiment and we measure the displacements at equally spaced points, say $h_0, h_1, h_2, \ldots, h_p$. *The recurrence formula Eq. 33 with* $n = 1, 2, 3, \ldots$ *allows us to obtain the value of the first integer* n *compatible with the* $h_0, h_1, h_2, \ldots, h_p$ *values and the coefficients of the difference Eq. 33. Then, from Eq. 35, the differential equation governing our problem can be easily obtained.*

As one example, let us assume that we know the exact values

$$h_i = \frac{e^{i\Delta x} - e^{-i\Delta x}}{e - e^{-1}} \; ; i = 0, 1, ..., 5 \tag{42}$$

Then, Eq. 33 for $n = 1$ *is not satisfied for* h_2, *but the same equation for* $n = 2$ *becomes*

$$h_{m+2} = (e^{\Delta x} + e^{-\Delta x}) h_{m+1} - h_m \tag{43}$$

which is satisfied for h_4 *and* h_5 *and then we can conclude that* $n = 2$ *and that Eq. 43 is the difference equation leading to exact values for all the discrete points.*

The characteristic roots of Eq. 43 are $e^{\Delta x}$ *and* $e^{-\Delta x}$ *and, according to Eq. 35, the characteristic roots of the associated differential equation are* $+1$ *and* -1. *Thus, Eq. 36 with* $k/T = 1$ *is implied.*

Note that Eq. 43 shows that Eq. 37 is one solution of the functional equation

$$h(x+2y) = (e^y + e^{-y}) h(x+y) - h(x) \tag{44}$$

Table 1 shows different exact and approximate solutions obtained by Eq. 37, Eq. 38 and Eq. 43 for $\Delta x = 0.4$.

x	Exact solution Eq. 37	Exact solution of difference Eq. 43	Approximate solution of difference Eq. 38
0.0	0.00000	0.00000	0.00000
0.4	0.34952	0.34952	0.34952
0.8	0.75571	0.75571	0.75496
1.2	1.28443	1.28443	1.28119
1.6	2.02141	2.02141	2.01241
2.0	3.08616	3.08616	3.06562
2.4	4.65131	4.65131	4.60933
2.8	6.97065	6.97065	6.89052

Table 1: Three different solutions of the example equation

The coincidence of the solutions of the differential and the difference equations at the common points is not casual. In fact, if we assume an infinitely differentiable solution of the functional equation

$$\sum_{i=0}^{n} a_i y^{(n-i)}(x) = 0, \tag{45}$$

using the Taylor expansion, we can write

$$y(x+\Delta_j) \approx \sum_{k=0}^{m} \frac{\Delta_j^k y^{(k)}(x)}{k} \; ; \; j = 1, ..., n \tag{46}$$

and by derivation of Eq. 45 $(m-n)$ times, we get

$$\sum_{i=0}^{n} a_i y^{(n-i+s)}(x) = 0 \; ; \; s = 0, 1, ..., m-n \tag{47}$$

The system (46)-(47), independently of the value of m, allows us to eliminate all m derivatives of y and to obtain a difference equation of order n. When m tends to infinity we get the coincidence of solutions. Note that the order of the difference equation remains constant when m increases.

On the other hand, it is well known that the general solution of the differential Eq. 45 can be written as

$$y(x) = \sum_{k=1}^{m} P_k(x) \exp\{w_k x\} \tag{48}$$

and the general solution of a difference equation as

$$y(x) = \sum_{k=1}^{m} P_k(x) w_k^x \tag{49}$$

where w_k $(k = 1, 2, \ldots, n)$ are the solutions of their characteristic equations. Note that both equations are of exactly the same form.

Until now, we have been working with linear differential equations of constant coefficients. But, could we do a similar thing if the coefficients of the equations are not constants?

4. From Differential Equations to Functional Equations

We start from a differential equation and we look for an equivalent functional equation. We only study the case of the following linear ordinary differential equation

$$\sum_{i=0}^{n} a_i(x) f^{(n-i)}(x) = h(x) \tag{50}$$

where f, h, a_i $(i = 0, \ldots, n)$ are infinitely differentiable functions in a certain domain D. Without loss of generality we can assume $a_0(x) = 1$.

We also assume that the value of f is known at n points of the domain

$$D : \{p_j = x + \Delta_j \,, j = 1, \ldots, n\} \,.$$

Using the Taylor expansion we have

$$f(x + \Delta_j) = \sum_{k=0}^{m} \frac{\Delta_j^k f^{(k)}(x)}{k!} + O(\Delta_j^{(m+1)}(x)) \quad ; \quad \forall j = 1, \ldots, n \tag{51}$$

where we assume $m > n$.
By differentiating $(m - n - 1)$ times Eq. 50, we get

$$\sum_{i=0}^{n+k-1} A_{ki}(x) f^{(n-i+k-1)}(x) = h^{(k-1)}(x); \; k = 1, 2, ..., m - n + 1 \tag{52}$$

where the upper index denote the order of derivation and the functions A_{ki} $(k = 1, 2, \ldots, m - n + 1)$ are given by :

$$A_{1i}(x) = a_i(x) \; ; \; i = 0, 1, \ldots, n$$

$$A_{(k+1)i}(x) = \begin{cases} A_{ki}(x) & if \;\; i = 0 \\ A_{ki}(x) + A'_{k(i-1)}(x) & if \;\; i = 1, 2, \ldots, n + k - 1 \\ A'_{k(n+k-1)}(x) & if \;\; i = n + k \end{cases}$$

It is worthwhile mentioning that because $a_0(x) = 1$, the first coefficient of all equations in expression 52 is equal to 1.

Eq. 51, without the complementary term, and Eq. 52 can be written, in matrix form, as

$$\begin{pmatrix} \mathbf{N} & \mathbf{D} \\ \mathbf{C} & \mathbf{B} \end{pmatrix} \cdot \mathbf{F} = \begin{pmatrix} \mathbf{H} \\ \Delta\mathbf{F} \end{pmatrix} \tag{53}$$

where $\mathbf{N}, \mathbf{D}, \mathbf{C}, \mathbf{B}, \mathbf{F}, \mathbf{H}$, and $\Delta\mathbf{F}$ are the following matrices:

$$\mathbf{F} = \begin{pmatrix} f^{(m)}(x) \\ \vdots \\ \vdots \\ \vdots \\ f^{(n)}(x) \\ f^{(n-1)}(x) \\ \vdots \\ \vdots \\ f(x) \end{pmatrix} \quad ; \quad \mathbf{H} = \begin{pmatrix} h(x) \\ h'(x) \\ \vdots \\ \vdots \\ \vdots \\ h^{(m-n)}(x) \end{pmatrix} \quad ; \quad \Delta\mathbf{F} = \begin{pmatrix} f(x+\Delta_n) \\ \vdots \\ \vdots \\ f(x+\Delta_1) \end{pmatrix}$$

$$\mathbf{N} = \begin{pmatrix} 0 & 0 & \cdots & \cdots & \cdots & \cdots & 1 \\ 0 & 0 & \cdots & \cdots & \cdots & 1 & A_{21} \\ \cdots & \cdots & \cdots & \cdots & \cdots & \cdots & \cdots \\ 0 & 0 & \cdots & 1 & A_{k1} & \cdots & A_{k(k-1)} \\ \cdots & \cdots & \cdots & \cdots & \cdots & \cdots & \cdots \\ 0 & 1 & \cdots & \cdots & \cdots & \cdots & A_{(m-n)(m-n-1)} \\ 1 & A_{(m-n+1)1} & \cdots & \cdots & \cdots & \cdots & A_{(m-n+1)(m-n)} \end{pmatrix}$$

$$\mathbf{D} = \begin{pmatrix} A_{11} & \cdots & A_{1n} \\ \cdots & \cdots & \cdots \\ A_{kk} & \cdots & A_{k(k+n-1)} \\ \cdots & \cdots & \cdots \\ A_{(m-n)(m-n)} & \cdots & A_{(m-n)(m-1)} \\ A_{(m-n+1)(m-n+1)} & \cdots & A_{(m-n+1)m} \end{pmatrix}$$

$$\mathbf{C} = \begin{pmatrix} \frac{\Delta_n^m}{m!} & \frac{\Delta_n^{(m-1)}}{(m-1)!} & \cdots & \frac{\Delta_n^n}{n!} \\ \cdots & \cdots & \cdots & \cdots \\ \frac{\Delta_1^m}{m!} & \frac{\Delta_1^{(m-1)}}{(m-1)!} & \cdots & \frac{\Delta_1^n}{n!} \end{pmatrix} \quad ; \quad \mathbf{B} = \begin{pmatrix} \frac{\Delta_n^{(n-1)}}{(n-1)!} & \cdots & \Delta_n & 1 \\ \cdots & \cdots & \cdots & \cdots \\ \frac{\Delta_1^{(n-1)}}{(n-1)!} & \cdots & \Delta_1 & 1 \end{pmatrix}$$

where the explicit dependence of matrices $\mathbf{N}$ and $\mathbf{D}$ on x has been omitted for the sake of clarity. From Eq. 53 we can eliminate all derivatives of the function f and get a functional equation. This is what we do in the following paragraphs.

First, we row-manipulate the matrix $(\mathbf{N}\ \mathbf{D})$ in order to transform the matrix $\mathbf{N}$ into an inverse unit diagonal matrix $\mathbf{P}$. These transformations produce some modifications in matrices **D** and **H**, which become $\mathbf{D}^*$ and $\mathbf{H}^*$:

$$\mathbf{M} = \begin{pmatrix} \mathbf{N} & \mathbf{D} \\ \mathbf{C} & \mathbf{B} \end{pmatrix} \approx \begin{pmatrix} \mathbf{P} & \mathbf{D}^* \\ \mathbf{C} & \mathbf{B} \end{pmatrix} \; ; \; \mathbf{G} = \begin{pmatrix} \mathbf{H} \\ \Delta\mathbf{F} \end{pmatrix} \approx \begin{pmatrix} \mathbf{H}^* \\ \Delta\mathbf{F} \end{pmatrix}$$

where

$$\mathbf{P} = \begin{pmatrix} 0 & 0 & \cdots & 0 & 1 \\ 0 & 0 & \cdots & 1 & 0 \\ \cdots & \cdots & \cdots & \cdots & \cdots \\ 0 & 1 & \cdots & 0 & 0 \\ 1 & 0 & \cdots & 0 & 0 \end{pmatrix}$$

Next, we transform matrix **C** into the null matrix by row-manipulations of matrices **M** and **G**. It is easy to check that this is equivalent to making the following transformation

$$\mathbf{B}^* = \mathbf{B} - \mathbf{CPD}^* \quad \text{and} \quad \Delta\mathbf{F}^* = \Delta\mathbf{F} - \mathbf{CPH}^*.$$

With this, the system (53) becomes equivalent to the system

$$\begin{pmatrix} \mathbf{P} & \mathbf{D}^* \\ \mathbf{0} & \mathbf{B}^* \end{pmatrix} \cdot \mathbf{F} = \begin{pmatrix} \mathbf{H}^* \\ \Delta\mathbf{F}^* \end{pmatrix} \tag{54}$$

From Eq. 54, we get

$$\mathbf{B}^* \cdot \begin{pmatrix} f^{(n-1)}(x) \\ \vdots \\ f(x) \end{pmatrix} = \begin{pmatrix} f(x+\Delta_n) \\ \vdots \\ f(x+\Delta_1) \end{pmatrix} + \mathbf{K} \tag{55}$$

where $\mathbf{K} = -\mathbf{CPH}^*$.
Now, from Eq. 55, we get

$$f(x) - \sum_{j=1}^{n} r^{n-j+1} \cdot f(x+\Delta_j) = \sum_{j=1}^{n} r^j \cdot k_j \tag{56}$$

where $(r^1 \ldots \quad r^n)$ is the last row of the matrix $\mathbf{B}^{*-1}$, that is,

$$r^j = \frac{(-1)^{n+j} \cdot b^{jn}}{\det \, \mathbf{B}^*} \quad , \text{ with } b^{\,jn} = \text{Adjoint}_{(j,n)} \text{ of } \mathbf{B}^*.$$

In this way, we obtain a difference Eq. 56 of order n, which approximate Eq. 50.

In addition, once the manipulations have been performed for a given value of m, if one wants to do the same process for $m+p$, one can start from the manipulated matrices $\mathbf{N}^*$ and $\mathbf{D}^*$ instead of starting from the initial **N** and **D** matrices, with the corresponding saving in computational time.

If we increase the value of m we shall get a better approximation. However, this can be done without increasing the value of n. In other words, by increasing m we get a sequence of difference equations of order n which approximate the initial differential equation. In the limit, we shall obtain a difference equation which is an exact replicate of the starting differential equation in the sense that it gives the same solutions at the grid points.

But we can go even further, because Eq. 56 can be interpreted as one functional equation in the variables $(x, \Delta_1, \Delta_2, ..., \Delta_n)$ and then we get a functional equation which is equivalent to Eq. 50.

Below, we give some examples.

EXAMPLE 2 .- *We apply the above method to the following differential equation*

$$xf'(x) - kf(x) = 0$$

where k is a given constant.

In this case, because $n = 1$, we take a single point $x + \Delta$.

With $m = 3$, the matrices in Eq. 53 are

$$\mathbf{N} = \begin{pmatrix} 0 & 0 & 1 \\ 0 & 1 & -\frac{k}{x} \\ 1 & -\frac{k}{x} & \frac{2k}{x^2} \end{pmatrix} \; ; \; \mathbf{D} = \begin{pmatrix} -\frac{k}{x} \\ \frac{k}{x^2} \\ -\frac{2k}{x^3} \end{pmatrix} \; ; \; \mathbf{H} = \begin{pmatrix} 0 \\ 0 \\ 0 \end{pmatrix}$$

$$\mathbf{C} = \begin{pmatrix} \frac{\Delta^3}{3} & \frac{\Delta^2}{2} & \Delta \end{pmatrix} \; ; \; \mathbf{B} = (1) \; ; \; \Delta\mathbf{F} = (f(x+\Delta))$$

After manipulating matrix **N** *for the first time we get*

$$\mathbf{D}^* = \begin{pmatrix} -\frac{k}{x} \\ -\frac{k(k-1)}{x^2} \\ -\frac{k(k-1)(k-2)}{x^3} \end{pmatrix}$$

and after making the matrix **C** *null, it results*

$$\mathbf{B}^* = \left(1 + k\frac{\Delta}{x} + \frac{k(k-1)}{2!}\left(\frac{\Delta}{x}\right)^2 + \frac{k(k-1)(k-2)}{3!}\left(\frac{\Delta}{x}\right)^3\right)$$

Due to the fact that we have an homogeneous differential equation, the matrices on the right hand side do not suffer any transformation and we get the difference equation

$$f(x) = \mathbf{B}^{*-1} f(x+\Delta)$$

It is easy to check that when increasing the value of m the added terms in matrix **B** *are of the form*

$$\frac{k(k-1)\dots(k-m+1)}{m\,!}\left(\frac{\Delta}{x}\right)^m$$

Thus, in the limit, we have

$$\mathbf{B}^* = (1 + \frac{\Delta}{x})^k$$

and then the functional equation equivalent to the initial differential equation is

$$f(x) = (1 + \frac{y}{x})^{-k} f(x+y)$$

and the difference equation

$$f(x) = (1 + \frac{\Delta}{x})^{-k} f(x+\Delta)$$

EXAMPLE 3 .- *We now apply the above method to the equation*

$$f'(x) - \frac{k}{x} f(x) = x^2$$

which is a complete equation associated with the homogeneous equation in example 2.

For $m = 3$, *all matrices are of the same form as before with the exception of matrix* $\mathbf{H}$, *which now becomes*

$$\mathbf{H} = \begin{pmatrix} x^2 \\ 2x \\ 2 \end{pmatrix}$$

Thus, after the first manipulation we get

$$\mathbf{H}^* = \begin{pmatrix} x^2 \\ (k+2)x \\ k^2+2 \end{pmatrix}$$

and after making the matrix $\mathbf{C}$ *null, we obtain*

$$\Delta\mathbf{F}^* = \Delta\mathbf{F} + \mathbf{K} \quad ; \quad \mathbf{K} = \left(-\Delta\, x^2 - \frac{\Delta^2}{2\,!}\,(k+2)\,x - \frac{\Delta^3}{3\,!}\,(k^2+2) \right)$$

where $\mathbf{D}^*$ *and* $\mathbf{B}^*$ *are the matrices indicated in the previous example.*

Thus, the approximate difference equation becomes

$$f(x) - \mathbf{B}^{*-1} f(x+\Delta) = \mathbf{B}^{*-1}\mathbf{K}$$

Finally, after some calculations, for m going to infinity we get the functional equation

$$f(x) - (1 + \frac{y}{x})^{-k} f(x+y) = \frac{x^3}{3-k} \left[1 - (1 + \frac{y}{x})^{3-k} \right]$$

and the difference equation

$$f(x) - (1 + \frac{\Delta}{x})^{-k} f(x+\Delta) = \frac{x^3}{3-k} \left[1 - (1 + \frac{\Delta}{x})^{3-k} \right]$$

EXAMPLE 4 .- *Now we deal with the constant coefficients linear equation*

$$f''(x) - f(x) = 0$$

We consider $n = 2$, *i.e.* $\{x + \Delta_1 , x + \Delta_2\}$, *and m=6. Thus, we have*

$$\mathbf{N} = \begin{pmatrix} 0 & 0 & 0 & 0 & 1 \\ 0 & 0 & 0 & 1 & 0 \\ 0 & 0 & 1 & 0 & -1 \\ 0 & 1 & 0 & -1 & 0 \\ 1 & 0 & -1 & 0 & 0 \end{pmatrix} \quad ; \quad \mathbf{D} = \begin{pmatrix} 0 & -1 \\ -1 & 0 \\ 0 & 0 \\ 0 & 0 \\ 0 & 0 \end{pmatrix}$$

$$\mathbf{C} = \begin{pmatrix} \frac{\Delta_2^6}{6!} & \frac{\Delta_2^5}{5!} & \cdots & \frac{\Delta_2^2}{2!} \\ \frac{\Delta_1^6}{6!} & \frac{\Delta_1^5}{5!} & \cdots & \frac{\Delta_1^2}{2!} \end{pmatrix} \quad ; \quad \mathbf{B} = \begin{pmatrix} \Delta_2 & 1 \\ \Delta_1 & 1 \end{pmatrix}$$

and **H** *is the column null matrix of dimension 5.*

After all the manipulations, we get

$$\mathbf{B}^* = \begin{pmatrix} \Delta_2 + \frac{\Delta_2^3}{3!} + \frac{\Delta_2^5}{5!} & 1 + \frac{\Delta_2^2}{2!} + \frac{\Delta_2^4}{4!} + \frac{\Delta_2^6}{6!} \\ \Delta_1 + \frac{\Delta_1^3}{3!} + \frac{\Delta_1^5}{5!} & 1 + \frac{\Delta_1^2}{2!} + \frac{\Delta_1^4}{4!} + \frac{\Delta_1^6}{6!} \end{pmatrix}$$

and when m goes to infinity we obtain

$$\mathbf{B}^* = \begin{pmatrix} \frac{e^{\Delta_2} - e^{-\Delta_2}}{2} & \frac{e^{\Delta_2} + e^{-\Delta_2}}{2} \\ \frac{e^{\Delta_1} - e^{-\Delta_1}}{2} & \frac{e^{\Delta_1} + e^{-\Delta_1}}{2} \end{pmatrix}$$

Thus, we get the functional equation

$$f(x) = \frac{-(e^y - e^{-y})\ f(x+z) + (e^z - e^{-z})\ f(x+y)}{e^{(z-y)} - e^{(y-z)}}$$

and the difference equation

$$f(x) = \frac{-(e^{\Delta_1} - e^{-\Delta_1})\ f(x+\Delta_2) + (e^{\Delta_2} - e^{-\Delta_2})\ f(x+\Delta_1)}{e^{(\Delta_2-\Delta_1)} - e^{(\Delta_1-\Delta_2)}}$$

EXAMPLE 5 .- *Finally we deal with the equation*

$$f''(x) - f(x) = x^3$$

whose homogeneous equation is that given in example 4.

We start by taking again $m = 6$ *and we get the same* **N**, **D** *and* **C** *matrices and*

$$\mathbf{H} = \begin{pmatrix} x^3 \\ 3x^2 \\ 6x \\ 6 \\ 0 \end{pmatrix} \quad ; \quad \mathbf{H}^* = \begin{pmatrix} x^3 \\ 3x^2 \\ 6x + x^3 \\ 6 + 3x^2 \\ 6x + x^3 \end{pmatrix}$$

*When m goes to infinity we observe the following : the manipulations on the matrix (**ND**) consist in adding to each row the row which is two places above it, that is,*

$$h^*_{2k} = \sum_{i=0}^{n-1} h_{2(k-i)} \qquad h^*_{2k+1} = \sum_{i=0}^{n} h_{2(k-i)+1}$$

and taking into account that $h_k = 0$ if $k > 4$, we have

$$h^*_j = \begin{cases} x^3 & \text{if } j = 1 \\ 3x^2 & \text{if } j = 2 \\ 6x + x^3 & \text{if } j = 2k+1\ ,\ k > 0 \\ 6 + 3x^2 & \text{if } j = 2k\ ,\ k > 1 \end{cases}$$

After making the matrix **C** *null, we get*

$$\Delta\mathbf{F}^* = \Delta\mathbf{F} - \mathbf{CPH}^* = \Delta\mathbf{F} + \begin{pmatrix} P(\Delta_2) \\ P(\Delta_1) \end{pmatrix}$$

where

$$P(\Delta) = -x^3\frac{\Delta^2}{2!} - 3x^2\frac{\Delta^3}{3!} - \sum_{n=2}^{\infty} x(6+x^2)\frac{\Delta^{2n}}{(2n)!} - \sum_{n=2}^{\infty}(6+3x^2)\frac{\Delta^{2n+1}}{(2n+1)!}$$

which can be written as

$$P(\Delta) = -x^3\left(\frac{e^{\Delta}+e^{-\Delta}}{2} - 1\right) - 3x^2\left(\frac{e^{\Delta}-e^{-\Delta}}{2} - \Delta\right) -$$
$$- 6x\left(\frac{e^{\Delta}+e^{-\Delta}}{2} - 1 - \frac{\Delta^2}{2}\right) - 6\left(\frac{e^{\Delta}-e^{-\Delta}}{2} - \Delta - \frac{\Delta^3}{3!}\right)$$

Finally, we get the difference equation

$$f(x) - E_1 f(x+\Delta_2) - E_2 f(x+\Delta_1) = E_1 P(\Delta_2) + E_2 P(\Delta_1)$$

where

$$E_1 = \frac{-(e^{\Delta_1} - e^{-\Delta_1})}{e^{(\Delta_2-\Delta_1)} - e^{(\Delta_1-\Delta_2)}} \quad ; \quad E_2 = \frac{e^{\Delta_2} - e^{-\Delta_2}}{e^{(\Delta_2-\Delta_1)} - e^{(\Delta_1-\Delta_2)}}$$

Acknowledgements

We thank the Dirección General de Investigación Científica y Técnica (DGICYT) (proyect PB90-0339), and the Subdirección General de Formación y Perfeccionamiento de Personal Investigador (ref. PG89-13770603), for partial support of this work.

References

[1] J. Aczél, *Lectures on functional equations and their applications*, volume **19** of Mathematics in Science and Engineering. Academic Press, 1966.

[2] E. Castillo, and R. Ruiz-Cobo. *Funcional Equations and Modelling in Science and Engineering.* Marcel Dekker, 1992.

[3] E. Castillo, M. Ruiz-Dávila, and M.R. Ruiz-Cobo. Resolución numérica de ecuaciones diferenciales mediante una ecuación funcional equivalente. In *XIV Jornadas Hispano-Lusas de Matemáticas. Tenerife. Spain*, pages 1143–1148, 1989.

[4] T. Levi-Civita, Sulle funzioni che ammettono una formula d'addizione del tipo $f(x+y) = \sum_{i=1}^{n} X_i(x)Y_i(y)$. *Atti. Accad. Nazl. Lincei*, Rend. **5-22**, Pt. 2, pages 181-183, 1913.

[5] P. Stäckel, Sulla equazione funzionale $f(x+y) = \sum_{i=1}^{n} X_i(x)Y_i(y)$. *Atti. Accad. Nazl. Lincei*, Rend. **5-22**, Pt. 2, pages 392-393, 1913.

[6] C. Stephanos, Sur une catégorie d'équations fonctionnelles. *Rend. Circ. Mat. Palermo*, **18**, pages 360-362, 1904.

[7] C. Stephanos, Sur une catégorie d'équations fonctionnelles. In *Math. Kongress Heidelberg*, pages 200-201, 1905.

Functional Analysis, Approximation Theory
and Numerical Analysis
Ed. John M. Rassias

ON SHAPE FROM SHADING PROBLEM

J. CHABROWSKI & KEWEI ZHANG

Centre for Mathematical Analysis - Department of Mathematics
The University of Queensland
Lucia 4067 Qld, Brisbane, Australia.

1. Introduction.

The purpose of this note is to study the variational approch to the eikonal equation

$$(1) \qquad (\frac{\partial u}{\partial x_1})^2 + (\frac{\partial u}{\partial x_2})^2 = \mathcal{E}(x_1, x_2)$$

in a bounded domain $\Omega \subset \mathbb{R}^2$. This equation arises in many branches of applied sciences. In particular, the equation (1) appears in an area of computer vision in the so called shape - from - shading problem in which one tries to solve the problem of how object shape can be recovered from image shading.

More precisely, one seeks a function $u(x_1, x_2)$ representing surface depth in the direction of z-axis, satisfying the image irradiance equation

$$R(\frac{\partial u}{\partial x_1}, \frac{\partial u}{\partial x_2}) = E(x_1, x_2).$$

Here R denotes the reflectance map (which is known) containing information about illumination and surface reflecting conditions, E is an image formed by projection of light along the z-axis onto a plane parallel to the x_1, x_2 - plane, and Ω is the image domain.

The equation (1) can be obtained in the case where the reflectance map corresponds to the situation in which an overhead, distant point - source illuminates a Lambertian surface. For a detailed discussion of this case we refer to papers [4] and [5]. We only mention here that if a small surface portion with normal direction $(-\frac{\partial u}{\partial x_1}, -\frac{\partial u}{\partial x_2}, 1)$ is illuminated by a distant, overhead point-source of unit power in direction $(0,0,1)$, then, according to Lambert's law, the emitted radiance and reflectance map are given by the cosine of the angle between the two directions, namely $\left((\frac{\partial u}{\partial x_1})^2 + (\frac{\partial u}{\partial x_2})^2 + 1\right)^{-\frac{1}{2}}$. Therefore, if $E(x_1, x_2)$ denotes the corresponding image, the image irradiance equation in this situation takes the form

$$\left[(\frac{\partial u}{\partial x_1})^2 + (\frac{\partial u}{\partial x_2})^2 + 1\right]^{-\frac{1}{2}} = E(x_1, x_2).$$

Since $0 < E(x_1, x_2) \leqq 1$, we set $\mathcal{E}(x_1, x_2) = E(x_1, x_2)^{-2} - 1$ and write the above equation in the form (1).

The first uniqueness result in class C^2-functions was obtained by Deift - Sylvester [8] in case $\mathcal{E}(x_1, x_2) = \frac{x_1^2 + x_2^2}{1 - x_1^2 - x_2^2}$ on a unit disc. This has been extended by Bruss [6] to $\mathcal{E}(x_1, x_2) = f(\sqrt{x_1^2 + x_2^2})$, with f satisfying some regularity assumptions. Solutions obtained in these papers are spherically symmetric. In paper [4], Brooks - Chojnacki - Kozera gave examples of solutions which are not spherically - symmetric. The question of the existence of C^1-unbounded solutions is discussed in paper Brooks - Chojnacki - Kozera [4]. A recent result by Kozera [9] solves the problem of recovering the shape of a smooth Lambertian surface from two images obtained by consecutive illumination of the surface by distant point light source in different directions.

Finally, we mention a paper by Horn and Brooks [11] suggesting a variational approach to the shape-from-shading problem. This paper contains some computational observation on variational aspect of this problem. However, this paper does not say anything about the existence of a solution (in some generalised sense) through the variational approach. We point out that our paper has been motivated by **[11]**. We show that, in general, the variational approach does not lead to an exact solution of the shape - from - shading problem. In particular, Theorem 2 shows that any function u satisfying the inequality $|Du(x)|^2 \leqq \mathcal{E}(x)$ can be regarded as a "minimum" of the energy functional associated with (1) in the sense that there exists a suitable sequence $\{u_n\}$, with $u_n |_{\partial\Omega}= u |_{\partial\Omega}$, such that

$$\lim_{n\to\infty} \int_\Omega \mid |Du_n(x)|^2 - \mathcal{E}(x) \mid \, dx = 0 \text{ and } u_n \rightharpoonup u \text{ weak} - * \text{ in } W^{1,\infty}(\Omega).$$

2. Observation on Young measures.

Let $\Omega \subset \mathbb{R}^2$ be bounded domain with a Lipschitz boundary $\partial\Omega$. For $x \in \Omega$ we set $x = (x_1, x_2)$. $W^{1,p}(\Omega)$, $1 \leqq p \leqq \infty$, denotes the usual Sobolev space. Since $\partial\Omega$ is Lipschitz, elements of $W^{1,p}(\Omega)$ admit traces on $\partial\Omega$. For basic information on Sobolev's spaces we refer to Adam's book **[1]**. Throughout this note we assume that $\mathcal{E}(x)$ is a nonnegative function in $C(\bar{\Omega})$. The weak convergence in $W^{1,p}(\Omega)$ is denoted by $\rightharpoonup$ and the strong convergence by $\to$. We associate with (1) a functional given by

$$I(u) = \int_\Omega \mid |Du(x)|^2 - \mathcal{E}(x) \mid \, dx.$$

This functional is not convex and consequently it is not lower semicontinuous. To avoid this difficulty we consider the relaxed problem (see **[7]** or **[9]**) to get some insight in I which will serve as a basis for the construction of a minimizing sequence.

We need the following result on Young measures (see [**2**], **[3]**).

THEOREM 1. *Let $\{z_j\}$ be bounded sequence in $L^1(\Omega; \mathbb{R}^s)$. Then there exist a subsequence $\{z_\nu\}$ of $\{z_j\}$ and a family $\{\nu_x\}$, $x \in \Omega$, of probability measures on $\mathbb{R}^s$, depending measurably on $x \in \Omega$, such that for any measurable subset $A \subset \Omega$*

$$f(\cdot, z_\nu) \rightharpoonup \langle \nu_x, f(x, \cdot) \rangle \text{ in } L^1(A)$$

for every Carathéodory function $f : \Omega \times \mathbb{R}^s \to \mathbb{R}$ *such that* $f(\cdot, z_\nu)$ *is sequentially weakly relatively compact in* $L^1(A)$.

We commence with a simple result on minimizing sequences of I.

PROPOSITION 1. . *Suppose that there exists a sequence* $\{u_j\}$ *in* $W^{1,2}(\Omega)$ *such that*

$$\lim_{j\to\infty} \int_\Omega \big|\, |Du_j|^2 - \mathcal{E}(x) \,\big| \, dx = 0.$$

Then up to a subsequence $u_j \rightharpoonup u$ *in* $W^{1,2}(\Omega)$ *and* $|Du(x)|^2 \leqq \mathcal{E}(x)$ *a.e. on* Ω. *If* $\{\nu_x\}$, $x \in \Omega$, *is family of Young measures corresponding to* $\{Du_j\}$, *then supp* $\nu_x \subset \{P;\, |P|^2 = \mathcal{E}(x)\}$.

PROOF: It is easy to see that $\{u_j\}$ is bounded $W^{1,2}(\Omega)$ and we may assume that $u_j \rightharpoonup u$ in $W^{1,2}(\Omega)$. We set $F(x,P) = \big|\, |P|^2 - \mathcal{E}(x) \,\big|$. Let $CF(x,P)$ denote a lower convex envelope of $F(x,P)$. Then a minimizing sequence for I is also a minimizing sequence for a functional I_C (see [7] or [9]) given by

$$I_C = \int_\Omega CF(x, Du)\, dx$$

and we have

$$0 = \lim_{j\to\infty} \int_\Omega F(x, Du_j)\, dx = \lim_{j\to\infty} CF(x, Du_j)\, dx = \int_\Omega CF(x, Du)\, dx.$$

On the other hand by Theorem 1 there exists a family of Young measures $\{\nu_x\}$, $x \in \Omega$, such that

$$\lim_{j\to\infty} \int_\Omega F(x, Du_j)\, dx = \int_\Omega \langle \nu_x(\cdot), F(x,\cdot)\rangle\, dx = 0,$$

so

$$\int_\Omega \langle \nu_x(\cdot), F(x,\cdot)\rangle\, dx = \int_\Omega CF(x, Du(x))\, dx,$$

and consequently

$$\operatorname{supp}\nu_x \subset \{P;\, F(x,P)=0\} = \{P;\, |P|^2 = \mathcal{E}(x)\}.$$

Noting that

$$CF(x,Du(x)) = \begin{cases} |Du(x)|^2 - \mathcal{E}(x) & \text{if } |Du(x)|^2 - \mathcal{E}(x) > 0 \\ 0 & \text{elsewhere}\,, \end{cases}$$

we see that $|Du(x)|^2 \leqq \mathcal{E}(x)$ a. e. on Ω.

This result will serve as a guide for the construction of a sequence mentioned at the end of the previous section. First of all, let $\phi_n = u_n - u$, then $\phi_n \rightharpoonup 0$ in $W^{1,2}(\Omega)$ and

$$0 = \lim_{n\to\infty} \int_\Omega \big|\, |Du(x) + D\phi_n(x)|^2 - \mathcal{E}(x) \,\big|\, dx = \int_\Omega \langle \bar{\nu}_x, \big|\, |Du(x)+\lambda|^2 - \mathcal{E}(x)\,\big| \rangle\, dx,$$

where $\{\bar{\nu}_x\}$, $x \in \Omega$, is a family of Young measures corresponding to the sequence $\{D\phi_n\}$. The last identity implies that supp $\bar{\nu}_x$ is contained in the boundary of a disc of radius $\mathcal{E}(x)$ with center at $-Du(x)$. The idea is, given a function u satisfying $|Du(x)|^2 \leqq \mathcal{E}(x)$ on Ω, to construct a sequence ϕ_n with these properties. Obviously, there might exist many such sequences. However, our aim is to construct the simplest sequence of this nature. On a microscopic level to satisfy the condition $\int \lambda\, d\bar{\nu}_x = 0$, in case $0 < |Du(x)|^2 < \mathcal{E}(x)$, we will construct our sequence in such a way that supp $\bar{\nu}_x$ will consist of two antipodal points. On the other hand we note that at points where $|Du(x)|^2 = \mathcal{E}(x)$, we have

$$\begin{aligned} 0 &= \langle \bar{\nu}_x, \big|\, |Du(x)+\lambda|^2 - \mathcal{E}(x)\,\big| \rangle \geqq \langle \bar{\nu}_x, |Du(x)+\lambda|^2 - \mathcal{E}(x) \rangle \\ &= \langle \bar{\nu}_x, |Du(x)|^2 + 2\lambda\cdot Du(x) + |\lambda|^2 - \mathcal{E}(x) \rangle = \langle \bar{\nu}_x, |\lambda|^2 \rangle. \end{aligned}$$

This implies that $\bar{\nu}_x = \delta_0$. Again on microscopic level, terms of our sequence should be equal to 0 around such a point.

The idea of using Young measures to examine structure of oscillations of weakly convergent sequences is not new. For detailed bibliographical comments on this subject we refer to [**7**] and [**10**]. A modern treatment of Young measures starts with paper by Tartar [**13**].

3. Main result.

We are now in a position to establish the main result of this note.

THEOREM 2. *Let $u \in C^1(\bar{\Omega})$ be a function such that $|Du(x)|^2 \leqq \mathcal{E}(x)$ on Ω. Then there exists a sequence $\{u_j\}$ in $W^{1,\infty}(\Omega)$ with $u_j\,|_{\partial\Omega} = u\,|_{\partial\Omega}$ and such that*

$$\lim_{j\to\infty} \int_\Omega \big|\, |Du_j(x)|^2 - \mathcal{E}(x) \,\big|\, dx = 0$$

and $u_j \rightharpoonup u$ weak-$*$ in $W^{1,\infty}(\Omega)$.

PROOF: Let $K > 0$ be a constant such that

$$\max_{\bar\Omega} |Du(x)| \leqq K \text{ and } \max_{\bar\Omega} \mathcal{E}(x) \leqq K^2.$$

We approximate Ω by a sequence of unions of squares $H_j = \bigcup_{k=1}^{I_j} D_k^j$ with $H_j \subset \Omega$ and $\lim_{j\to\infty} |\Omega - H_j| = 0$. We assume that edges of D_k^j are parallel to the coordinate axis with the length $d(D_k^j) = \frac{1}{2^j}$. Let $M = \max(1, K)$. For each integer n, by the uniform continuity of $\mathcal{E}$ and Du, we can find an integer $j_n > n$ such that

$$|Du(x) - Du(x_k^{j_n})| \leqq \frac{1}{j_n} < \frac{1}{16nM|\Omega|} \tag{2}$$

$$|\mathcal{E}(x) - \mathcal{E}(x_k^{j_n})| \leqq \frac{1}{j_n} < \frac{1}{4n|\Omega|} \tag{3}$$

for all $x \in D_k^{j_n}$, where $x_k^{j_n}$ is the centre of the square $D_k^{j_n}$ and

$$|\Omega - H_{j_n}| \leqq \frac{1}{8Mj_n}. \tag{4}$$

We now distinguish three cases: *i*) $Du(x_k^{j_n}) = 0$, $\mathcal{E}(x_k^{j_n}) \neq 0$, *ii*) $|Du(x_k^{j_n})|^2 = \mathcal{E}(x_k^{j_n})$, *iii*) $0 < |Du(x_k^{j_n})|^2 < \mathcal{E}(x_k^{j_n})$.

Case *i*). Let $x_k^{j_n} = (x_{1,k}^{j_n}, x_{2,k}^{j_n})$. We define the function $\phi_k^{j_n}$ on $D_k^{j_n}$ by

$$\phi_k^{j_n}(x) = \begin{cases} (x_1 - x_{1,k}^{j_n} + \frac{1}{2^{j_n+1}})\mathcal{E}(x_k^{j_n}) \text{ for } \begin{cases} x_2 - x_{2,k}^{j_n} \geqq x_1 - x_{1,k}^{j_n} \\ x_2 - x_{2,k}^{j_n} \leqq -x_1 + x_{1,k}^{j_n} \\ x_1 \geqq x_{1,k}^{j_n} - \dfrac{1}{2^{j_n+1}} \end{cases} \\ (x_{1,k}^{j_n} + \frac{1}{2^{j_n+1}} - x_1)\mathcal{E}(x_k^{j_n}) \text{ for } \begin{cases} x_1 \leqq x_{1,k}^{j_n} + \dfrac{1}{2^{j_n+1}} \\ x_2 - x_{2,k}^{j_n} \leqq x_1 - x_{1,k}^{j_n} \\ -x_1 + x_{1,k}^{j_n} \leqq x_2 - x_{2,k}^{j_n} \end{cases} \\ (x_2 - x_{2,k}^{j_n} + \frac{1}{2^{j_n+1}})\mathcal{E}(x_k^{j_n}) \text{ for } \begin{cases} x_2 \geqq x_{2,k}^{j_n} - \dfrac{1}{2^{j_n+1}} \\ x_2 - x_{2,k}^{j_n} \leqq x_1 - x_{1,k}^{j_n} \\ -x_1 + x_{1,k}^{j_n} \geqq x_2 - x_{2,k}^{j_n} \end{cases} \\ (x_{2,k}^{j_n} + \frac{1}{2^{j_n+1}} - x_2)\mathcal{E}(x_k^{j_n}) \text{ for } \begin{cases} x_2 \leqq x_{2,k}^{j_n} + \dfrac{1}{2^{j_n+1}} \\ x_2 - x_{2,k}^{j_n} \geqq x_1 - x_{1,k}^{j_n} \\ -x_1 + x_{1,k}^{j_n} \geqq x_2 - x_{2,k}^{j_n} \end{cases} \end{cases}$$

It is easy to see that $\phi_k^{j_n} \in \overset{\circ}{W}{}^{1,\infty}(D_k^{j_n})$ and

$$(5) \qquad \|\phi_k^{j_n}\|_{L^\infty(D_k^{j_n})} = \frac{1}{2^{j_n+1}} < \frac{1}{8nM|\Omega|} \text{ and } |D\phi_k^{j_n}(x)| = \mathcal{E}(x_k^{j_n}) \text{ a.e. on } D_k^{j_n}.$$

Case *ii*). We set $\phi_k^{j_n}(x) \equiv 0$ on $D_k^{j_n}$ end extend this function by zero outside.

Case *iii*). We set

$$\psi_k^{j_n}(x) = \left(\frac{\partial u(x_k^{j_n})}{\partial x_2} x_1 - \frac{\partial u(x_k^{j_n})}{\partial x_1} x_2\right) \frac{\sqrt{\mathcal{E}(x_k^{j_n}) - |Du(x_k^{j_n})|^2}}{|Du(x_k^{j_n})|}$$

and define a function $\gamma_k^{j_n}(x)$ defined on the strip between lines $\psi_k^{j_n}(x) = 1$ and $\psi_k^{j_n}(x) = -1$ by

$$\gamma_k^{j_n}(x) = \begin{cases} 1 - \psi_k^{j_n}(x) \text{ if } 0 \leqq \psi_k^{j_n}(x) \leqq 1 \\ 1 + \psi_k^{j_n}(x) \text{ if } -1 \leqq \psi_k^{j_n}(x) \leqq 0. \end{cases}$$

We see that $\gamma_k^{j_n}(x) = 0$ on the lines $\psi_k^{j_n}(x) = 1$ and $\psi_k^{j_n}(x) = -1$. Moreover, we have $|D\gamma_k^{j_n}(x)|^2 = \mathcal{E}(x_k^{j_n}) - |Du(x_k^{j_n})|^2$ and $D\gamma_k^{j_n}(x_k^{j_n}) = 0$ a.e. on the strip $-1 \leqq \psi_k^{j_n}(x) \leqq 1$. We now extend $\gamma_k^{j_n}$ periodically into $\mathbb{R}^2$ and denote this extension again by $\gamma_k^{j_n}$. We now set $\omega_{k,m}^{j_n}(x) = \frac{1}{m}\gamma_k^{j_n}(mx)$. Then we have $|\omega_{k,m}^{j_n}(x)| \to 0$, as $m \to \infty$, uniformly in $x \in \mathbb{R}^2$. Let us denote the restriction of $\omega_{k,m}^{j_n}$ to $D_k^{j_n}$ by $g_{k,m}^{j_n}$. It is clear that

$$\|g_{k,m}^{j_n}\|_{L^\infty(D_k^{j_n})} \leqq \frac{1}{m} \text{ and } \|Dg_{k,m}^{j_n}\|_{L^\infty(D_k^{j_n})} = \sqrt{\mathcal{E}(x_k^{j_n}) - |Du(x_k^{j_n})|^2}.$$

Let $U_{k,m}^{j_n}$ denote a square contained in $D_k^{j_n}$ with edges parallel to coordinate axes and of length $\frac{1}{2^{j_n}} - 2\|g_{k,m}^{j_n}\|_{L^\infty(D_k^{j_n})}$ and such that

$$\text{dist}\,(\partial D_k^{j_n}, U_{k,m}^{j_n}) = \|g_{k,m}^{j_n}\|_{L^\infty(D_k^{j_n})}.$$

We now define a function $h_{k,m}^{j_n}$ on $\partial D_k^{j_n} \cup U_{k,m}^{j_n}$ by

$$h_{k,m}^{j_n}(x) = \begin{cases} 0 & \text{for } x \in \partial D_k^{j_n} \\ g_{k,m}^{j_n}(x) & \text{for } x \in U_{k,m}^{j_n} \\ 0 & \text{for } x \in \Omega - D_k^{j_n}. \end{cases}$$

The function $h_{k,m}^{j_n}$ is Lipschitz continuous on its domain of definition and its Lipschitz constant does not exceed $\max\left(1, \sqrt{\mathcal{E}(x_k^{j_n}) - |Du(x_k^{j_n})|^2}\right)$. Let $\phi_{k,m}^{j_n}$ be a Lipschitz extension of $h_{k,m}^{j_n}$. We now choose m_n such that $m_n > n$, $m_n > j_n$ and

$$|D_k^{j_n} - U_{k,m_n}^{j_n}| \leqq \frac{1}{4nI_{j_n}(1+2M^2)}. \tag{6}$$

We set $\phi_n(x) = \sum_{k=1}^{I_{j_n}} \beta_k^n(x)$, where

$$\beta_k^n(x) = \begin{cases} \phi_k^{j_n}(x) \text{ if } x_k^{j_n} \text{ satisfies } i) \\ 0 \quad \text{if } x_k^{j_n} \text{ satisfies } ii) \\ \phi_{k,m_n}^{j_n}(x) \text{ if } x_k^{j_n} \text{ satisfies } iii). \end{cases}$$

It then follows from the construction of $\phi_k^{j_n}$ and $\phi_{k,m_n}^{j_n}$ (see (5)) that

$$\|\phi_n\|_{L^\infty(\Omega)} \leqq \frac{1}{n} \text{ and } \|D\phi_n\|_{L^\infty(\Omega)} \leqq M.$$

Let $u_n = u + \phi_n$ and we write $H_j = H_j^1 \cup H_j^2 \cup H_j^3$, where H_j^1, H_j^2 and H_j^3 are unions of squares from cases i), ii) and iii), respectively. In the next step of the proof we show that

$$\lim_{n\to\infty} \int_\Omega \big|\, |Du_n(x)|^2 - \mathcal{E}(x) \,\big| \, dx = 0.$$

We have by (4) the following estimate

$$\begin{aligned} \int_\Omega \big|\, |Du_n(x)|^2 - \mathcal{E}(x) \,\big| \, dx &\leqq \int_{H_{j_n}} \big|\, |Du_n(x)|^2 - \mathcal{E}(x) \,\big| \, dx + M^2|\Omega - H_{j_n}| \\ &\leqq \frac{1}{4n} + \int_{H_{j_n}} \big|\, |Du(x) + D\phi_n(x)|^2 - \mathcal{E}(x) \,\big| \, dx \qquad (7) \\ &\leqq \frac{1}{4n} + a_n + b_n + c_n, \end{aligned}$$

where

$$a_n = \sum_{k=1}^{I_{j_n}} \int_{D_k^{j_n}} \big|\, |Du(x) + D\phi_n(x)|^2 - |Du(x_k^{j_n}) + D\phi_n(x)|^2 \,\big| \, dx,$$

$$b_n = \sum_{k=1}^{I_{j_n}} \int_{D_k^{j_n}} \big|\, |Du(x_k^{j_n}) + D\phi_n(x)|^2 - \mathcal{E}(x_k^{j_n}) \,\big| \, dx,$$

$$c_n = \sum_{k=1}^{I_{j_n}} \int_{D_k^{j_n}} |\mathcal{E}(x) - \mathcal{E}(x_k^{j_n})| \, dx.$$

It follows from the uniform continuity of Du and $\mathcal{E}$ (see (2) and (3)) that

$$a_n \leqq \frac{4M}{n}|\Omega| \text{ and } c_n \leqq \frac{|\Omega|}{j_n}.$$

To estimate b_n we write

$$\begin{aligned} b_n = & \sum_{D_k^n \in H_{j_n}^1} \int_{D_k^{j_n}} \left| |Du(x_k^{j_n}) + D\phi_n(x)|^2 - \mathcal{E}(x_k^{j_n}) \right| dx \\ & + \sum_{D_k^{j_n} \in H_{j_n}^2} \int_{D_k^{j_n}} \left| |Du(x_k^{j_n}) + D\phi(x)|^2 - \mathcal{E}(x_k^{j_n}) \right| dx \\ & + \sum_{D_k^{j_n} \in H_{j_n}^3} \int_{D_k^{j_n}} \left| |Du(x_k^{j_n}) + D\phi_n(x)|^2 - \mathcal{E}(x_k^{j_n}) \right| dx \\ & = I_1 + I_2 + I_3. \end{aligned}$$

It follows from the constructions in cases *i*) and *ii*) that

$$I_1 = 0 \quad \text{and} \quad I_2 = 0.$$

Using (6) we estimate I_3 as follows

$$\begin{aligned} I_3 = & \sum_{D_k^{j_n} \in H_{j_n}^3} \left[\int_{D_k^{j_n} - U_{k,m_n}^{j_n}} 2M^2 \, dx \right. \\ & \left. + \int_{U_{k,m_n}^{j_n}} \left| |Du(x_k^{j_n})|^2 + 2Du(x_k^{j_n}) D\phi_{k,m_n}^{j_n}(x_k^{j_n}) + |D\phi_{k,m_n}|^2 - \mathcal{E}(x_k^{j_n}) \right| \right] dx \\ \leqq & \frac{1}{4n} + \sum_{D_k^{j_n} \in H_{j_n}^3} \int_{U_{k,m_n}^{j_n}} \left| |Du(x_k^{j_n})|^2 + \left(\sqrt{\mathcal{E}(x_k^{j_n}) - |Du(x_k^{j_n})|^2}\right)^2 - \mathcal{E}(x_k^{j_n}) \right| dx \\ & = \frac{1}{4n}. \end{aligned}$$

The last inequality follows from the fact that

$$|D\phi_{k,j_n}(x)|^2 = \mathcal{E}(x_k^{j_n}) - |Du(x_k^{j_n})|^2.$$

Consequently, combining (7), (8) and the last estimate for I_3 we get

$$\int_\Omega \big|\, |Du_n(x)|^2 - \mathcal{E}(x) \,\big| \, dx \leqq \frac{1}{n}.$$

Finally, we observe that since $u_n = u + \phi_n$ with $\|\phi_n\|_{L^\infty(\Omega)} \leqq \frac{1}{n}$ and $\|D\phi_n\|_{L^\infty(\Omega)} \leqq M$, we have $\phi_n \rightharpoonup 0$ weak-$*$ in $\overset{\circ}{W}{}^{1,\infty}(\Omega)$. Hence $u_n \rightharpoonup u$ weak-$*$ in $W^{1,\infty}(\Omega)$ and $u_n\, |_{\partial\Omega} = u\, |_{\partial\Omega}$.

In Theorem 2 we have chosen an L^1-norm to measure the difference between $\mathcal{E}(x)$ and $Du(x)$. The same result continues to hold if we choose the L^p norm $(p > 1)$. The difference between $\mathcal{E}$ and Du can only be measured in an average sense which makes the so called "noise" appear.

We close this paper with two examples illustrating Theorem 2.

Example 1. If $u \equiv 0$ on Ω, then given a nonnegative function $\mathcal{E} \in C(\bar{\Omega})$, there exists a sequence u_n in $\overset{\circ}{W}{}^{1,\infty}(\Omega)$ such that

$$\lim_{n\to 0} \int_\Omega \big|\, |Du_n(x)|^2 - \mathcal{E}(x) \,\big| \, dx = 0 \text{ and } u_n \rightharpoonup 0 \text{ weak } -* \text{ in } W^{1,\infty}(\Omega).$$

Here $\overset{\circ}{W}{}^{1,\infty}(\Omega)$ denotes a Sobolev space whose elements have zero trace on $\partial\Omega$.

Example 2. Suppose that $\Omega = \{x;\ |x|^2 < R\}$. Let f and g be in $C[0, R]$ with $0 \leqq g(t) \leqq f(t)$ on $[0, R]$ and $g(0) = 0$. Then the function $u(|x|) = \int_0^{|x|} g(t)\, dt$ satisfies the inequality $|Du(|x|)|^2 \leqq f(|x|)^2$ on Ω. By Theorem 2 there exists a sequence u_n in $W^{1,\infty}(\Omega)$ such that $u_n\, |_{\partial\Omega} = u\, |_{\partial\Omega}$ and

$$\lim_{n\to\infty} \int_\Omega \big|\, |Du_n(x)|^2 - f(|x|)^2 \,\big| \, dx = 0 \text{ and } u_n \rightharpoonup u \text{ weak } -* \text{ in } W^{1,\infty}(\Omega).$$

Functions $u \in C^1(\bar{\Omega})$ with $Du(x)|^2 \leqq \mathcal{E}(x)$ on Ω, due to the aproximation property from Theorem 2, are called in the current literature "noisy" solutions to the equation (1).

REFERENCES

1. R.A. Adams, "Sobolev spaces," Academic Press, New York, 1975.
2. J.M. Ball, *A version of the fundamental theorem for Young measures*, Lecture Notes in Physics - Springer Verlag **344** (1988), 207–215.
3. H. Berliochi & J.M. Lasry, *Intégrandés normales et mesures paramétrées en calcul des variations*, Bull. Soc. Math. France **101** (1973), 129–184.
4. M.J. Brooks, W. Chojnacki & R. Kozera, *Shading without shape*, Quarterly of Appl. Math., in press.
5. M.J. Brooks, W. Chojnacki & R. Kozera, *Circularly - symmetric eikonal equations and non-uniqueness in computer vision*, J. Math. Analysis and Appl. in press.
6. A.R. Bruss, *The eikonal equation: some results applicable to computer vision*, J. Math. Phys **23(5)** (1982), 890–896.
7. B. Dacorogna, "Continuity and weak lower semicontinuity of non-linear functionals," Lecture Notes in Maths 992, Springer Verlag, 1982.
8. P. Deift & J. Sylvester, *Some remarks on the shape - from - shading problem in computer vision*, J. Math. Anal. Appl. **84(1)** (1981), 235–248.
9. I. Ekeland & R. Temam, "Analyse convexe et problémes variationnels," Dunod, Paris, 1974.
10. Lawrence C. Evans, "Weak convergence methods for nonlinear partial differential equations," CBMS Regional Conference Series in Mathematics (74), Am. Math. Soc.
11. Berthold K.P. Horn & Michael J. Brooks, *The variational approach to shape from shading*, Computer Vision, Graphics and Image Processing **33** (1986), 174– 208.

12. R. Kozera, *Existence and uniqueness in photometric stereo*, School of Information Science and Technology, Flinders University of South Australia, preprint TR 90 - 14.

13. L. Tartar, "Compensated compactness," Heriot - Watt Symposium, Vol. 4, Pitman, 1978.

Functional Analysis, Approximation Theory
and Numerical Analysis
Ed. John M. Rassias

ON THE APPROXIMATIONS TO ANALYTIC FUNCTIONS

Chiu-Cheng Chang
School of Accountancy and Business
Nanyang Technological University
Nanyang Avenue
Singapore 2263

Abstract

In this paper we prove a theorem which shows that the uniqueness problem for entire functions of exponential type is equivalent to the approximation problem for analytic functions. This theorem is then combined with theorems on uniqueness to produce a number of results on the approximation of analytic functions. Using the concept of Pólya property and examples of functions which are known to have such property, we give additional results in approximation. We also extend our results to approximate functions in $L^p(B)$ where B is a Carathéodory domain.

INTRODUCTION

In this paper we prove a theorem which shows that the uniqueness problem for entire functions of exponential type [3, 4, 6, 7, 8] is equivalent to the approximation problem for analytic functions. Thus to each uniqueness theorem for entire functions of exponential type there corresponds an approximation theorem for analytic functions and conversely. This result is then combined with the known uniqueness theorems of DeMar [6, 7, 8], Child [4], Strenk [13] and Chang [3] to yield many approximation theorems. Using examples of functions which are known to have the Pólya property [3], we are able to give more specific results in approximation. Among them Runge's theorem on approximation by polynomials, a strengthened version of approximation by rational functions, approximation by trigonometric polynomials and approximation by exponential functions are simple examples. We also extend these results, using Farrell's method [9], to approximate functions in $L^p(B)$ where B is a Carathéodory domain and $L^p(B)$ is the space of all functions f analytic on B such that $\int\int_B |f(z)|^p dxdy < \infty$ where $z = x + iy$. A by-product of this extention is an improvement of a theorem of Davis [5].

MAIN RESULT

Let Ω be a simply connected domain in the complex plane. Let $K[\Omega]$ denote the class of all entire functions f of exponential type such that the Borel transform of f, denoted by F, is analytic on Ω^c, the complement of Ω. Let $H(\Omega)$ denote the space of analytic functions on Ω with the topology of uniform convergence on compact subsets of Ω. Let $\{L_n\}$ be a sequence of linear functionals defined on $K[\Omega]$ by

$$L_n(f) = \frac{1}{2\pi i}\int_\Gamma g_n(z)F(z)dz \tag{1}$$

where g_n is in $H(\Omega)$, $n = 0, 1, 2, \cdots$, and $\Gamma \subseteq \Omega$ is a simple closed contour such that F is analytic outside and on Γ. A class $K[\Omega]$ is said to be a uniqueness class for $\{L_n\}$ if the zero function is the only function f in $K[\Omega]$ with the property that $L_n(f) = 0$ for $n = 0, 1, 2, \cdots$.

THEOREM 1. *A necessary and sufficient condition for $K[\Omega]$ to be a uniqueness class for $\{L_n\}$ defined by (1) is that the linear span $\langle g_n\rangle$ of $\{g_n\}$ be dense in $H(\Omega)$.*

Proof. (Sufficiency) Let $\{L_n\}$ be defined on $K[\Omega]$ by (1) and the linear span $\langle g_n\rangle$ of the generating functions $\{g_n\}$ for $\{L_n\}$ be dense in $H(\Omega)$. Let $f \in K[\Omega]$ be such that $L_n(f) = 0$ for all $n = 0, 1, 2, \cdots$. We must show that $f \equiv 0$. Let g be any function in $H(\Omega)$. Since $\langle g_n \rangle$ is dense in $H(\Omega)$, there exists a sequence $\{h_n\}$ of linear combinations of functions in $\{g_n\}$, $h_n = \sum_{k=0}^{N(n)} a_{nk} g_k$, converging uniformly to g on compact subsets of Ω. It follows from Köthe's duality theorem [10] that to the Borel transform F of f there corresponds a unique continuous linear functional L defined on $H(\Omega)$ by

$$L(g) = \frac{1}{2\pi i} \int_\Gamma g(z)F(z)dz$$

where $\Gamma \subseteq \Omega$ is a simple closed contour such that F is analytic outside and on Γ. Since $L_n(f) = \frac{1}{2\pi i} \int_\Gamma g_n(z)F(z)dz = 0$ for $n = 0, 1, 2, \cdots$ and Γ is a compact subset of Ω, $L(g) = \frac{1}{2\pi i} \int_\Gamma g(z)F(z)dz = \frac{1}{2\pi i} \int_\Gamma (\lim_{n\to\infty} h_n(z))F(z)dz = \lim_{n\to\infty} \frac{1}{2\pi i} \int_\Gamma h_n(z)F(z)dz = \lim_{n\to\infty} \frac{1}{2\pi i} \int_\Gamma (\sum_{k=0}^{N(n)} a_{nk} g_k(z))F(z)dz = \lim_{n\to\infty} \frac{1}{2\pi i} \sum_{k=0}^{N(n)} \int_\Gamma a_{nk} g_k(z)F(z)dz = 0$. Since g is arbitrary, L is the null functional on $H(\Omega)$. Hence $F \equiv 0$ which implies $f \equiv 0$.

(Necessity) Assume that $\langle g_n \rangle$ is not dense in $H(\Omega)$. We shall show that there exists an f in $K[\Omega]$ such that $L_n(f) = 0$ for all $n = 0, 1, 2, \cdots$, but $f \not\equiv 0$, so $K[\Omega]$ is not a uniqueness class for $\{L_n\}$. By assumption, the closure $\overline{\langle g_n \rangle}$ of $\langle g_n \rangle$ is a proper subset of $H(\Omega)$. Hence there exists a g in $H(\Omega)$ but $g \notin \overline{\langle g_n \rangle}$. Since the topology on $H(\Omega)$ is defined by the countable collection of norms $\|f\|_n < \varepsilon$, $n = 1, 2, \cdots$, this means there exist a $\delta > 0$ and a positive integer N such that $\|g - h\|_n > \delta$ for all $n \geq N$ and for all h in $\overline{\langle g_n \rangle}$. Let G be the subspace generated by $\overline{\langle g_n \rangle}$ and g; that is, the subspace consisting of all elements of the form $h + \alpha g$ with α in C and $h \in \overline{\langle g_n \rangle}$. Define L on G by $L(h + \alpha g) = \alpha\delta$. We first show that L is well defined. Let $h_1 + \alpha g$ and $h_2 + \beta g$ be in G and $h_1 + \alpha g = h_2 + \beta g$. Then $h_1 - h_2 = (\beta - \alpha)g$. If $\beta - \alpha$ were not zero, we would have $g = \frac{h_1 - h_2}{\beta - \alpha}$ which is in $\overline{\langle g_n \rangle}$. Since $g \notin \overline{\langle g_n \rangle}$, $\beta - \alpha = 0$. Hence we have $h_1 = h_2$ and $\alpha = \beta$ which implies that L is well defined. It is clear that L is linear. Thus L is a linear functional on

G. Since $|\alpha|\delta \leq |\alpha| \; \|g - (\frac{-h}{\alpha})\|_n = \|h + \alpha g\|_n$ for all $n \geq N$, we have $|L(s)| \leq \|s\|_n$ for all s in G and for all $n \geq N$. Since $\| \quad \|_n$ is a norm, the hypothesis in the complex version of Hahn-Banach theorem is satisfied by L and we may extend L to all of $H(\Omega)$ such that $|L(t)| \leq \|t\|_n$ for all t in $H(\Omega)$ and for all $n \geq N$. Now we show that L is continuous at the zero function. Let $\varepsilon > 0$ be given. Let $V = \{x \mid x \in C \text{ and } |x| < \varepsilon\}$. Then U defined by $U = \{t \mid t \in H(\Omega) \text{ and } \|t\|_N < \varepsilon\}$ is a neighborhood of the zero function in $H(\Omega)$ with the property that $L(U) \subseteq V$. Since $L(0) = 0$ and V is an arbitrary neighborhood of $0, L$ is continuous at the zero function. Let t_0 be in $H(\Omega)$. Then $L(t_0 + U) = L(t_0) + L(U) \subseteq L(t_0) + V$. Since $L(t_0) + V$ is an arbitrary neighborhood of $L(t_0)$ and $t_0 + U$ is a neighborhood of t_0 in $H(\Omega)$, we conclude that L is a continuous linear functional on $H(\Omega)$ with $L(g) = \delta$ and $L(h) = 0$ for all h in $\overline{\langle g_n \rangle}$. By Köthe's duality theorem there exists a unique F locally analytic on Ω^c with $F(\infty) = 0$ such that $L(t) = \frac{1}{2\pi i} \int_\Gamma t(z)F(z)dz$ for all t in $H(\Omega)$ and $\Gamma \subseteq \Omega$ is a simple closed contour so chosen that F is analytic outside and on Γ. Let f be the function in $K[\Omega]$ such that its Borel transform is F. Then $L_n(f) = \frac{1}{2\pi i} \int_\Gamma g_n(z)F(z)dz = L(g_n) = 0$ for $n = 0, 1, 2, \cdots$. But $L(g) = \delta$, so $F \not\equiv 0$. Therefore, $f \not\equiv 0$ and so $K[\Omega]$ is not a uniqueness class for $\{L_n\}$.

APPLICATIONS OF THEOREM 1

We can combine Theorem 1 with each of the known uniqueness theorems to yield results on approximation. The following definition [3] is included for ready reference:

DEFINITION 2. Let Ω_w and Ω_z be simply connected domains in C. Let $m(w, z)$ be holomorphic on $\Omega_w \times \Omega_z$. Then m has the *Pólya property with respect to z on* Ω_z if and only if for all simple closed contours $\Gamma \subseteq \Omega_z$, if f is analytic on Γ and if $\int_{\Gamma} m(w,z)f(z)dz \equiv 0$, then f has an analytic continuation to the Jordan region enclosed by Γ.

The combination of Theorem 1 and the uniqueness theorems of Chang [3] yields the following approximation results (Theorems 3, 4 and 5):

THEOREM 3. *Let Ω_w be a domain and Ω_z a simply connected domain in C and let $w_0 \in \Omega_w$. Let $m(w,z)$ be holomorphic on $\Omega_w \times \Omega_z$. Let $g_n(z) = D_w^{(n)}[m(w,z)]_{w=w_0}$, $n = 0,1,2,\cdots$ where $D_w^{(n)}[m(w,z)]_{w=w_0}$ is the n-th partial derivative of $m(w,z)$ with repect to w evaluated at w_0. Then $m(w,z)$ has the Pólya property on Ω_z if and only if the linear span $\langle g_n \rangle$ of $\{g_n\}$ is dense in $H(\Omega_z)$.*

THEOREM 4. *Let Ω_w be a domain and Ω_z a simply connected domain in C. Let $m(w,z)$ be holomorphic on $\Omega_w \times \Omega_z$. Let $\{w_n\}$ be a sequence of points in Ω_w with a limit point in Ω_w and $m(w_n,z) = g_n(z)$ for $n = 0,1,2,\cdots$. Then $m(w,z)$ has the Pólya property on Ω_z if and only if $\langle g_n \rangle$ is dense in $H(\Omega_z)$.*

THEOREM 5. *Let Ω_w be a domain and Ω_z a simply connected domain in C. Let $m(w,z)$ be holomorphic on $\Omega_w \times \Omega_z$. Let $\{w_n\}$ be a sequence of points in Ω_w such that $\sum_{n=0}^{\infty} |w_n - w_{n+1}|$ converges and $w_n \to \delta_0 \in \Omega_w$. Let $D_w^{(n)}[m(w,z)]_{w=w_n} = g_n(z)$ for $n = 0,1,2,\cdots$ where $D_w^{(n)}[m(w,z)]_{w=w_n}$ is the n-th partial derivative of $m(w,z)$ with respect to w evaluated at w_n. Then $m(w,z)$ has the Pólya property on Ω_z if and only if $\langle g_n \rangle$ is dense in $H(\Omega_z)$.*

Combining Theorem 1 with a theorem of DeMar [6] yields the following:

THEOREM 6. *Let Ω be a simply connected domain in the complex plane. Let W be analytic on Ω. Then $\langle W^n \rangle$ is dense in $H(\Omega)$ if and only if W is univalent on Ω.*

Combining Theorem 1 and a theorem of Child [4, p.61], which says that if $K[\Omega]$ is a uniqueness class for $\{L_n\}$ defined by (1), then $K[\Omega]$ is also a uniqueness class for $\{L_n^*\}$ defined as in (1) with g_n' as generating functions, we have

THEOREM 7. *Let Ω be a simply connected domain in the complex plane. Let g_n be analytic on Ω, $n = 0, 1, 2, \cdots$. If $\langle g_n \rangle$ is dense in $H(\Omega)$, then $\langle g_n' \rangle$ is dense in $H(\Omega)$.*

A consequence of Theorem 1 taken together with a theorem of DeMar [8] is the following:

THEOREM 8. *Let p be a positive integer and α a primitive p-th root of 1. Let Ω be a simply connected domain and let W and $h_k : k = 0, 1, \cdots, p-1$, be analytic on Ω with $W(\Omega)$ a p-symmetric domain. Let $\Omega_1 \subseteq \Omega$ be a simply connected domain such that W is univalent on Ω_1 and $W(\Omega_1) = W(\Omega)$. Let $Z : W(\Omega_1) \to \Omega_1$ be the inverse of W. Let $\Delta(z)$ be $\det(h_{kq}(z))$ where $h_{kq}(z) = h_k(Z(\alpha^q W(z))) : k, q = 0, 1, \cdots, p-1$. Then the linear span $\langle W^{pn} h_k \rangle$ of $\{[W(z)]^{pn} h_k(z)\}_{n=0}^{\infty}$, $k = 0, 1, \cdots, p-1$, is dense in $H(\Omega)$ if and only if W is univalent on Ω and $\Delta(z) \neq 0$ for all $z \in \Omega$ such that $W(z) \neq 0$.*

In the special case where $\Omega = \{z = x + iy \mid |y| < \pi\}$, $W(z) = e^z - 1$ and $h_k(z) = (e^z - 1)^k e^{a_k z}$; a_k real, Strenk's main theorem [13, p.31] shows that the hypothesis that $W(\Omega)$ be p-symmetric in Theorem 8 is superfluous. Hence we have the following

THEOREM 9. *Let p be a positive integer and α a primitive p-th root of 1. Let $\Omega = \{z = x + iy \mid |y| < \pi\}$, $W(z) = e^z - 1$ and $h_k(z) = (e^z - 1)^k e^{a_k z}$; a_k real, $k = 0, 1, \cdots, p-1$. Let $\Delta(z)$ be $\det(h_{kq}(z))$ where $h_{kq}(z) = \alpha^{kq}(e^z - 1)^k (1 + \alpha^q (e^z - 1))^{a_k}$,*

$k, q = 0, 1, \cdots, p-1$. *Then the linear span* $\langle (e^z - 1)^{pn+k} e^{a_k z} \rangle$ *is dense in* $H(\Omega)$ *if and only if* $\Delta(z) \neq 0$ *for all* z *in* Ω *such that* $z \neq 0$.

SPECIFIC RESULTS IN APPROXIMATION.

We are now in the position to use above results and functions which are known to have the Pólya property to obtain more specific results in approximation.

APPROXIMATION BY POLYNOMIALS. Since $W(z) = z$ is univalent on any simply connected domain Ω, $\langle z^n \rangle$ is dense in $H(\Omega)$ by Theorem 6. This is Runge's theorem for simply connected domains. We restate it in a theorem.

THEOREM 10. *Every function analytic on a simply connected domain* Ω *can be uniformly approximated on compact subsets of* Ω *by polynomials.*

APPROXIMATION BY RATIONAL FUNCTIONS. We shall only illustrate some interesting results as it will become obvious how to obtain similar ones. Our first theorem in the following is an improvement of a theorem of Rubel and Taylor [12], in the case of simply connected domain.

THEOREM 11. *Let* Ω_z *be a simply connected domain properly contained in* C. *Let* $\{w_n\}$ *be a convergent sequence in* Ω_z^c *with each* w_n *occurring at most finitely many times. Let* k *be a fixed positive integer. Then* $\langle \frac{1}{(w_n - z)^k} \rangle$ *is dense in* $H(\Omega_z)$.

Proof. Since $\frac{1}{(w-z)^k}$ has the Pólya property on Ω_z, the assertion follows from Theorem 4.

THEOREM 12. *Let* Ω_z *be a simply connected domain properly contained in* C. *Let* w_0 *be a point in* Ω_z^c. *Then the linear span* $\langle \frac{1}{(w_0 - z)^n} \rangle$ *of the sequence* $\{ \frac{1}{(w_0 - z)^n} \}$ *of rational functions is dense in* $H(\Omega_z)$.

Proof. Since $\frac{1}{w-z}$ has the Pólya property on Ω_z and the n-th partial derivative of $\frac{1}{w-z}$ with respect to w evaluated at w_0 is $(-1)^n \frac{n}{(w_0-z)^{n+1}}$, the conclusion then follows from Theorem 3.

COROLLARY 13. *Let Ω_z be a simply connected domain properly contained in C. Let w_0 be a point in Ω_z^c. Let k be any fixed positive integer. Then the linear span $\langle \frac{1}{(w_0-z)^{n+k}} \rangle$ of the sequence $\{\frac{1}{(w_0-z)^{n+k}}\}_{n=1}^{\infty}$ is dense in $H(\Omega_z)$.*

THEOREM 14. *Let Ω_z be a simply connected domain properly contained in C. Let $\{w_n\}$ be a sequence of points in Ω_z^c such that $\sum_{n=0}^{\infty} |w_n - w_{n+1}|$ converges. Then $\langle \frac{1}{(w_n-z)^n} \rangle$ is dense in $H(\Omega_z)$.*

Proof. Theorem 5 and the fact that $\frac{1}{(w-z)^n}$ has the Pólya property on Ω_z.

Since $\frac{1}{(w-z)^k}$, k a fixed positive integer, is not the only rational function which has the Pólya property, we can obtain many more similar results on rational approximation of analytic functions by applying Theorems 3, 4 and 5 to rational functions which have the Pólya property.

APPROXIMATION BY TRIGONOMETRIC POLYNOMIALS. Again we shall only illustrate some interesting results as it will become evident that the key to more approximation results is to find more functions which have the Pólya property. But first we shall apply Theorem 6 to some functions univalent on particular domains.

DEFINITION 15. A complex *trigonometric polynomial* is a finite sum of the form $\sum_{n=1}^{N} a_n \cos nz + b_n \sin nz$ where a_n and b_n are complex numbers and z in C.

THEOREM 16. *The set of all trigonometric polynomials is dense in $H(\Omega)$ where $\Omega = \{z = x + iy \mid |x| < \pi\}$*

Proof. Since e^{iz} is univalent on Ω, $\langle e^{inz} \rangle$ is dense in $H(\Omega)$ by Theorem 6. It follows from Euler's identity that $\langle \cos nz + i \sin nz \rangle$ is dense in $H(\Omega)$.

THEOREM 17. *Let* $\Omega_1 = \{z = x + iy \mid |x| < \frac{\pi}{2}\}$, $\Omega_2 = \{z = x + iy \mid 0 < x < \pi\}$, $\Omega_3 = \{z = x + iy \mid |y| < \frac{\pi}{2}\}$ *and* $\Omega_4 = \{z = x + iy \mid 0 < y < \pi\}$. *Then* $\langle \sin^n z \rangle$ *is dense in* $H(\Omega_1)$, $\langle \cos^n z \rangle$ *is dense in* $H(\Omega_2)$, $\langle \sin h^n z \rangle$ *is dense in* $H(\Omega_3)$, *and* $\langle \cos h^n z \rangle$ *is dense in* $H(\Omega_4)$.

Proof. Theorem 6.

THEOREM 18. *Let* $\{w_n\}$ *be any sequence in* C *converging to a point in* C. *Let* Ω *be any simply connected domain in* C. *Then* $\langle \sin w_n z + \cos w_n z \rangle$ *is dense in* $H(\Omega)$.

Proof. Since $\sin wz + \cos wz$ has the Pólya property with respect to z on Ω, the assertion follows from Theorem 4.

APPROXIMATION BY EXPONENTIAL POLYNOMIALS. We first use Theorem 6 to obtain the following result:

THEOREM 19. *Let* $\Omega = \{z = x + iy \mid |y| < \pi\}$. *Then* $\langle e^{nz} \rangle$ *is dense in* $H(\Omega)$.

This result can be improved by applying a theorem of Rubel [11]. Let A be a set of positive integers. Let $\Delta(A)$ denote the upper density of A, defined by $\Delta(A) = \limsup_{t \to \infty} \frac{A(t)}{t}$, where $A(t)$, the counting function of A, is defined as the number of integers n in A for which $n \leq t$. Then we can state Rubel's theorem in the following form.

THEOREM 20. *Let* $\Omega = \{z = x + iy \mid |y| < \pi\}$. *Let* A *be a set of positive* *integers. Let $\{L_n\}$, $n \in A$, be the sequence of linear functionals defined on $K[\Omega]$ as in (1) with e^{nz}, $n \in A$, as generating functions for $\{L_n\}$. Then a necessary and sufficient condition for $K[\Omega]$ to be a uniqueness class for $\{L_n\}$ is that $\Delta(A) = 1$.*

Then Theorem 1 and Theorem 20 jointly imply the following:

THEOREM 21. *Let $\Omega = \{z = x + iy \mid |y| < \pi\}$. Let A be a set of positive integers. Then a necessary aand sufficient condition for $\langle e^{nz} \rangle$, $n \in A$, to be dense in $H(\Omega)$ is that $\Delta(A) = 1$.*

THEOREM 22. *Let Ω be any simply connected domain in C. Let $\{w_n\}$ be a convergent sequence in C with a limit point in C. Then $\langle e^{w_n z} \rangle$ is dense in $H(\Omega)$.*

Proof. Theorem 4 and the fact that e^{wz} has the Pólya property with respect to z on Ω.

APPROXIMATION BY OTHER ANALYTIC FUNCTIONS. We shall only illustrate some result again as it is evident how to obtain similar ones.

THEOREM 23. *Let $f \in H(C)$. Then $\langle f^n \rangle$ is dense in $H(C)$ if and only if f is of the form $f(z) = az + b$ where $a \neq 0$.*

Proof. The functions $az + b$, $a \neq 0$, are the only entire functions univalent in the whole plane.

APPROXIMATION IN $L^p(B)$. For any $p > 0$, we let $L^p(B)$ be the class of functions $f(z)$ which are analytic on a Carathéodory domain B [9] such that $\int\int_B |f(z)|^p dxdy < \infty$ where $z = x + iy$. Since the set of all polynomials is dense in $H(B)$, the question as to whether it is also dense in $L^p(B)$ arises naturally. Here density in $L^p(B)$ is, of course, measured by surface integral. Farrell [9] proves that the answer to this question is affirmative. We point out here that Farrell's proof does not depend on the given sequence being a sequence of polynomials, but only on its span being dense in the space $H(B)$ where B is a Carathéodory domain. Thus, he actually proved that any sequence whose span is dense in $H(B)$ also has a dense span in $L^p(B)$, $p > 0$. We list some consequences of this for the purpose of illustration.

THEOREM 24. *Let B be a Carathéodory domain. Let $p > 0$. Let $f \in L^p(B)$. Then there exists a sequence $\{q_n\}$ of rational functions with poles in B^c such that $\lim_{n\to\infty} \int\int_B |f(z) - q_n(z)|^p dxdy = 0$.*

THEOREM 25. *Let $\Omega = \{z = x + iy \mid |x| < \pi\}$. Let $B \subset \Omega$ be a Carathéodory domain. Let $p > 0$ and $f \in L^p(B)$. Then there exists a sequence $\{q_n\}$ of trigonometric polynomials such that $\lim_{n\to\infty} \int\int_B |f(z) - q_n(z)|^p dxdy = 0$.*

THEOREM 26. *Let $\Omega = \{z = x + iy \mid |y| < \pi\}$. Let $B \subset \Omega$ be a Carathéodory domain. Let $p > 0$. Let A be a set of positive integers such that the upper density $\Delta(A)$ of A is equal to 1. Then $\langle e^{nz} \rangle$, $n \in A$, is dense in $L^p(B)$.*

The case $p = 2$ is of great interest. We note that $L^2(B)$ is a Hilbert space under the norm $\|f\|^2 = \int\int_B |f(z)|^2 dxdy$. We now compare Theorem 26 with a theorem of Davis [5]. By a theorem of Rubel [11], $\Delta(A) = 1$ if and only if $\Delta_M(A) = 1$ where $\Delta_M(A)$, the upper mean density of A, is defined by $\Delta_M(A) = \limsup_{t\to\infty} \frac{1}{t} \int_0^t \frac{A(x)}{x} dx$. Davis shows for sequences $\{\lambda_n\}$ of distinct complex numbers, $\lambda_n \to \infty$, that if $\Delta_M(A) > \frac{c(\widehat{B})}{2\pi}$ where $c(\widehat{B})$ is the circumference of the convex hull of B, then $\{e^{\lambda_n z}\}$, $\lambda_n \in A$, is complete in $L^2(B)$. Hence for $\{\lambda_n\}$ an increasing sequence of positive integers, if $\Delta(A) = 1$, then the linear span $\langle e^{nz} \rangle$ of $\{e^{nz}\}$, $n \in A$, is dense in $L^2(B)$ for all B with $c(\widehat{B}) < 2\pi$. But Theorem 26 says that if $\Delta(A) = 1$, then

$\langle e^{nz} \rangle$, $n \in A$, is dense in $L^2(B)$ for all B such that $B \subset \Omega$. This is an improvement of Davis result for $\{\lambda_n\}$ an increasing sequence of positive integers.

BIBLIOGRAPHY

[1] Boas, R. P. *Entire Functions.* Academic Press, Inc., New York, 1954.

[2] Buck, R. C. "Interpolation Series," *Transactions of the American Mathematical Society,* vol. 64 (1948), pp. 283-298.

[3] Chang, C. C. "On the Uniqueness Problem for Entire Functions of Exponential Type". To appear in Soochow Mathematical Journal.

[4] Child, J. D. *Uniqueness and Interpolation Results for Entire Functions of Exponential Type.* Ph. D. Thesis, University of Cincinnati, 1972.

[5] Davis, P. "On the Applicability of Linear Differential Operators of Infinite Order to Functions of Class $L^2(B)$," *American Journal of Mathematics,* vol. 75 (1952), pp. 475-491.

[6] DeMar, R. F. "A Uniqueness Theorem for Entire Functions," *Proceedings of the American Mathematical Society,* vol. 16 (1965), pp. 69-71.

[7] DeMar, R. F. "Uniqueness Classes for Difference Functionals," *Canadian Journal of Mathematics,* vol. 18 (1966), pp. 593-607.

[8] DeMar, R. F. "Uniqueness Classes for Periodic-Type Functionals," *Duke Mathematical Journal,* vol. 37 (1970), pp. 29-40.

[9] Farrell, O. J. "On the Approximation to an Analytic Function by Polynomials," *Bulletin of the American Mathematical Society,* vol. 40 (1934), pp. 908-914.

[10] Köthe, G. "Dualität in der Funktionentheorie," *J. Reine Angew. Math.,* vol. 191 (1953), pp. 30-49.

[11] Rubel, L. A. "Necessary and Sufficient Conditions for Carlson's Theorem on Entire Functions," *Transactions of the American Mathematical Society*, vol. 83 (1956), pp. 417-429.

[12] Rubel, L. A. and Taylor, B. A. "Functional Analysis Proofs of Some Theorems in Function Theory," American Mathematical Monthly, vol. 76 (1969), pp. 483-489.

[13] Strenk, R. E. *Uniqueness Classes for Periodic-Type Difference Functionals.* Ph. D. Thesis, University of Cincinnati, 1972.

Functional Analysis, Approximation Theory
and Numerical Analysis
Ed. John M. Rassias

Error Estimate in Non-equi-mesh Spline Finite Strip Method for Thin Plate Bending Problem

Wu Ciqian Wang Zhehui
Department of Computer Science
Zhongshan University
Guang Zhou, PR China

ABSTRACT

Spline finite strip method based on equi-mesh for structure analysis has been studied by C.Q. Wu , Y.K. Cheung and S.C. Fan in 1981 [1]. In this paper we present spline finite strip method based on non-equi-mesh. For a model problem, we give error estimate analysis. As for non-equi-mesh processing generalized trapezoid element method and its error estimate analysis on abitrary area, we shall discussed in [2].

1. Assumptions

As in [3], suppose that D_x, D_y, D_1, D_{xy} are positive constants. D_x , $D_y > D_1$, $W(x,y)$ is displacement function on the bounded area Ω. Let $T = \partial\Omega$ be boundary of Ω and $W \in H^2(\Omega), f \in H^0(\Omega)$.

Definition

$$J(W) = \iint_\Omega \{\frac{1}{2}[D_x W_{xx}^2 + 2D_1 W_{xx} W_{yy} + D_y W_{yy}^2 + 4D_{xy} W_{xy}^2] - Wf\} dxdy$$

$$a(u,v) = \iint_\Omega [D_x u_{xx} v_{xx} + D_1 u_{xx} v_{yy} + D_1 u_{yy} v_{xx} + D_y u_{yy} v_{yy} + 4D_{xy} u_{xy} v_{xy}] dxdy$$

$$(f,v) = \iint_\Omega fv\, dxdy$$

$$\forall\ W, u, v \in H^2(\Omega),\ \forall\ f \in H^0(\Omega)$$

$$\|u\|_{k,\Omega} = \{ \iint_\Omega [\sum_{0\le\alpha+\beta\le k} (\frac{\partial^k u}{\partial x^\alpha \partial y^\beta})^2] dxdy \}^{1/2}$$

$$|u|_{k,\Omega} = \{ \iint_\Omega [\sum_{\alpha+\beta=k} (\frac{\partial^k u}{\partial x^\alpha \partial y^\beta})^2] dxdy \}^{1/2}$$

$$\forall\ u \in H^k(\Omega),\ k = 0,1,2,\ldots,$$

By $\|u\|_k$, $|u|_k$ denote $\|u\|_{k,\Omega}$, $|u|_{k,\Omega}$ respectively, if there is no confusion.

$$Lu = D_x u_{xxxx} + (2D_1 + 4D_{xy}) u_{xxyy} + D_y u_{yyyy} \qquad \forall\ u \in C^4(\Omega)$$

$$H_E^k(\Omega) = \{ u;\ u \in H^k(\Omega),\ u = u_x = u_y = 0,\ (x,y) \in \Gamma \}$$

$$C_E^k(\overline{\Omega}) = \{ u;\ u \in C^k(\overline{\Omega}),\ u = u_x = u_y = 0,\ (x,y) \in \Gamma \}$$

Thin plate bending problem in structure analysis becomes the following problems

find $u^* \in H^2(\Omega)$ **such that**

$$J(W^*) = \min_{W \in H^2(\Omega)} J(W) \tag{1.1}$$

Lemma (1.1)

For all $u,v \in H^2(\Omega)$**, there exist constants** B_1 B_2 **such that**

$$a(u,v) \le B_1 |u|_2 |v|_2 \tag{1.2}$$

$$a(u,u) \ge B_2 (|u|_2)^2 \tag{1.3}$$

Lemma (1.2)

Let $u \in c^4(\overline{\Omega})$ **then u satisfies**

$$Lu = f \qquad (x,y) \in \Omega \tag{1.4}$$

$$u = u_x = u_y = 0 \qquad (x,y) \in \Gamma \tag{1.5}$$

iff $u \in H_E^2(\Omega)$ **and**

$$a(u,v) = (f,v) \qquad \forall\ v \in H_E^2(\Omega) \tag{1.6}$$

holds.

Say $\Gamma \in C^k$ and $\Omega \in C^k$ if Γ can be expressed as

$$\Gamma = \{ (x,y) \mid x=x(t),\ y=y(t),\ \alpha \leq t \leq \beta \}$$

where $x(t), y(t) \in C^k [\alpha,\beta]$ are single value functions of $t \in [\alpha,\beta]$.

Lemma (1.3)

Let $\Omega \in C^{4+k}$ or Ω be a convex polygonal domain and $f \in H^k(\Omega)$ then solution u of (1.4) and (1.5) belongs to $H_E^{4+k}(\Omega)$. Furthermore,

(i) if $\Omega \in C^{4+k}$ then

$$\| u \|_{4+k} \leq C_0' \| f \|_k$$

(ii) if Ω be a convex polygonal domain then

$$\| u \|_{3+k} \leq C_0 \| f \|_k$$

where C_0', C_0 just depend on Ω.

2. Numerical method and error estimate

In the following description, we assume that Ω be regular domain: $\Omega = [a,b]x[c,d]$.

Let π_x: $a = x_0 < x_1 < \ldots < x_n = b$

π_y: $y_{-3} < \ldots < y_0 = c < y_1 < \ldots < y_m = b < \ldots < y_{m+3}$

be divisions along x-direction and y-direction, respectively.

Let $l_k = x_{k+1} - x_k$, $k = 0(1)n-1$

$h_j = y_j - y_{j-1}$, $j = -2(1)m+3$

$h_x = \max \{l_k\}$, $h_y = \max \{ h_j \}$, $h = \max\{h_x, h_y\}$

Assume that there exists a constant $\tau > 0$ such that

$$\min \{ \min\{l_k\}, \min\{h_j\} \} \geq \tau h \tag{2.1}$$

lemma (2.1)

$$\tau h \le l_k, h_j \le h, \quad h^{-1} \le l_k^{-1}, \quad h_j^{-1} \le h^{-1}/\tau$$

$$k = 0(1)n-1, \quad j=-2(1)m+3 \tag{2.2}$$

Let $\Phi_j(y) = (y_{j+2}-y_{j-2})[y_{j-2}, y_{j-1}, y_j, y_{j+1}, y_{j+2}](y-.)_+^3$

be the well-known cubic spline, where $j=-1(1)m+1$. For an element $E_k=[x_k, x_{k+1}]\times[c,d]$

$$\Theta_H W = \sum_{i=1}^{4} \mu_{ik}(x)\lambda_{ik}(y) \qquad (x,y) \in E_k \tag{2.3}$$

is a Hermite interpolation of the displacement function W(x,y) along x-direction.

Here $\mu_{1k}(x) = 1-3t^2 + 2t^3$, $\mu_{2k}(x) = (t-2t^2+t^3)l_k$

$\mu_{3k}(x) = 3t^2 - 2t^3$, $\mu_{4k}(x) = (-t^2+t^3)l_k$ $\quad t = (x - x_k)/l_k$

$\lambda_{1k}(y) = W(x_k, y)$, $\lambda_{2k}(y)=W_x(x_k, y)$

$\lambda_{3k}(y)=W(x_{k+1}, y)$, $\lambda_{4k}(y)=W_x(x_{k+1}, y)$

$k=0(1)n-1$

Let $\overline{\psi} = \{\psi_{-1}, \psi_0, \cdots, \psi_m, \psi_{m+1}\}$

$= \{\Phi_{-1}, \Phi_0-\alpha\Phi_{-1}, -\beta\Phi_1+\Phi_0-\gamma\Phi_{-1}, \Phi_2, \cdots, \Phi_{m-2},$

$-\gamma'\Phi_{m+1}+\Phi_m-\beta'\Phi_{m-1}, \Phi_m-\alpha'\Phi_{m+1}, \Phi_{m+1}\}$

where $\alpha=\Phi_0(y_0)/\Phi_{-1}(y_0), \alpha'=\Phi_m(y_m)/\Phi_{m+1}(y_m)$

$\beta=(h_2+h_1)/h_0, \quad \gamma=(h_0+h_{-1})/h_1$

$$\beta'=(h_m+h_{m-1})/h_{m+1}, \gamma'=(h_{m+1}+h_{m+2})/h_m \tag{2.4}$$

Then non-equi-mesh spline finite strip approximation of W(x,y) is

$$\tilde{W}(x,y) = \sum_{i=1}^{4} \mu_{ik}(x) \sum_{j=-1}^{m+1} C_{ijk}\psi_j(y)$$
$$= \sum_{j=-1}^{m+1} \left(\sum_{i=1}^{4} C_{ijk}\mu_{ik}(x)\right)\psi_j(y)$$

where $C_{i,-1,k}, C_{i,0,k}, C_{i,m,k}, C_{i,m+1,k}$ **are suitable numbers depending on boundary conditions.**

In the following description, we just study the error analysis for a special -clamped boundary -case i.e.

$$W = W_x = W_y = 0 \qquad (x,y)\in\Gamma$$

In this case

$$C_{i,-1,k} = C_{i,m+1,k} = 0 \qquad i=1,2,3,4$$
$$C_{i0k} = C_{imk} = 0 \qquad i=1,3 \tag{2.5}$$

and we write

$$\tilde{W}(x,y)= \sum_{j=0}^{m} \left(\sum_{i=1}^{4} C_{ijk}\mu_{ik}(x)\right)\psi_j(y) \tag{2.6}$$

where C_{ijk} **satisfies**

$$C_{3jk}=C_{1,j,k+1}, C_{4jk}=C_{2,j,k+1} \qquad j=0(1)m, k=0(1)n-1$$
$$C_{1j0} = C_{2j0}= C_{1jn} =C_{2jn} = 0 \qquad j=0(1)m$$
$$C_{10k} =C_{1mk} = C_{30k} = C_{3mk} = 0 \qquad k=0(1)n \tag{2.7}$$

Denote $V = \{ \tilde{W}(x,y)= \sum_{j=0}^{m} \left(\sum_{i=1}^{4} C_{ijk}\mu_{ik}(x)\right)\psi_j(y),\ (x,y)\in E_k,\ k=0(1)n-1 \ |C_{ijk}$ **satisfy(2.7)**$\}$

Lemma (2.2)

$$V\subset H_E^2(\Omega)$$

By $\hat{W}(x,y)$ **we denote the numerical solution of (1.1) in V.**

$$\hat{W}(x,y)= \sum_{j=0}^{m} \left(\sum_{i=1}^{4} \hat{C}_{ijk}\mu_{ik}(x)\right)\psi_j(y) \qquad (x,y)\in E_k,$$

which satisfies

$$J(\hat{W}) = \min_{\tilde{W} \in V} J(\tilde{W}) \tag{2.9}$$

Theorem (2.1)

For all $v \in V \subset H_E^2(\Omega)$ we have

$$a(\hat{W}, v) = (f, v)$$

$$a(W^*, v) = (f, v)$$

and then
$$a(W^* - \hat{W}, v) = 0 \tag{2.10}$$

Let $\Theta_s \lambda_{ik}(y)$ be the cubic B-spline interpolation of type I (fixed-support boundary condition) of function $\lambda_{ik}(y)$. For all $w \in C_E^4(\overline{\Omega})$ we define the tensor product approximation of W as

$$\Theta_{HS} W = \sum_{i=1}^{4} \mu_{ik}(x) \Theta_s \lambda_{ik}(y) \qquad (x,y) \in E_k, \quad k=0(1)n-1$$

Obviously $\Theta_{HS} W \in V$

Lemma(2.3) Let $I_k = [x_k, x_{k+1}]$, $f(x) \in c^\alpha(I_k)$, $\alpha=2,3,4$.
By $\Theta_H f$ denote Hermite interporation of f(x) on I_k, let $R(x) = f(x) - \Theta_H f(x)$. Then from [4] we have

$$\left[\int_{x_k}^{x_{k+1}} R^{(j)}(x)^2 dx \right]^{1/2} = O(l_k^{\alpha-j})$$

and

$$\max_{x \in I_k} |R^{(j)}(x)| = O(l_k^{\alpha-j})$$

where $j=0,1,2$, $\alpha=2,3,4$

Lemma(2.4) Let $g(y) \in c^\alpha[c,d]$, $\alpha=2,3,4$. By $\Theta_S g$ denote the I-type cubic spline interpolation of g(y) on the division π_y let $R(y) = g(y) - \Theta_S g(y)$ from

[4] we have

$$\left[\int_c^d R^{(j)}(y)^2 dy \right]^{1/2} = O(h_y^{\alpha-j})$$

and

$$\max_{y\in[c,d]} |R^{(j)}(y)| = O(h_y^{\alpha-j})$$

where $j=0,1,2,\ \alpha=2,3,4$

Theorem (2.2)

Let $W(X,Y) \in C_E^4(\Omega)$, then there exists a positive constant $C_1(W)$ independing on h such that

$$|W-\Theta_{HS}W|_{2,\Omega} \le C_1(W)h^2$$

Proof:

Suppose $\alpha+\beta=2,\ \alpha,\beta\ge 0$ For all $W \in C_E^4(\Omega)$

$$(W-\Theta_{HS}W)^{(\alpha,\beta)} = W^{(\alpha,\beta)} - \sum_{i=1}^{4} \mu_{ik}^{(\alpha)}(x)\lambda_{ik}^{(\beta)}(y)$$

$$+ \sum_{i=1}^{4} \mu_{ik}^{(\alpha)}(x)[\lambda_{ik}^{(\beta)}(y)-(\Theta_s\lambda_{ik}(y))^{(\beta)}]$$

$$\therefore \sum_{k=0}^{n-1} \int_c^d \int_{x_k}^{x_{k+1}} \{(W-\Theta_{HS}W)^{(\alpha,\beta)}\}^2 dxdy$$

$$\le 2\sum_{k=0}^{n-1} \int_c^d\int_{x_k}^{x_{k+1}}\{(W^{(0,\beta)}-\Theta_H W^{(0,\beta)})^{(\alpha,0)}\}^2 dx$$

$$+8\sum_{k=0}^{n-1}\sum_{i=1}^{4} \max|\mu_{ik}^{(\alpha)}(x)|^2 \int_{x_k}^{x_{k+1}}\int_c^d\{[\lambda_{ik}(y)-\Theta_s\lambda_{ik}(y)]^{(0,\beta)}\}^2 dy \qquad (2.11)$$

Lemma (2.1) (2.3) (2.4) estimate RHS of (2.11) as $O(h^4)$

then there exists a positive constant $C_1(W)$ independing on h such that

$$|W-\Theta_{HS}W|^2_{2,\Omega} \leq C_1(W)^2 h^4$$

$$\therefore \quad |W-\Theta_{HS}W|_{2,\Omega} \leq C_1(W)h^2$$

Theorem (2.3)

For all $W(X,Y) \in C^4_E(\Omega)$,there exists positive constants C_2, C_3 independing on h and W such that

$$|W-\Theta_{HS}W|_{2,\Omega} \leq C_2 h|W|_{3,\Omega} + C_3 h^2 |W|_{4,\Omega}$$

Proof:

As in Theorem (2.2), we estimate the RHS of (2.11) to proof this theorem.

Theorem (2.4)

Let W^* and $\hat{W}$ be solution of variational problem (1.1) and (2.9), respectively. Then there exists a positive constant d_2 independing on h such that

$$|W^* - \hat{W}|_{2,\Omega} \leq d_2 h^2$$

Proof: By Theorem (2.1) and the fact that $\Theta_{HS}W^* - \hat{W} \in V \subset H^2_E(\Omega)$ We know that

$$a(W^* - \hat{W}, \Theta_{HS}W^* - \hat{W}) = 0$$

$$a(W^* - \hat{W}, W^* - \hat{W}) = a(W^* - \hat{W}, W^* - \Theta_{HS}W^*) + a(W^* - \hat{W}, \Theta_{HS}W^* - \hat{W})$$

$$= a(W^* - \hat{W}, W^* - \Theta_{HS}W^*)$$

By Lemma (1.1) and Theorem (2.2) we have

$$B_2(|W^* - \hat{W}|_{2,\Omega})^2 \leq a(W^* - \hat{W}, W^* - \hat{W})$$

$$\leq B_1 |W^* - \hat{W}|_{2,\Omega} |W^* - \Theta_{HS}W^*|_{2,\Omega}$$

$$\leq B_1 |W^* - \hat{W}|_{2,\Omega} C_1(W^*)h^2$$

$$\therefore \quad |W^* - \hat{W}|_{2,\Omega} \leq (B_1/B_2)C_1(W^*)h^2 \overset{df}{=} d_2 h^2 \tag{2.12}$$

Theorem (2.5) There exists positive constant d_0, d_1 independing on h such that

$$|W^* - \hat{W}|_{0,\Omega} \leq d_0 h^3$$

$$|W^* - \hat{W}|_{1,\Omega} \leq d_1 h^{5/2}$$

Proof: By z we have denote the solution of the boundary problem

$$Lz = W^* - \hat{W} \qquad (x,y) \in \Omega$$

$$z = z_x = z_y = 0 \qquad (x,y) \in \Gamma$$

by Lemma (1.3) , embedding theorem and Lemma (1.2) we have

$$a(z,v) = (W^* - \hat{W}, v) \qquad \forall\ v \in H_E^2(\Omega) \tag{2.13}$$

From Theorem (2.1) we have

$$a(W^* - \hat{W}, \Theta_{HS} z) = 0 \tag{2.14}$$

Replace v by $W^* - \hat{W}$ in (2.13) we have

$$(|W^* - \hat{W}|_{0,\Omega})^2 = (W^* - \hat{W}, W^* - \hat{W}) = a(z, W^* - \hat{W})$$

$$= a(W^* - \hat{W}, z - \Theta_{HS} z) + a(W^* - \hat{W}, \Theta_{HS} z)$$

$$= a(W^* - \hat{W}, z - \Theta_{HS} z)$$

$$\leq B_1 |W^* - \hat{W}|_{2,\Omega} |z - \Theta_{HS} z|_{2,\Omega} \tag{2.15}$$

By Theorem (2.3) and Lemma (1.3) we have

$$|z - \Theta_{HS} z|_{2,\Omega} \leq C_2 h |z|_{3,\Omega} + C_3 h^2 |z|_{4,\Omega}$$

$$\leq C_2 h C_0 |W^* - \hat{W}|_{0,\Omega} + C_3 h^2 C_0 |W^* - \hat{W}|_{1,\Omega}$$

$$\because (|W^* - \hat{W}|_{1,\Omega})^2 = \iint_\Omega [(W^* - \hat{W})_x^2 + (W^* - \hat{W})_y^2] dxdy$$

$$= -\iint_\Omega \Delta(W^*-\hat{W})(W^*-\hat{W})dxdy + 0$$

$$\leq |W^*-\hat{W}|_{2,\Omega}\, |W^*-\hat{W}|_{0,\Omega}$$

$$\leq d_2 h^2 |W^*-\hat{W}|_{0,\Omega}\,.$$

$$\therefore |W^*-\hat{W}|_{1,\Omega} \leq d_2^{1/2} h\, |W^*-\hat{W}|_{0,\Omega}^{1/2} \tag{2.17}$$

By (2.12) (2.15) (2.16) (2.17) solve the inequition we have

$$|W^*-\hat{W}|_{0,\Omega} \leq d_0 h^3$$

By (2.17) we have

$$|W^*-\hat{W}|_{1,\Omega} \leq d_1 h^{5/2}$$

Theorem (2.6) There exists positive constant μ_0 independing on h such that

$$\max_{\Omega} \left| \frac{\partial^{\alpha+\beta}(W^*-\hat{W})}{\partial x^\alpha\, \partial y^\beta} \right| \leq \mu_0 h^{(6-\alpha-\beta)/2}$$

where $\alpha+\beta = 0,1,2,\ \alpha,\beta \geq 0$

Proof:

We can write $\tilde{W} \in V$ as

$$\tilde{W}(x,y) = \sum_{j=0}^{m} \left(\sum_{i=1}^{4} C_{ijk}\,\mu_{ik}(x)\right)\psi_j(y),\quad (x,y) \in E_k,$$

Orthogonalize $\{\mu_{ik}(x))\psi_j(y)\}_{i=1(1)4, j=1(1)m}$ to $\{\tilde{\psi}_j\}_{j=1(1)m}$ where $\iint_\Omega \psi_i \psi_j\, dxdy = \delta_{ij}$, $M=4m$ then we rewrite $\tilde{W}$ as

$$\tilde{W} = \sum_{j=1}^{M} e_j \psi_j \qquad (x,y) \in E_k$$

we have

$$\iint_\Omega \tilde{W}^2 dxdy = \sum_{ij=1}^{M} \iint_\Omega e_i e_j \psi_i \psi_j\, dxdy = \sum_{i=1}^{M} e_i^2 = \|\tilde{W}\|_{0,\Omega}^2$$

$$\therefore \max_{E_k} |\widetilde{W}| \le \left(\sum_{i=1}^{M} e_i^2\right)^{1/2} \left(\sum_{i=1}^{M} \tilde{\psi}_i^2\right)^{1/2} \le C' \|\widetilde{W}\|_{0,\Omega} \tag{2.18}$$

Suppose that $|W^* - \hat{W}|$ reaches its maximum in E_p then

$$\max_{\Omega} |W^* - \hat{W}| \le \max_{E_p} |W^* - \hat{W}|$$

$$\le \max_{E_p} |W^* - \Theta_{HS} W^*| + \max_{E_p} |\Theta_{HS} W^* - \hat{W}| \tag{2.19}$$

For $\Theta_{HS} W^* - \hat{W} \in V$ by (2.18) we have

$$\max_{E_p} |\Theta_{HS} W^* - \hat{W}| \le C' \|\Theta_{HS} W^* - \hat{W}\|_{0,\Omega}$$

$$\le C' \|W^* - \Theta_{HS} W^*\|_{0,\Omega} + C' \|W^* - \hat{W}\|_{0,\Omega} \tag{2.20}$$

From Lemma (2.3) (2.4) we have

$$\max_{E_p} |W^* - \Theta_{HS} W^*| = O(h^4),\ \|W^* - \Theta_{HS} W^*\|_{0,\Omega} = O(h^3)$$

and $\|W^* - \hat{W}\|_{0,\Omega} = O(h^3)$ by (2.19) and (2.20) we know that there exists a positive constant μ_1 such that

$$\max_{\overline{\Omega}} |W^* - \hat{W}| \le \mu_1 h^3$$

Similarly there exists positive constants μ_2 and μ_3 such that

$$\max_{\overline{\Omega}} |(W^* - \hat{W})^{(\alpha,\beta)}| \le \mu_2 h^{5/2} \qquad \alpha + \beta = 1$$

$$\max_{\overline{\Omega}} |(W^* - \hat{W})^{(\alpha,\beta)}| \le \mu_3 h^2 \qquad \alpha + \beta = 2$$

Set $\mu_0 = \max\{ \mu_1 \mu_2 \mu_3 \}$ we conclude that

$$\max_{\Omega} \left| \frac{\partial^{\alpha+\beta} (W^* - \hat{W})}{\partial x^{\alpha} \partial y^{\beta}} \right| \leq \mu_0 h^{(6-\alpha-\beta)/2}$$

Reference

[1] C.Q.Wu , Y. K. Cheung and S. C. Fan,"Spline finite stripe in structure analysis", Guangdong Science and Technology Publishing House, 1986

[2] C.Q. Wu and Wang Zhehui, " Generalized trapezoid element method and its error analysis for structure ayalysis", to be published

[3] Wang Zhehui, Wu Ciqian, "The Uniqueness Existence of Solution And Normal Boundary Condition for Thin Bending Problem", to be published

[4] Martin H. Schultz, "Spline Analysis", Prentice-Hall, Inc. 1973

Functional Analysis, Approximation Theory
and Numerical Analysis
Ed. John M. Rassias

The Hyers-Ulam stability of a functional equation containing partial difference operators

Zbigniew Gajda

Institute of Mathematics, Silesian University,

Bankowa 14, PL-40-007 Katowice, Poland

Abstract

We examine the stability of a functional equation containing difference operators of higher orders. It appears in a problem raised by J. Schwaiger in 1984. We show that this equation is stable in the Hyers-Ulam sense, which means that any function satisfying the equation with a certain accuracy only is uniformly approximated by its exact solution.

1. Introduction

Let $(G,+)$ and $(H,+)$ be two Abelian Groups and let X be a linear space over the rationals. We consider the following functional equation

$$\Delta^m_{1,a}\Delta^n_{2,b}\ F(x,y) = 0, \quad a,x \in G, \quad b,y \in H \tag{1}$$

in which $F : G \times H \to X$ is regarded as an unknown function, whereas $\Delta^m_{1,a}$ and $\Delta^n_{2,b}$ stand for the m-th and n-th iterates of the partial difference operators defined by

$$\Delta_{1,a}F(x,y) := F(x+a,y) - F(x,y)$$

and

$$\Delta_{2,b}F(x,y) := F(x,y+b) - F(x,y).$$

To describe solutions of Eq. 1 we recall that a function $p : G \to X$ is said to be a polynomial function of order $m-1$ if and only if it satisfies the following Fréchet functional equation:

$$\Delta^m_a p(x) = 0, \qquad a,x \in G. \tag{2}$$

Here, Δ_a^m denotes the m-th iterate of the ordinary difference operator defined by

$$\Delta_a p(x) := p(x+a) - p(x).$$

More information about algebraic properties of polynomial functions can be found in [3] and [6].

In 1984, during the 22nd International Symposium on Functional Equations in Oberwolfach, Prof. J. Schwaiger conjectured that any function $F : G \times H \to X$ satisfying Eq. 1 can be represented in the form

$$F = P + Q,$$

where $P(\cdot, y)$ is a polynomial function of order $m-1$ for each fixed $y \in H$ and $Q(x, \cdot)$ is a polynomial function of order $n-1$ for each fixed $x \in G$ (conversely, every function of this form fulfils Eq. 1). The same conjecture appears among open problems collected at the end of [7]. The author of the present article proved it to be right in his previous paper [4].

Once we know the general form of solutions of Eq. 1 it is natural to ask if the equation is stable in the sense introduced by S. Ulam in [8]. Originally, Ulam was interested in the stability of the linear functional equation. What he had in mind was the question whether any function satisfying such an equation with some bounded error is uniformly closed to an exact solution of the equation. It was D. H. Hyers [5] who first answered Ulam's question in affirmative. Since then the concept of stability has been studied thoroughly in relation to various functional equations. At present it is usually referred to as the Hyers-Ulam stability. The aim of the present paper is to prove that, under the assumption that X is a Banach space, Eq. 1 is stable in the Hyers-Ulam sense.

2. Auxiliary results

Our further considerations are based on the following result of M. Albert and J. A. Baker [1] concerning the stability of Eq. 2:

THEOREM A. *Let $(G, +)$ be an Abelian group and let $(X, \|\cdot\|)$ be a Banach space. With every positive integer m one can associate a non-negative constant k_m with the following property: if for some $\varepsilon > 0$ a function $f : G \to X$ satisfies the inequality*

$$\|\Delta_a^m f(x)\| \le \varepsilon, \qquad a, x \in G,$$

then there exists a polynomial function $p : G \to X$ of order $m-1$ such that

$$\|f(x) - p(x)\| \le k_m \varepsilon, \qquad x \in G.$$

REMARK 1. In a less rigorous formulation the above theorem states that any function $f : G \to X$ for which the transformation

$$G^2 \ni (a, x) \to \Delta_a^m f(x) \in X$$

is bounded, can be decomposed into a sum $f = p + r$, where $p : G \to X$ is a polynomial function of order $m - 1$ and $r : G \to X$ is bounded. Moreover, one may show that this decomposition is unique up to a constant function.

The next lemma is due to K. Baron (cf. [2]) and its proof was enclosed in our paper [4].

LEMMA 1. *Let $(G, +)$ be an Abelian group, let Y be a linear space over the rational field and assume that L is a linear subspace of Y. If $f : G \to Y$ is a function such that*

$$\Delta_a^m f(x) \in L, \quad a, x \in G,$$

then $f = p + l$, where $p : G \to Y$ is a polynomial function of order $m - 1$ and

$$l(x) \in L, \quad x \in G.$$

We shall also need the following

LEMMA 2. *Let $(G, +)$ be an Abelian group, let Y be a real or complex linear space and let V be a linear subspace of Y which splits into the direct sum of two subspaces W and Z. Suppose that Z is endowed with a norm $\|\cdot\|_Z$ such that $(Z, \|\cdot\|_Z)$ becomes a Banach space and denote by proj_Z the projection operator mapping V onto its direct component Z. Furthermore, assume that a function $f : G \to Y$ satisfies the following conditions:*

$$\Delta_a^m f(x) \in V, \quad a, x \in G \qquad (i)$$

and for some $\varepsilon > 0$,

$$\|\mathrm{proj}_Z (\Delta_a^m f(x))\|_Z \le \varepsilon, \quad a, x \in G. \qquad (ii)$$

Then f can be expressed in the form

$$f = p + q + r,$$

where $p : G \to Y$ is a polynomial function of order $m - 1$, $q : G \to W$, $r : G \to Z$ and

$$\|r(x)\|_Z \le k_m \varepsilon, \quad x \in G,$$

k_m being the constant from the assertion of Theorem A.

Proof. Condition (i) jointly with Lemma 1 imply that there exist a polynomial function $p_0 : G \to Y$ of order $m-1$ and a function g mapping G into V such that

$$f = p_0 + g.$$

Since V is the direct sum of subspaces W and Z, the function g admits the unique decomposition

$$g = q + r_0$$

with a function q assuming values in W and a function r_0 assuming values in Z. Taking into account the linearity of difference operators we obtain

$$\Delta_a^m f(x) = \Delta_a^m g(x) = \Delta_a^m q(x) + \Delta_a^m r_0(x), \quad a, x \in G.$$

It is also clear that $\Delta_a^m q(x) \in W$ and $\Delta_a^m r_0(x) \in Z$, which yields the identity

$$\text{proj}_Z \left(\Delta_a^m f(x)\right) = \Delta_a^m r_0(x), \quad a, x \in G.$$

Consequently, by condition (ii), we have

$$\|\Delta_a^m r_0(x)\|_Z \le \varepsilon, \quad a, x \in G.$$

Applying Theorem A we may represent r_0 in the form $r_0 = p^* + r$, where $p^* : G \to Z$ is a polynomial function of order $m-1$ and $r : G \to Z$ is such that

$$\|r(x)\|_Z \le k_m \varepsilon, \quad x \in G.$$

Putting $p := p_0 + p^*$ we get a polynomial function of order $m-1$ and moreover,

$$f = p_0 + g = p_0 + q + r_0 = p_0 + q + p^* + r = p + q + r.$$

The functions p, q and r have all the desired properties and the proof is finished.

3. Main results

THEOREM 1. *Let $(G,+)$ and $(H,+)$ be two Abelian groups, suppose $(X, \|\cdot\|)$ to be a Banach space and let us fix two positive integers m, n and a positive real constant ε. Assume that $F : G \times H \to X$ is a function satisfying the inequality*

$$\|\Delta_{1,a}^m \Delta_{2,b}^n F(x,y)\| \le \varepsilon, \quad a, x \in G, \quad b, y \in H. \tag{3}$$

Then for an arbitrarily small positive constant δ there exist functions $P, Q : G \times H \to X$ with the following properties:

$$\bigwedge_{y \in H} P(\cdot, y) \text{ is a polynomial function of order } m-1, \tag{a}$$

$$\bigwedge_{x\in G} Q(x,\cdot) \textit{ is a polynomial function of order } n-1 \tag{b}$$

and

$$\|F(x,y)-P(x,y)-Q(x,y)\| \le k_m k_n \varepsilon + \delta, \quad (x,y)\in G\times H, \tag{c}$$

where k_m, k_n are the constants occurring in the assertion of Theorem A.

Proof. Let Y denote the space of all functions mapping H into X. We define the following linear subspaces of Y:

$V := \{\varphi \in Y :$ the transformation $H^2 \ni (b,y) \to \Delta_b^n \varphi(y) \in X$ is bounded $\}$,

$W := \{\varphi \in Y : \ \Delta_b^n \varphi(y) = 0$ for all $b, y \in H \}$,

$Z := \{\varphi \in Y : \ \varphi$ is bounded $\}$,

$L := \{\varphi \in Y : \ \varphi$ is constant $\}$.

In the space Z one may introduce the uniform convergence norm $\|\cdot\|_\infty$ defined by

$$\|\varphi\|_\infty := \sup_{y\in H} \|\varphi(y)\|, \quad \varphi \in Z.$$

Together with this norm Z becomes a Banach space and L is its closed subspace. Factorizing Y, V, W and Z by the subspace L we obtain the quotient spaces

$$\tilde{Y} := Y/L, \quad \tilde{V} := V/L, \quad \tilde{W} := W/L, \quad \tilde{Z} := Z/L.$$

In what follows the coset of L determined by a function $\varphi \in Y$ will be designated by $[\varphi]$. It is an immediate consequence of Remark 1 that $\tilde{V}$ coincides with the direct sum of $\tilde{W}$ and $\tilde{Z}$. The space $\tilde{Z}$ turns into a Banach space when we equip it with the quotient norm $\|\cdot\|_{\tilde{Z}}$ related to $\|\cdot\|_\infty$ as follows:

$$\|[\varphi]\|_{\tilde{Z}} := \inf\{\|\psi\|_\infty : \ \psi \in [\varphi]\}, \ \ [\varphi] \in \tilde{Z}.$$

Now we define a function $\tilde{f} : G \to \tilde{Y}$ by the formula

$$\tilde{f}(x) := [F(x,\cdot)], \quad x \in G.$$

Then we have

$$\Delta_a^m \tilde{f}(x) = \left[\Delta_{1,a}^m F(x,\cdot)\right], \quad a, x \in G.$$

Putting

$$\varphi_{a,x} := \Delta_{1,a}^m F(x,\cdot), \quad a, x \in G$$

and calling to mind inequality (3) we derive

$$\|\Delta_b^n \varphi_{a,x}(y)\| = \|\Delta_{2,b}^n \Delta_{1,a}^m F(x,y)\| = \|\Delta_{1,a}^m \Delta_{2,b}^n F(x,y)\| \le \varepsilon$$

for arbitrarily chosen $a, x \in G$, $b, y \in H$. Hence it follows, in particular, that for all $a, x \in G$ one has

$$\varphi_{a,x} \in V, \text{ i.e. } \Delta_a^m \tilde{f}(x) \in \tilde{V}.$$

Moreover, on account of Theorem A, for any fixed $a, x \in G$ the function $\varphi_{a,x}$ admits a representation

$$\varphi_{a,x} = \pi_{a,x} + \rho_{a,x},$$

where $\pi_{a,x} : H \to X$ is a polynomial function of order $n-1$ and $\rho_{a,x} : H \to X$ is bounded with an estimation of its supremum norm given by

$$\|\rho_{a,x}\|_\infty \leq k_n \varepsilon.$$

On the other hand,

$$\Delta_a^m \tilde{f}(x) = [\varphi_{a,x}] = [\pi_{a,x}] + [\rho_{a,x}], \quad a, x \in G.$$

The first term of the above sum belongs to $\tilde{W}$, whereas the second is an element of $\tilde{Z}$, which implies that

$$\text{proj}_{\tilde{Z}} \left(\Delta_a^m \tilde{f}(x) \right) = [\rho_{a,x}], \quad x \in G.$$

Hence,

$$\left\| \text{proj}_{\tilde{Z}} \left(\Delta_a^m \tilde{f}(x) \right) \right\|_{\tilde{Z}} = \|[\rho_{a,x}]\|_{\tilde{Z}} \leq \|\rho_{a,x}\|_\infty \leq k_n \varepsilon, \quad a, x \in G.$$

Applying Lemma 2 to the function $\tilde{f}$ we receive a decomposition

$$\tilde{f} = \tilde{p} + \tilde{q} + \tilde{r},$$

where $\tilde{p} : G \to \tilde{Y}$ is a polynomial function of order $m-1$, $\tilde{q} : G \to \tilde{W}$, $\tilde{r} : G \to \tilde{Z}$ and

$$\|\tilde{r}(x)\|_{\tilde{Z}} \leq k_m k_n \varepsilon, \quad x \in G. \tag{4}$$

Let us represent the functions $\tilde{p}$, $\tilde{q}$ and $\tilde{r}$ in the form

$$\tilde{p}(x) = [p(x)], \quad \tilde{q}(x) = [q(x)], \quad \tilde{r}(x) = [r(x)], \quad x \in G$$

with some $p : G \to Y$, $q : G \to W$ and $r : G \to Z$. Then we have

$$[\Delta_a^m p(x)] = \Delta_a^m \tilde{p}(x) = [0], \quad x \in G,$$

which means that

$$\Delta_a^m p(x) \in L, \quad a, x \in G.$$

From Lemma 1 it follows that $p = p_0 + l$, where $p_0 : G \to Y$ is a polynomial function of order $m-1$ and

$$l(x) \in L, \quad x \in G.$$

Consequently, for every $x \in G$ the cosets $[p(x)]$ and $[p_0(x)]$ coincide. From this place on, the constant δ will be treated as fixed (possibly very small, nevertheless positive). In view of (4), to each $x \in G$ one can assign an element $r_0(x) \in [r(x)]$ such that

$$\|r_0(x)\|_\infty \leq k_m k_n \varepsilon + \delta.$$

Evidently, the cosets $[r(x)]$ and $[r_0(x)]$ are identical for every $x \in G$. As a result,

$$[F(x,\cdot)] = [p_0(x)] + [q(x)] + [r_0(x)], \quad x \in G$$

or equivalently,

$$F(x,\cdot) = p_0(x) + q(x) + r_0(x) + l_0(x), \quad x \in G$$

with some $l_0 : G \to L$. If we now put

$$q_0(x) := q(x) + l_0(x), \quad x \in G,$$

then the function q_0 assumes values in $W + L \subset W$. Next we define functions P, Q and R mapping $G \times H$ into X by the formulae

$$P(x,y) := p_0(x)(y), \quad Q(x,y) := q_0(x)(y), \quad R(x,y) := r_0(x)(y)$$

for all $(x,y) \in G \times H$. Then

$$F(x,y) = P(x,y) + Q(x,y) + R(x,y), \quad (x,y) \in G \times H$$

and it is readily seen that P and Q have the properties (a) and (b) from the assertion of our theorem. Finally, notice that

$$\sup_{(x,y)\in G\times H} \|R(x,y)\| = \sup_{x\in G}\sup_{y\in H} \|r_0(x)(y)\| = \sup_{x\in G} \|r_0(x)\|_\infty \leq k_m k_n \varepsilon + \delta,$$

which ensures property (c) and completes the proof.

By induction based on similar arguments one can prove the following generalization of Theorem 1:

THEOREM 2. *Let $(G_1,+),\ldots,(G_m,+)$ be Abelian groups and suppose that $(X,\|\cdot\|)$ is a Banach space. Moreover, fix a system of positive integers $n_1,\ldots,n_m$ and a positive real constant ε. Assume that $F : G_1 \times \ldots \times G_m \to X$ is a function fulfilling the inequality*

$$\left\|\Delta^{n_1}_{1,a_1}\ldots\Delta^{n_m}_{m,a_m} F(x_1,\ldots,x_m)\right\| \leq \varepsilon$$

for all $a_i, x_i \in G$, $i = 1,\ldots,m$. Then for arbitrarily small positive δ there exist functions $P_i : G_1 \times \ldots \times G_m \to X$ $(i = 1,\ldots,m)$ such that P_i is a polynomial function *of order $n_i - 1$ with respect to the i-th variable while the remaining variables are fixed and*

$$\left\| F(x_1, ..., x_m) - \sum_{i=1}^{m} P_i(x_1, ..., x_m) \right\| \le \left(\prod_{i=1}^{m} k_{n_i} \right) \varepsilon + \delta$$

for all $(x_1, ..., x_m) \in G_1 \times ... \times G_m$. As usual k_{n_i} $(i = 1, ..., m)$ are the constants appearing for the first time in Theorem A.

REMARK 2. The paper [1] provides some estimations of the constants k_n $(n = 1, 2, ...)$, but it is not known whether they are sharp. Therefore, there is not much practical sense in trying to improve Theorems 1 and 2 by eliminating from them the constant δ which only imperceptibly weakens their assertions. Nonetheless such improvement would be desirable from the aesthetic point of view. We do not know if it can be achieved without imposing additional assumptions on the space X. However, if X is finite-dimensional, the constant δ can be avoided as is easily seen from the proof of Theorem 1 and from the following

PROPOSITION. *If* $\dim X < \infty$, *then keeping the meaning of symbols used in the proof of Theorem 1, we have*

$$\bigwedge_{[\varphi]\in\tilde{Z}} \quad \bigvee_{\psi\in[\varphi]} \quad \|[\varphi]\|_{\tilde{Z}} = \|\psi\|_\infty .$$

Proof. Identifying elements of the space X with constant functions mapping H into X, we may write

$$\|[\varphi]\|_{\tilde{Z}} = \inf\{\|\varphi + c\|_\infty : \ c \in X\}.$$

Fix a coset $[\varphi] \in \tilde{Z}$ and put $\eta := 2\|\varphi\|_\infty$. If $c \in X$ is chosen from the complement of the closed ball $B(0, \eta)$ with the centre 0 and the radius η, then

$$\|\varphi + c\|_\infty \ge \|c\| - \|\varphi\|_\infty > \|\varphi\|_\infty \ge \|[\varphi]\|_{\tilde{Z}} .$$

Consequently,

$$\|[\varphi]\|_{\tilde{Z}} = \inf\{\|\varphi + c\|_\infty : \ \ c \in B(0, \eta)\}.$$

We may select a sequence $\{c_n\}_{n\in\mathbf{N}}$ of elements of $B(0, \eta)$ such that

$$\|[\varphi]\|_{\tilde{Z}} = \lim_{n\to\infty} \|\varphi + c\|_\infty .$$

Since $B(0, \eta)$ is compact, the sequence $\{c_n\}_{n\in\mathbf{N}}$ contains a subsequence $\{c_{n_k}\}_{k\in\mathbf{N}}$ convergent to a $c_0 \in B(0, \eta)$. Hence

$$\|[\varphi]\|_{\tilde{Z}} = \lim_{k\to\infty} \|\varphi + c_{n_k}\|_\infty = \|\varphi + c\|_\infty ,$$

which ends the proof, because $\psi := \varphi + c_0 \in [\varphi]$.

References

1. M. Albert and J. A. Baker, *Functions with bounded differences*, Ann. Polon. Math. **43** (1983), p.93-103.

2. K. Baron, *Functions with differences in subspaces.* In: Proceedings of the 18th International Symposium on Functional Equations, University of Waterloo, Faculty of Mathematics, Waterloo, Ontario, Canada, 1980.

3. D. Z. Djoković, *A representation theorem for* $(X-1)(X-2)...(X-n)$ *and its applications*, Ann. Polon. Math. **22** (1969), p. 189-198.

4. Z. Gajda, *A solution to a problem of J. Schwaiger*, Aequationes Math. **32** (1987), p. 38-44.

5. D. H. Hyers, *On the stability of the linear functional equation*, Proc. Nat. Acad. Sci. U.S.A. **27** (1941), p. 222-224.

6. S. Mazur and W. Orlicz, *Grundlegenden Eigenschaften der polynomischen Operationen* I, II, Studia Math. **5** (1934), p. 50-68, 179-189.

7. Th. M. Rassias (Editor), *Nonlinear Analysis*, World Scientific, Singapore - New Jersey - Hong Kong, 1987.

8. S. Ulam, *Problems in modern mathematics*, Chapter VI, Science Editions, Wiley, New York, 1960.

Functional Analysis, Approximation Theory
and Numerical Analysis
Ed. John M. Rassias

ASYMPTOTIC BEHAVIOR OF DYNAMICAL SYSTEM AND PROCESSES ON BANACH INFINITE DIMENSIONAL SPACES

A. F. IZÉ

Universidade Federal de São Carlos,
Departmento de Matemática
Caixa Postal 676, 13560-São Carlos-SP, Brasil

Abstract

This paper proves a fixed point index theorem for non-compact maps and gives an application to prove the existence of positive solutions of evolutionary equations in infinite dimensional Banach sequence spaces.

1. INTRODUCTION

In [10] we presented a topological method that can be applied to study the asymptotic behavior of differential equations in Banach spaces. We prove here a fixed point index theorem for non-compact maps and give an application on the existence of positive solutions of evolutionary equations in infinite dimensional Banach sequence spaces. The results generalize Theorem of Hartman and Wintner [7] and [8] to an infinite dimensional spaces and also the results proved in [11]. The proof shows the generality and simplicity of the method developed in [10]. Infinite systems of ordinary differential equations arise in the theory of branching process, [2] semidiscretization of partial differential equations, [3], [5], degradation of polymers, [3] and perturbation theory of quantum mechanics [1].

2. PRELIMINARIES

We begin by recalling a few definitions and results from [10].

Suppose X is a Banach space, $\mathbb{R}^+ = [0, \infty)$, $u : \mathbb{R} \times X \times \mathbb{R}^+ \to X$ is a given mapping and define $U(\sigma, t) : X \to X$ for $\sigma \in \mathbb{R}^+$ by $U(\sigma, t)x = u(\sigma, x, t)$. A process on X is a mapping $u : \mathbb{R} \times X \times \mathbb{R}^+ \to X$ satisfying the following properties:

i) u is continuous,
ii) $U(\sigma, \sigma) = I$ (identity),
iii) $U(\sigma + s, t)U(\sigma, s) = U(\sigma, s + t)$.

A process is said to be an autonomous process or a semidynamical system if $U(\sigma, t)$ is independent of σ, i.e., $U(\sigma, t) = U(0, t)$ for each $\sigma \in \mathbb{R}$ and $t \geq 0$. When this is the case, define $T(t) = U(0, t)$ and note $T(t)x$ is continuous at each $(t, x) \in \mathbb{R} \times X$.

Definition. Suppose u is a process on X. The trajectory $\tau^+(\sigma, x)$ through $(\sigma, x) \in \mathbb{R} \times X$ is the set on $\mathbb{R} \times X$ defined by

$$\tau^+(\sigma, x) = \{(\sigma + t, U(\sigma, t)x) | t \in \mathbb{R}^+\} .$$

The orbit $\gamma^+(\sigma, x)$ through (σ, x) is the set in X defined by

$$\gamma^+(\sigma, x) = \{U(\sigma, t)x | t \in \mathbb{R}^+\} .$$

If there exists a backward continuation for the process u, we define for some $t^- \in [-\infty, 0]$

$$\tau^-(\sigma, x) = \{(\sigma + t, U(\sigma, t)x) | t \in (t^-, 0)\} .$$
$$\gamma^-(\sigma, x) = \{U(\sigma, t)x | t \in (t^-, 0)\} .$$

An integral of the process in $\mathbb{R}$ is a continuous functions $y : \mathbb{R} \to X$ such that for any $\sigma \in \mathbb{R}$, $\tau^+(\sigma, y(\sigma)) = \{(\sigma + t, y(\sigma + t)) | t \geq 0\}$. An integral y is an integral through $(\sigma, x) \in \mathbb{R} \times X$, if $y(\sigma) = x$.

We assume in the following that the integral through each $(\sigma, x) \in \mathbb{R} \times X$ is unique. We define $\tau^{-1}(x) = \{(\sigma, y) \in \mathbb{R} \times X |, \exists t > 0$ such that $U(\sigma, t)y = x\}$. If $P_o = (\sigma, x) \in \mathbb{R} \times X$ and $z \in \gamma^+(\sigma, x)$ we define

$$t_z = \inf\{t \geq 0 | U(\sigma, t)x = z\} ,$$
$$Q_z = \{(\sigma + t_z, U(\sigma, t_z)x)\} ,$$
$$[P_o, Q_z] = \{(\sigma + t, U(\sigma, t)x) | 0 \leq t \leq t_z\} .$$

Let Ω be an open set of $\mathbb{R} \times X$, ω an open set of Ω, $\omega \neq \emptyset$ and $\partial\omega = \overline{\omega} \cap (\Omega - \omega)$ the boundary of ω with respect to Ω. We put

$$S^\circ = \{P_\circ = (\sigma, x) \in \partial\omega | \exists z \in \gamma^+(\sigma, x)$$
$$\text{with } (P_\circ, Q_z) \neq \emptyset \text{ and } (P_\circ, Q_z) \cap \overline{\omega} = \emptyset\}.$$
$$S = \{P_\circ = (\sigma, x) \in \partial\omega | \exists z \in \gamma^+(\sigma, x)$$
$$\text{with } (P_\circ, Q_z) \neq \emptyset \text{ and if there exists } \tau^-(P_\circ), \exists Q \in \omega$$
$$\text{such that } (Q, P_\circ] \subset \overline{\omega}\}$$
$$S^* = \{Q \in S^\circ | \exists P_\circ = (\sigma, x) \in \omega - S$$
$$\text{with } Q \in \tau^+(\sigma, x) \text{ and } [P_\circ, Q) \subset \omega - S\}$$

The points of S° are called demiegress points, the points of S are called egress points. The points of S^* are called strict egress points.

The above definition applies even if ω has an empty interior. For example, if $X = l^2$ and ω is the positive cone in l^2, $\omega = \{x \in l^2 | x \geq 0\}$, $S = \{x = (x_1, x_2, \ldots) | x \geq 0$ and $x_i > 0$ for at least one $i\}$. $\omega = \partial\omega$ and the trajectory through a point $P_\circ = (\sigma, x) \in \omega$ either leaves the set ω at some time $\overline{t} > \sigma$ or $\overline{t} = \sigma$ and then $\tau^+(P_\circ) \cap \overline{\omega} = \emptyset$ and $P_\circ \in S^\circ$. A point $P_\circ \in S^\circ$ is an egress point, that is, $P_\circ \in S$, if there exists an backward continuation of the process and a small piece of the trajectory is contained in ω. If $P_\circ \in S^\circ$ and there is a small piece of the left trajectory through $P_\circ$ contained in $\omega - S$ then $P_\circ \in S^*$, that is, $P_\circ$ is a strict egress point. If all points of S are strict egress points, that is, $S = S^*$, S^* is closed with respect to $\overline{\omega}$.

Given a point $P_\circ = (\sigma, x) \in \omega - S$, if the trajectory $\tau^+(\sigma, x)$ of the process is contained in $\omega - S$ for every $t > 0$ we say that the trajectory is asymptotic with respect to $\omega - S$. If the trajectory is not asymptotic with respect to $\omega - S$ then there is a $t > 0$ such that $(\sigma + t, U(\sigma, t)x) \in S$. Taking:

$$t_{P_\circ} = \inf\{t > 0 | (\sigma + t, U(\sigma, t)x) \in S\}$$
$$Q = (\sigma + t_{P_\circ}, U(\sigma, t_{P_\circ})x) = C(P_\circ)$$

we have

$$[P_\circ, Q] \subset \overline{\omega}\ .$$

The point $C(P_\circ)$ is called the consequent of $P_\circ$. Define G to be the set of all $P_\circ = (\sigma, x) \in \omega - S$ such that there are $C(P_\circ)$ and $C(P_\circ) \in S^*$. G is called the left shadow of ω. Consider the mapping, the consequent operator:

$$K : S^* \cup G \to S^*$$

$K(P_o) = C(P_o)$ if $P_o \in \omega$ and $K(P_o) = P_o$ if $P_o \in S^*$.

Lemma 1. *If $S = S^*$, the consequent operator $K : S^* \cup G \to S^*$ is continuous.*

Lemma 2. *If $S = S^*$ and the solution operator $U(t, \sigma)$ is a conditional condensing map for $t > \sigma$ then $K : G \to S$ is a conditional condensing map.*

Following Nussbaum [14] we say that a subset A of a Banach space X is admissible if A has a locally finite covering $\{A_j : j \in J\}$ by closed convex sets $A_j \in X$.

Let ω be a non-empty subset of Ω, and with S and S^* denoting respectively the set of egress and strict egress points of ω. Assume that there exists a non-empty closed set Z where $Z \subset \omega \cup S$ and the following conditions are satisfied:

i) $S = S^*$
ii) Z is admissible
iii) there exists a continuous map $\Phi : S \to S$ such that $\Phi(P) \neq P$ for every $P \in S$.
iv) ΦK is a condensing map
v) $i_{\bar{\omega}}(\Phi K, Z - S) \neq 0$.

Then there exists at least one point $P_o = (\sigma, x) \in Z - S$ such that the trajectory $\tau^+(\sigma, x)$ through $P_o = (\sigma, x)$ is contained in $\omega - S$.

Proof. Assume that the theorem is not true. Then $C(P_o) \in S$ for every $P_o \in Z - S$ and then $Z - S \subset G$. Then $Z = (Z \cap S) \cup (Z - S) \subset S \cup G$. From iv) ΦK is a condensing map and from iii) $\Phi(P) \neq P$ for every $P \in S$, that is $\Phi K(P_o) \neq P_o$ for every $P \in Z = S$. Hence $i_{\bar{\omega}}(\Phi K, Z - S) = 0$ which is a contradiction. Then there exists at least one point $P_o \in Z - S$ such that the trajectory $\tau^+(\sigma, P_o)$ is asymptotic with respect to $\omega - S$.

Corollary 1. *Let ω be a non-empty, subset of Ω, and with S and S^* denoting the set of egress and strict egress points of ω respectively. Assume that there exists a closed set Z, $\phi \neq Z \subset \omega \cup S$ and the following conditions are satisfied:*

i) $S = S^*$
ii) *Z is admissible*
iii) *there exists a continuous map $\Phi : S \to S$ such that $\Phi(P) \neq P$ for every $P \in S$*
iv) *ΦK is a condensing map and ΦK has a fixed point in Z.*

Then there exists at least one point $P_0 = (\sigma, x) \in Z - S$ such that the trajectory $\tau^+(\sigma, x)$ through $P_0(\sigma, x)$ is contained in $\omega - S$.

We will need the following Lemma [4], [14].

Lemma 3. *Let X be Banach space, $K \subset X$ a cone and $F : \overline{K_R} \to K$ a condensing map, $K_R = K \cap B_R(0)$. Suppose that*

a) *$Fx \neq \lambda x$ for $\|x\| = R$ and $\lambda > 0$.*
b) *There exists a smaller radius $r \in (0, R)$ and an $e \in K|\{0\}$ such that $x - Fx \neq \lambda e$ for $\|x\| = r$ and $0 < \lambda \leq 1$.*

Then F has a fixed point in $\{x \in K | r \leq \|x\| \leq R\}$.

3. MAIN RESULTS

Let X be a real Banach sequence space and consider the system of ordinary differential equations defined in X:

$$\begin{aligned} &\dot{x}_i + \sum_{j=1}^{\infty} a_{ij}(t) x_j = 0, \qquad i, 1, 2, \dots \\ &x_i(0) = x_i^0 \end{aligned} \tag{1}$$

where $a_{ij}(t)$ are continuous functions of the real variable t for $0 \leq t < \infty$, $x^0 = (x_1^0, x_2^0, \dots) \in X$.

System (1) can be written in the form

$$\begin{aligned} &\dot{x} + A(t)x = 0 \\ &x(0) = x^0 \ . \end{aligned} \tag{2}$$

We assume that for each $t \in [0, T]$, $-A(t)$ is the infinitesimal generator of a C^0 semigroup on the space X, the domain $D(A(t)) = D$ is independent of t, is dense in X and that the initial value problem (1) has an unique classical solution defined in $[0, \infty)$. Assume also continuity with respect to initial conditions for the solutions of (1). See [8] and [11]. Our purpose here is to apply Theorem 1 to prove the existence of a positive solution of system (1).

Theorem 2. *Assume the hypotheses*

i). *The solution operator*

$$U(t, \sigma)x^0 = x^0 - \int_\sigma^t A(s)x(s)ds$$

is a conditional condensing map for $t > \sigma$

ii) $\sum_{j=1}^{\infty} a_{ij}(t)x_j > 0$ *for every* $i = 1, 2, \ldots, x_j \geq 0, j = 1, 2, \ldots$.

Then system (1) *has a monotone decreasing solution* $x(t) = (x_1(t), x_2(t), \ldots)$, $x(t) \not\equiv 0$ *such that* $x_i(t) > 0$ *and* $\dot{x}_i(t) < 0$ *for every* $i = 1, 2, \ldots, t \geq 0$, *and consequently* $x_i(t)$ *are monotone decreasing for* $t \geq 0$.

If the solution operator $K(t, 0)x^\circ = x^\circ - \int_\sigma^t A(s)x(s)ds$ *is compact for* $t > \sigma$ *and* $\sum_{j=1}^{\infty} a_{ij}(t)x_j \geq 0$ *for every* $i = 1, 2, \ldots$, $x_j \geq 0$, $j = 1, 2, \ldots$. *Then* (1) *has as monotone non-increasing solution* $x(t) = (x_1(t), x_2(t), \ldots), x(t) \not\equiv 0$, *such that* $\dot{x}_i(t) \leq 0$ *for every* $i = 1, 2, \ldots$, $t \geq 0$ *and consequently* $x_i(t)$ *are monotone non-increasing for* $t \geq \sigma$.

Proof. Let us assume first that

$$\sum_{j=1}^{\infty} a_{ij}(t)x_j > 0, \qquad \text{for every } i, \quad x_j \geq 0, \quad j = 1, 2, \ldots$$

or equivalently, that $A(t)$ is a strongly positive operator.

For $\sigma > 0$ let

$$\omega = \{x = (x_1, x_2, \ldots) \in X | x_i > 0, \quad i = 1, 2, \ldots\} .$$

For $0 < r < R$ let

$$Z = \{x = (x_1, x_2, \ldots) \in X | x_i \geq 0, r \leq \|x\| \leq R\}$$
$$S = \{x = (x_1, x_2, \ldots) \in X | x_i \geq 0, x_i = 0 \text{ and } x_j \neq 0 \text{ for at least one } i \text{ and one } j\} .$$

The closure of $\omega, \overline{\omega}$ is a cone in the space X and Z is a conic sector. If $x = c$ or $x = l^\infty$, $\overline{\omega}$ is a solid cone in X.

From hypothese ii), $\dot{x}_i < 0$, then the derivatives along the solutions of (1) on the points of S are negative, then the points of S are strict egress points. At the origin $\dot{x}_i = 0$ for every i, whence the origin is not an egress point. The derivative along the solutions of (1) on the points of $Z_R = \{x \in Z | \|x\| = R\}$ and $Z_r = \{x \in Z | \|x\| = r\}$ are negative, $\dot{x}_i < 0$. Then the points of Z_r are strict egress points and the points of $Z_R - S$ are strict ingress points. The points of $Z_R \cap S$ are not egress points. The continuous function $\Phi : S \to S$ defined by $\Phi(x_1, x_2, \dots) = (0, x_1, x_2, \dots)$ satisfies $\Phi(x) \neq x$ for every $x \in S$.

Assume that for every $x^\circ \in Z - S$, the solution through x° does not remain in ω for $t > 0$. Since the points of S are strict egress points $Z - S \subset G$. The consequent operator $K : Z \to S$ defined by

$$K(t, o)(x^\circ)_i = x_i^\circ - \int_o^t \sum_{j=1}^{\infty} a_{ij}(s) x_j(s) ds, \qquad i = 1, 2, \dots , \tag{4}$$

is defined in $Z \subset \overline{\omega}$. Since the points of S are strict egress, K is defined in $Z \cap S$ then in $\overline{Z}$, that is, $\overline{Z} \subset G \cup S$. For Lemma 2, the consequent operator $K : G \to S$ is a conditional condensing map.

Since Φ is continuous, $\Phi_o K$ is continuous, and since $\dot{x}_i < 0$ for every i, K takes Z into $\overline{\omega} \cap \{x \in \overline{\omega} | \; ||x|| < R\}$ and then K is a condensing map. Since Φ is an isometry $\Phi_o K$ is a condensing map.

Finally to prove the Theorem, we have to prove that $\Phi_o K$ has a fixed point in Z. If we prove that the conditions a) and b) of Lemma 3, are satisfied then $\Phi_o K : Z \to \overline{\omega}$ has a fixed point in Z.

a) If $||x^\circ|| = R$ for $\lambda > 1$, $||\lambda x^\circ|| > R$. Since we are assuming that $\sum_{j=1}^{\infty} a_{ij}(t) x_j > 0$, x is decreasing, $||\Phi K(x^\circ)|| = ||K(x^\circ)|| \leq ||x^\circ|| < ||\lambda(x^\circ)||$ then $||\lambda x^\circ - \Phi K(x^\circ)|| \geq ||\lambda x^\circ|| - ||\Phi K(x^\circ)|| > 0$ and $\lambda x^\circ \neq \Phi K(x^\circ)$

b) Fix $e = (0, 1, 0, 0, \dots)$. If $x_1^\circ \neq 0$, $x^\circ - \Phi_o K(x^\circ) = (x_1^\circ, x_2^\circ, \dots) - (0, x_1^\circ, x_2^\circ, \dots) + \left(0, \int_\sigma^t \sum_{j=1}^{\infty} a_{1j}(s) x_j(s) ds, \int_\sigma^t \sum_{j=1}^{\infty} a_{2j}(s) x_j(s) ds, \dots \right) \neq \lambda(0, 1, 0, 0, \dots)$, since the first coordinate $x_1^\circ \neq 0$.

If $x_1^\circ = 0$, $x^\circ \in S$ and $K x^\circ = x^\circ$ then

$$x^\circ - \phi_o K(x^\circ) = (0, x_2^\circ, x_3^\circ, \dots) - (0, 0, x_2^\circ, x_3^\circ, \dots) = (0, x_2^\circ, x_3^\circ - x_2^\circ, x_4^\circ - x_3^\circ, \dots)$$

Let us assume by contradiction that

$$x^\circ - \phi_o K(x^\circ) = \lambda e$$

that is

$$(0, x_2^\circ, x_3^\circ - x_2^\circ, x_4^\circ - x_3^\circ, \dots) = (0, \lambda e_1, \lambda e_2, \dots)$$

If $x_2^\circ = 0$ obviously $x^\circ - \phi_o k(x^\circ) \neq \lambda e$ since $\lambda e_1 \neq 0$.

If $x_2^\circ \neq 0$ we have

$$(0, x_2^\circ, x_3^\circ - x_2^\circ - x_3^\circ, \dots) = (0, \lambda e_1, \lambda e_2 + \lambda e_3, \dots)$$

that is

$$x_2^\circ = \lambda e_1, x_3^\circ = x_2^\circ + \lambda e_2 = \lambda e_1 + \lambda e_2 \ ,$$
$$x_4^\circ = x_3^\circ + \lambda e_3 = \lambda e_1 + \lambda e_2 + \lambda e_3 + \dots$$

Then

$$x^\circ = \lambda \left(0, e_1, e_1 + e_2, e_1 + e_2 + e_3, \dots \sum_{i=1}^{n} e_i, \dots \right)$$

and x° can not belong to l^P for any $p \geq 1$ and x° can not belong also to $c_\circ$. If $x = c$ or l^∞ since $\sum_{i=1}^{\infty} e_i < r$ and $\lambda < 1$

$$r = \|x^\circ\| = \lambda \sum_{i=1}^{\infty} e_i < r$$

a contradiction.

Therefore all conditions of Corollary 1 are satisfied and there exists at least one point $x^\circ \in Z - S$ such that the solution of (1) $x(t) = (x_1(t), x_2(t), \dots)$ through x° stays in $\omega - S$ for every $t \geq 0$ and since $\dot{x}_i < 0$, $i = 1, 2, \dots$, each $x_i(t)$ decreases monotonically to zero as $t \to \infty$.

If the solution operator $K(t, t_\circ)x^\circ = x^\circ - \int_{t_\circ}^t A(s)x(s)ds$, is compact for $t > t_\circ$ and $\sum_{j=1}^{\infty} a_{ij}(t)x_j \geq 0$, consider the system

$$\begin{aligned} \dot{Y}_i + \sum_{j=1}^{\infty} (a_{ij}(t) + \varepsilon_{ij}) Y_i = 0 \\ Y_i(0) = Y_i^\circ, \qquad i =, 1, 2 \dots . \end{aligned} \tag{4}$$

From the proof above, there exists for each ε_{ij}, a positive solution $Y_n(t)$ of (4) through some $Y_n^\circ = (Y_{1n}^\circ, Y_{2n}^\circ, \dots)$ and $Y_n(t)$ decreases monotonically as $t \to \infty$. When $E_{ij} \to 0$, there is a sequence of positive solutions $Y_n(t)$ of (4) through Y_n°. Let $E = \{Y_n^\circ\}$. If $K(t, 0)$ is compact for $t > t_\circ$, for $\bar{t} > 0$, the set $\{K(\bar{t}, 0)Y_{in}^\circ\} = \{Y_{in}(\bar{t})\}$ is compact and there exists a convergent subsequence $\{Y_{ink}(\bar{t})\}$, $Y_{ink}(\bar{t}) \to Y_\circ(\bar{t})$ and the solutions $Y_{ink}(t) \to Y(t)$, $Y(\bar{t}) = Y_\circ(\bar{t})$ on every interval $\bar{t} \leq t \leq T < \infty$.

The system

$$\dot{x}_i + \sum_{j=1}^{\infty} a_{ij}(t) = -f_i(t,x)$$
$$x_i(0) = x^\circ, \quad i = 1,2,\dots, \tag{5}$$

can be written in the form

$$\dot{x} + A(t)x = -f(t,x)$$
$$x(0) = x^\circ . \tag{6}$$

We assume that for each $t \in [0,T]$, $-A(t)$ is the infinitesimal generator of a C°-semigroup on the space X, the domain $D(A(t)) = D$ is independent of t, is dense in X, $f : [0,\infty) \times U \to X$ is continuous, $U \subset X$, open, $f(t,0) = 0$ and we assume existence of a unique classical solution of (4) in $[0,\infty)$, as well as continuity with respect to initial conditions; see [9] and [12].

Theorem 3. *Assume the hypotheses*

i) *The solution operator*

$$U(t,\sigma)x^\circ = x^\circ - \int_\sigma^t A(s)x(s)ds - \int_\circ^t f(t,x)$$

is a conditional condensing map for $t > \sigma$.

ii) $\sum_{j=1}^{\infty} a_{ij}(t)x_j + f_i(t,x) > 0$ *for every* $i = 1,2,\dots, x_j \geq 0,\ j = 1,2,\dots$.

Then system (5) has a solution $x(t) = (x_1(t), x_2(t), \dots)$, $x(t) \not\equiv 0$, such that $x_i(t) > 0$, $\dot{x}_i < 0$ for every $i = 1,2,\dots$, $t \geq 0$, and consequently the $x_i(t)$ are monotone decreasing.

If the solution operator $K(t,0)x^\circ = x^\circ - \int_\circ^t A(s)x(s) - \int_\circ^t f(s,x(s))ds$ is conditionally compact for $t > \sigma$ and $\sum_{j=1}^{\infty} a_{ij}(t)x_j \geq 0$ for every $i = 1,2,\dots$, $x_j \geq 0$, $j = 1,2,\dots$, then system (5) has solution $x(t) = (x_1(t), x_2(t), \dots)$ such that $x(t) \not\equiv 0$, $x_i(t) \geq 0$ and $x_i(t) \leq 0$ for every $i = 1,2,\dots$, $t \geq 0$ and consequently $x_i(t)$ are monotone non-increasing for $t \geq 0$.

The proof follows as in Theorem 2, assuming that $\sum_{j=1}^{\infty} a_{ij}(t)x_j + f_i(t,x) > 0$ and the conclusion is that there exists a positive solution of (5) through some point $x_o = (x_1^o, x_2^o, \dots) \in \omega - S$. If the solution operator

$$U(t,t_o)x^o = x^o - \int_{t_o}^{t} A(s)x(s)ds - \int_{t_o}^{t} f(s,x(s))ds$$

is conditionally compact and $\sum_{j=1}^{\infty} a_{ij}x_j + f_i(t,x) \geq 0$, from Lemma 2 and since this implies that K is compact, consider the system

$$\begin{aligned} &\dot{Y}_i + \sum (a_{ij}(t) + \varepsilon_{ij})x_j + f_i(t,x) = 0 \\ &Y_i(0) = Y_i^o \end{aligned} \tag{7}$$

and a subsequence $Y_{o,k_j}(t)$, $Y_{o,k_j}(0) = Y_i^o$ which converges, as $\varepsilon_{ij} \to 0$, uniformly on every interval $0 \leq t \leq t < \infty$.

For system (7) the solution $x(t)$ can become zero after a finite time. Theorem 3 generalizes a result of Hartman and Wintner [7].

Example 1. Let $\{a_i\} \in c$ the space of convergent sequences with norm $\|a\| = \sup_i |a_i|$. Assume that $\lim_{i\to\infty} a_i = a_\infty \neq 0$ and define $\alpha^i = (0,0,\dots, a_i, 0, \dots)$. Define $T(t)\alpha^i = \{e^{\lambda_i}\alpha^i\}$, $-\infty < \text{Re}\,\lambda_i \leq W < \infty$. $T(t)$ is a strongly continuous semigroup with infinitesimal generator A given by $A\alpha^i = \{\lambda_i\alpha^i\}$. $T(t)$ is compact if and only if $\underline{\lim}\,\text{Re}\,\lambda_i = -\infty$. Consider the system (8) $\dot{x}_i = -\lambda_i x_i - \sum_{j=1}^{\infty} g_{ij}(t)x_j$, $x_i(t_o) = x_i^o$ with $\lambda_i > 0$, $\sum_{i,j} \|g_{ij}\| < \infty$. This system can be written in the form $\dot{x} = Ax + G(t)x$, $x(t_o) = x^o$. Since $T(t)$ is compact and $G(t)$ is bounded, the consequent operator

$$K(t,t_o)x^o = x^o + \int_{t_o}^{t} (A + G(x))ds$$

is compact. From Theorem 2, there exists at least one monotone solution $x(t) = (x_1(t), x_2(t), \dots)$, of (8) such that $\lim_{t\to\infty} x(t) = 0$, $x(t) \geq 0$ and $-\dot{x}(t) \geq 0$.

Example 2. Let X be a Banach sequence space and consider the system (9) $\dot{x} = (A + B)x$. Assume that B from X to X is compact. A is the infinitesimal generator of a C^o–semigroup e^{At}, $t \geq 0$ in X and there exist constants $M, \gamma > 0$ such that $|e^{At}| \leq e^{-\gamma t}$, $t \geq 0$. If all solutions of system (9) are defined for $t \geq 0$ then the semigroup defined by (9) satisfies $T(t) = e^{At} + U(t)$ where $U(t)$ is conditionally compact (conditionally completely continuous) and therefore $T(t)$ is a conditional α- contraction. [6, pp. 123] and Webb [15]. If $A + B < 0$

there exist a solution $x(t)$ of (7) such that $x(t) > 0$, $\lim_{t \to \infty} x(t) = C$ and $\dot{x}(t) < 0$, $t > 0$.

References

1. N. Arley and V. Borchsenius, *On the theory of infinite systems of differential equations and their applications to the theory of stochastic processes and the perturbation theory of quantum mechanics*, Acta Math. 76-261-322 (1945).
2. D. R. Cox and H. D. Miller, *The theory of stochastic processes*, Methuen, London (1965).
3. K. Deimling, *Ordinary Differential Equations in Banach Spaces*, Lectures Notes **596**, Springer-Verlag (1977).
4. K. Deimling, *Nonlinear Functional Analysis*, Springer-Verlag, Berlin Heidelberg (1985).
5. R. W. Dickey, *Infinite systems of nonlinear oscillation equation related to the string*, Proc. Amer. Math. Soc. 23-459-468 (1969).
6. J. K. Hale, *Asymptotic behavior of dissipative systems. Mathematical surveys and monographs*, Amer. Math. Soc., **25** (1988).
7. P. Hartman and A. Wintner, *Linear differential and difference equation with monotone solutions*, Amer. J. Math. **75** (1953), 731–743.
8. P. Hartman and A. Wintner, *On monotone solutions of systems of nonlinear differential equations*, Amer. J. Math. (1954), 860–866.
9. D. Henry, *Geometry theory of semilinear parabolic equation*, Lecture Notes in Mathematics, **840** Springer-Verlag.
10. A. F. Izé, *On a fixed point index method for the analysis of the asymptotic behavior and boundary value problems of infinite dimensional dynamical systems and processes*, J. Differential Equations, **52** (2) (April 1984).
11. A. F. Izé, *An infinite dimensional extension of theorems of Hartman and Wintner on monotone positive solutions of ordinary differential equations*, Proc. Amer. Math. Soc.
12. T. Kato, *Quasilinear Equation of Evolution, with Applications to Partial Differential Equations*, Lecture Notes in Mathematics **448** ed. W. N. Everitt, Springer-Verlag, Berlin (1974).
13. M. A. Krasnoselskii, *Positive Solutions of Operator Equations*, Noordhoff, Groningen (1964).
14. R. G. Nussbaum, *Periodic solutions of some nonlinear, autonomous functional differential equations II.*, J. **14** (1973), 380–394.
15. G. Webb, *Compactness of bounded trajectories of dynamical systems in infinite dimensional spaces*, Proc. Roy. Soc. Edinburg Sect. A. (1979a), 19–33.

Functional Analysis, Approximation Theory
and Numerical Analysis
Ed. John M. Rassias

BANACH SPACES IN BERGMAN OPERATOR THEORY

Erwin Kreyszig

Carleton University, Ottawa, Canada, K1S 5B6

Abstract

We show how Banach space theory can be used for certain aspects of integral operators of Bergman type, which serve as transformations for translating methods and results on analytic functions into analogs for solutions of linear partial differential equations.

1. The Creation of Banach Space Theory: F. Riesz, S. Banach

The introduction of normed spaces and, in particular, Banach spaces, was an important landmark in the early development of *functional analysis*, which is usually considered to have begun in 1887 with five Notes on functionals by Volterra [*Opere* **1**, 294-328]., followed in 1906 by the appearance of three important papers, first, Fréchet's famous thesis containing the modern axiomatic definition of metric space, second, Hilbert's most important (the fourth) of his six *Mitteilungen* on integral equations, the earliest truly functional-analytic theory of those equations, and third, the paper by F. Riesz [*Oeuvres* I, 110-154] containing an axiomatic definition of topological space based on "Verdichtungsstelle" (we now say *Häufungspunkt*, accumulation point, *not* condensation point!), along with many modern ideas of general topology. This paper, obscurely published and included only in part in Riesz's 1908 paper given at the International Congress of Mathematicians, Rome [*Atti* 2, 18-24], was also influential on the basic work of 1921 by L. Vietoris, culminating with the introduction of the Vietoris separation axiom. Its German translation finally became more widely known in 1960 by the publication of Riesz's *Oeuvres*. Riesz's axioms, in modern formulation, were as follows.

For each subset S and point p, it is defined whether p is an accumulation point of S or not (is "isolated"), and this relation satisfies the axioms 1.-4.:

1. If S is finite, it has no accumulation points.
2. An accumulation point p of S is also an accumulation point of any set containing S.
3. If S is partitioned into S_1 and S_2, then any accumulation point of S is also an accumulation point of S_1 or S_2 (or both).
4. For an accumulation point p of S and a point $q \neq p$ there is a subset $\tilde{S}$ of S such that p is an accumulation point of $\tilde{S}$, but q is not.

Based on this, Riesz then defined neighborhood, interior point, boundary point, open set, connectedness, etc. Clearly, that was an early (not fully successful) attempt to define topological space, without using any distance concept.

With Hilbert space theory well under way, one recognized soon, certainly around 1909 when Riesz introduced his famous representation of bounded linear functionals on $C[a, b]$, that for numerous applications, Hilbert spaces are not general enough. One was thus looking for *suitable* more general spaces, and it seems that the idea of combining vector space structure with metric (missed by Fréchet at that early time) that led to the breakthrough in the form of normed spaces was somehow "in the air". And it was again Riesz who did the first step in 1915 or 1916 in his famous *Acta mathematica* paper (submitted 1916, publishing date of the volume 1918, delayed by the War) on compact operators (abstract Fredholm theory). There he phrased matters in terms of $C[a, b]$, but his axioms were those of what we now call a Banach space, because once he had defined the norm on $C[a, b]$, in that paper he never used anything else but the axioms of a complete normed space. And he emphasized clearly that he had much more in mind than $C[a, b]$:

> "The restriction to continuous functions made in this paper is not essential... [and] the...case treated here may be regarded as a test case (*Prüfstein*) for the general applicability [of the method]."

As Bourbaki [4], 268, phrased it, it seems that "only the scruples of a careful analyst to go away too far from classical mathematics" kept Riesz from a totally abstract formulation of his theory, as it is now common (see [10], Chap. 8).

Fours years later, in 1922, the appearance of Banach's thesis marked the beginning of a systematic theory of normed spaces. It is interesting that short before, papers by Helly and by Hahn also contained the axioms of normed space, and in 1922, N. Wiener published another independent paper in which he also stated equivalent axioms and advocated the use of *complex* scalars, which Banach had not used, for unknown reasons. It took Banach, Hahn, Steinhaus and others only ten years to fully develop the "elementary" theory of normed and Fréchet spaces, as it appeared in 1932 in Banach's classic [1], a book of great influence on the further development of functional analysis.

2. *HB* Spaces in the Theory of Operators of Bergman Type

$HB(\Omega)$ denotes the Banach space of holomorphic and bounded functions on a domain $\Omega \subset \mathbb{C}^n$, taken with the maximum norm. We need the case $n = 2$ and choose $\Omega = \Omega_1 \times \Omega_2 \subset \mathbb{C}^2$, where

$$\Omega_1 = \{z \mid |z| < \rho\}, \quad \Omega_2 = \{z^* \mid |z^*| < \rho\},$$

where $\rho > 0$ is fixed. We shall be concerned with partial differential equations of the form

$$Lu = u_{zz^*} + b(z, z^*)u_{z^*} + c(z, z^*)u = 0, \quad (z, z^*) \in \Omega. \tag{2.1}$$

Here, u_z has been eliminated in the usual fashion, without restricting generality.

We now take $X = HB(\Omega_{1,R})$ and $Y = HB(\Omega_R)$; here,

$$\Omega_{1,R} = \{z \mid |z| \leq R,\ R = \frac{1}{2}(\rho - \eta),\quad 0 < \eta < \frac{\rho}{2}\}, \tag{2.2}$$

the norm being defined by

$$\|f\|_R = \max_{z \in \Omega_{1,R}} |f(z)|, \tag{2.3}$$

and

$$\Omega_R = \Omega_{1,R} \times \Omega_{2,R},\quad \Omega_{2,R} = \{z^* \mid |z^*| \leq R,\ R \text{ as above}\}, \tag{2.4}$$

the norm being defined by

$$\|u\|_R = \max_{(z,z^*) \in \Omega_R} |u(z, z^*)|. \tag{2.5}$$

Our "retreat" from Ω to $\Omega_R \subset \Omega$, necessitated by the convergence of the Bergman series for the kernel of the operator to be defined, is notationally slightly more practical than assuming holomorphy of b and c on, say, $|z| < 3\rho,\ |z^*| < 3\rho$, to avoid it.

We then define a linear operator $T : X \to Y,\ f \mapsto u = Tf$ by

$$u(z, z^*) = Tf(z, z^*) = \int_{-1}^{1} k(z, z^*, t) f\left(\frac{z}{2}(1 - t^2)\right)(1 - t^2)^{-\frac{1}{2}}\, dt, \tag{2.6}$$

integrating from -1 to 1 along a C^1-arc C in $D = \{t \mid |t| \leq 1\} \subset \mathbb{C}$.

Condition *(A)*. The *kernel* is a holomorphic solution of

$$(1 - t^2)k_{z^*t} - t^{-1}k_{z^*} + 2ztLk = 0 \tag{2.7}$$

on $\Omega \times D$ such that $(1 - t^2)k_{z^*} \to 0$ as $t \to \pm 1$ uniformly in a neighborhood N of $0 \in \Omega$ and if C passes through the origin, then k_{z^*}/zt is continuous in $N \times D$.

Condition *(B)*. $b,\ c \in \mathbb{C}^\omega(\Omega)$.

Theorem 2.1. *(A), (B) imply* $Lu = LTf = 0$ *on* Ω. ([3], 10)

T is then called a *Bergman operator* for (2.1) on Ω. It is called *of first kind* and denoted by T_1 if its kernel satisfies

Condition *(C)*. $k|_{z=0} = k|_{z^*=0} = 1$.

This condition holds if

$$k(z, z^*, t) = 1 + \sum_{n=0}^{\infty} q_n(z, z^*) z^n t^{2n}, \quad q_n(z, 0) = 0. \tag{2.8}$$

Lemma 2.2. *(A), (B) imply absolute and uniform convergence of* (2.8) *on* $\frac{1}{2}\bar{\Omega} \times D$. ([9], 21).

In addition to T_1 we can also consider $\tilde{T}_1 : X \to Y$, $g \mapsto u = \tilde{T}_1 g$, defined by $\tilde{T}_1 g(f) = T_1 f$, where

$$g(z) = \int_{-1}^{1} f\left(\frac{z}{2}(1-t^2)\right)(1-t^2)^{-\frac{1}{2}}\, dt, \tag{2.9}$$

the contribution of the term 1 in (2.8) to (2.6).

Lemma 2.3. *The Maclaurin series of* $f(\frac{1}{2}z)$ *and* $g(z)$ *have the same radius of convergence.*

PROOF. This follows from

$$f(z) = \sum_{n=0}^{\infty} a_n z^n, \quad g(z) = \sum_{n=0}^{\infty} b_n z^n, \quad b_n = \frac{1}{2^n} B\left(\frac{1}{2}, n + \frac{1}{2}\right) a_n.$$

Theorem 2.4. *If* b, $c \in HB(\Omega)$ *in* (2.1), *then the Bergman operator of the first kind,*

$$\tilde{T}_1 : HB(\Omega_{1,R}) \to HB(\Omega_R), \quad g \mapsto u = \tilde{T}_1 g$$

is bounded.

PROOF. u in (2.6) with $T = \tilde{T}_1$ satisfying *(A)* and *(C)* can be represented (see (4b) i [3], 15)

$$\begin{aligned} u(z, z^*) &= \tilde{T}_1 g(z, z^*) \\ &= g(z) + \sum_{n=1}^{\infty} \frac{Q^{(n)}(z, z^*)}{2^{2n} B(n, n+1)} \int_0^z (z-\zeta)^{n-1} g(\zeta)\, d\zeta, \end{aligned} \tag{2.10}$$

where

$$Q^{(n)}(z, z^*) = \int_0^{z^*} P^{(2n)}(z, \zeta^*)\, d\zeta^*. \tag{2.11}$$

If $(z, z^*) \in \Omega_R$, then

$$|g(z)| \leq \|g\|_R, \quad |Q^{(n)}(z, z^*)| \leq \|Q^{(n)}\|_R.$$

Hence at any (z, z^*) where the series (2.10) converges,

$$|u(z,z^*)| \leq \|g\|_R + \sum_{n=1}^{\infty} \frac{\|Q^{(n)}\|}{2^{2n}B(n,n+1)} \left| \int_0^z (z-\zeta)^{n-1} g(\zeta)\, d\zeta \right|.$$

Clearly, the integral does not exceed $R^n \|g\|_R / n$ in absolute value. By the definition of dominants (see (14) in [3], 14),

$$|P^{(2n)}(z,z^*)| \leq \frac{2^{n+1}\Lambda_n K}{\left(1-\frac{|z|}{\rho}\right)^n \rho^{n-1} 1 \cdot 3 \cdots (2n-1)}, \tag{2.12}$$

where

$$\Lambda_n = \prod_{j=1}^{n-1}(A+j) = \frac{\Gamma(A+n)}{\Gamma(A+1)}, \quad A = 8K\rho(1+\rho).$$

ρ is fixed and $4K$ is an upper bound for $|b(z,z^*)|$ and $|c(z,z^*)|$ in Ω_R, resulting from

$$|b(z,z^*)| \leq \frac{K}{\left(1-\frac{|z|}{\rho}\right)\left(1-\frac{|z^*|}{\rho}\right)} < 4K \text{ when } |z|,\ |z^*| \leq R < \frac{\rho}{2}$$

and similarly for $c(z, z^*)$. Now if $|z| \leq R < \rho/2$, then $(1 - |z|/\rho)^{-n} < 2^n$, so that from (2.11) and (2.12),

$$\|Q^{(n)}\|_R \leq \frac{2^{2n}\Lambda_n K}{\rho^{n-2} 1 \cdot 3 \cdots (2n-1)}.$$

Together, with $1 \cdot 3 \cdots (2n-1)\ B(n, n+1) = (n-1)!/2^n$ we thus obtain

$$\|u\|_R \leq \|g\|_R \left[1 + K\rho^2 \sum_{n=2}^{\infty} \frac{\Lambda_n}{n!} \left(\frac{R}{R+\eta}\right)^n\right]. \tag{2.13}$$

The series on the right converges by the ratio test. Denoting its sum by M and taking the supremum over all g of norm one, we have

$$\|\tilde{T}_1\| \leq 1 + K\rho^2 M \tag{2.14}$$

and the theorem is proved.

In applications, a more explicit form of (2.14) is often practical:

Lemma 2.5. *For the operator in Theorem* 2.4,

$$\|\tilde{T}_1\|_R \leq 1 + \frac{1}{2}S^2 K\rho^2 (A+1)\ {}_2F_1(A+1, 1, 1, S),\ S = \frac{R}{R+\eta}. \tag{2.14*}$$

PROOF. Using the previous notations, we have from (2.11) and (2.12),

$$\|\tilde{T}_1\|_R \leq 1 + K\rho^2(A+1)H(S),$$

where

$$H(S) = \sum_{n=2}^{\infty} \frac{\tilde{\Lambda}_n}{n!} S^n$$

with

$$\tilde{\Lambda}_2 = 1, \quad \tilde{\Lambda}_n = \prod_{j=2}^{n-1} (A+j), \qquad n = 3, 4, \ldots.$$

so that

$$|H(S)| \leq \frac{1}{2} S^2 \,{}_2F_1(A+1, 1, 1, S)$$

and the result follows.

As a basic application of Theorem 2.4 we note

Theorem 2.6. *Let G be a total subset of associated $g \in HB(\Omega_{1,R})$. Then for a given equation (2.1) the set $\tilde{T}_1(G)$ is total in the set of solutions $\mathcal{S} \subset HB\ (\Omega_R)$ of (2.1).*

The proof is standard. Note that $S < 1$ by (2.2), whereas for $S = 1$ the hypergeometric series diverges because $\alpha + \beta - \gamma = A + 1 \geq 1$ (see [7], I, 57).

3. Hardy Spaces in the Theory of Operators of Bergman Type

In this section we choose $X = H^p(\Omega_1)$, $1 \leq p < \infty$, the Banach space of holomorphic functions on Ω_1 such that

$$M_p(r, f) = \left(\frac{1}{2\pi} \int_0^{2\pi} |f(re^{i\theta})|^p \, d\theta \right)^{1/p} \qquad (z = re^{i\theta}) \tag{3.1}$$

is bounded as $r \to \rho$, with norm defined by

$$\|f\|_p = \sup_{r \in J} M_p(r, f), \qquad J = (0, \rho).$$

We further choose $Y = H^p(\Omega)$, $1 \leq p < \infty$, the Banach space of holomorphic functions on $\Omega = \Omega_1 \times \Omega_2$ such that

$$M_p(\tilde{R}, u) = \left(\frac{1}{4\pi^2} \int_0^{2\pi} \int_0^{2\pi} |u(r_1 e^{i\theta_1}, r_2 e^{i\theta_2})|^p \, d\theta_1 \, d\theta_2 \right)^{1/p} \tag{3.2}$$

is bounded as $\tilde{R} = (r_1,\ r_2) \to (\rho, \rho),\ z = r_1 e^{i\theta_1},\ z^* = r_2 e^{i\theta_2}$, with

$$\|u\|_p = \sup_{\tilde{R} \in K} M_p(\tilde{R}, u), \qquad K = J \times J.$$

We then introduce $\tilde{T}_1 :\ X \to Y,\ g \mapsto \tilde{T}_1 g = u$, as defined by (2.6), (2.9), but now considered on the new space X into Y; for simplicity we use the same notation for the operator.

Theorem 3.1. *Conditions (A)-(C) in Sec. 2 imply that $\tilde{T}_1$ is bounded.*

PROOF. We use Bergman's standard notation from [3] and ideas from [13]. From (2.6) and (2.8) we have $u = g + F$, where

$$F(z, z^*) = \sum_{n=1}^{\infty} 2^{-2n} [B(n, n+1)]^{-1} q_n(z, z^*) \int_0^z (z - s)^{n-1} g(s)\ ds.$$

From this, the Minkowski inequality and θ_2-integration in the g-term,

$$\|u\|_p \le \sup_{r \in J} I_1 + \sup_{\tilde{R} \in K} I_2,$$

where

$$I_1 = \left(\frac{1}{2\pi} \int_0^{2\pi} |g(r_1 e^{i\theta_1})|^p\ d\theta_1 \right)^{1/p},$$

$$I_2 = \left(\frac{1}{4\pi^2} \int_0^{2\pi} \int_0^{2\pi} |F(r_1 e^{i\theta_1}, r_2 e^{i\theta_2})|^p\ d\theta_1\ d\theta_2 \right)^{1/p}.$$

From $|z - s| \le \rho$ and Lemma 2.2,

$$|F(z, z^*)| \le A \left| \int_0^z g(s)\ ds \right|$$

with a suitable constant A. By θ_2-integration,

$$I_2 \le \left(\frac{1}{2\pi} \int_0^{2\pi} A^p \left| \int_0^z g(s)\ ds \right|^p\ d\theta_1 \right)^{1/p}.$$

By Zygmund's formula (9.12) in [14], 19,

$$I_2 \le \int_0^{|z|} \left(\frac{1}{2\pi} \int_0^{2\pi} A^p |g(s)|^p\ d\theta_1 \right)^{1/p} |ds| \le A\rho \|g\|_p.$$

Now $\sup I_1 = \|g\|_p$, so that the completion of the proof, with a remarkably simple constant, is seen from

$$\|u\|_p \le (1 + \rho A)\ \|g\|_p.$$

As an important consequence of this theorem, we show next that the space of solutions of (2.1) generated by $\tilde{T}_1$ is complete.

Theorem 3.2. *$Y_1 = \tilde{T}_1(H^p(\Omega_1)) \subset Y = H^p(\Omega)$, $1 \leq p < \infty$, fixed, is a Banach space.*

PROOF. Since Y is a Banach space, it suffices to show that Y_1 is closed in Y, which follows by familiar arguments if one observes that $u(z,0) = g(z)$.

We claim that Y_1 can be made into a Banach algebra by defining

$$u_1 * u_2 = \tilde{T}_1(g_1 * g_2),$$

$$(g_1 * g_2)(z) = \sum_{n=0}^{\infty} \frac{c_n}{n!} z^n, \quad c_n \sum_{m=0}^{n} g_m^{(1)} g_{n-m}^{(2)}, \tag{3.3}$$

$$g_j(z) = \sum_{n=0}^{\infty} \frac{g_n^{(j)}}{n!} z^n, \quad j = 1, 2.$$

Theorem 3.3. *Let $u_j = \tilde{T}_1 g_j \in H^p(\Omega)$, $1 \leq p < \infty$, $j = 1,2$, be solutions of (2.1) satisfying (B). Then $u = u_1 * u_2 \in H^p(\Omega)$ and these solutions form a Banach algebra Y_1 with multiplication $*$, the identity being $\tilde{T}_1 1$.*

PROOF. $u_j \in H^p(\Omega)$ implies that $\sup_{\tilde{R} \in K} M_p(\tilde{R}, u_j) < \infty$. Hence, letting $r_2 = 0$, integrating over θ_2 and using $u_j(z,0) = g_j(z)$, we get $g_j \in H^p(\Omega_1)$. Now (3.3) converges on Ω_1, absolutely and uniformly on closed subsets of Ω_1, because it is majorized by the Cauchy product. By direct integration it follows that (3.3) is equivalent to

$$(g_1 * g_2)(z) = \frac{d}{dz} \int_0^z g_1(z-t) g_2(t)\, dt, \tag{3.4}$$

and Wigley [12] has shown that in this way one obtains a Banach algebra structure for $H^p(\Omega_1)$, with $g_1 = 1$ being the identity. Thus $g_1 * g_2 \in H^p(\Omega_1)$, and Theorem 3.1 implies that

$$u = u_1 * u_2 = \tilde{T}_1(g_1 * g_2) \in H^p(\Omega).$$

Hence Y_1 in Theorem 3.2 is an algebra with multiplication defined by (3.3), and is a Banach space by that theorem. To conclude that Y_1 is a Banach algebra, we map Y_1 into the Banach algebra $B(Y_1)$ of all bounded linear operators on Y_1. Denote this mapping by S and define it by $u \mapsto Su = \tilde{T}_{1u}$, where

$$\tilde{T}_{1u} v = u * v.$$

The mapping S is an isomorphism of Y_1 onto its range, the subalgebra $S(Y_1)$ of $B(Y_1)$. From the closed graph theorem it follows that S is also a homeomorphism, as is proved in

[5], 861. Hence the algebra Y_1 is algebraically and topologically equivalent to the Banach algebra $S(Y_1)$. This completes the proof.

Our results formulated for g, as a technical convenience, can readily by reformulated in terms of f by using

Lemma 3.4. *$f \in H^p(\Omega_1)$ implies $g \in H^p(\Omega_1)$ and*

$$\|g\|_p \leq 4^{1/p}\pi \|f\|_p \qquad (p \geq 1). \tag{3.5}$$

PROOF. For $z \in \Omega_1$ the inequality in the lemma in [6], 36, can be generalized to

$$|f(z)| \leq (2\rho)^{1/p}(\rho - r)^{-1/p}\|f\|_p \qquad (r = |z|).$$

By using this in $M_p(r, g)$, integrating over θ_1, setting $t = \cos\alpha$ and noting that $\rho - \frac{1}{2}r \sin^2\alpha \geq \frac{1}{2}\rho$ we obtain the result.

We prove next a multiplication theorem for associated functions.

Theorem 3.5 *If*

$$u_j = \tilde{T}_1 g_j \in H^2(\Omega), \qquad j = 1, 2, \tag{3.6}$$

with Ω as before, then

$$u = u_1 \times u_2 := \tilde{T}_1(g_1 g_2) \in H^1(\Omega).$$

PROOF. From (3.6) and the definition we have

$$\sup_{0<r_1,r_2<\rho} \int_0^{2\pi}\int_0^{2\pi} |u_j(z, z^*)|^2 \, d\theta_1 \, d\theta_2 < \infty, \quad j = 1, 2.$$

Setting $z^* = 0$, using $u_j(z, 0) = g_j(z)$ and integrating over θ_2, we obtain

$$\sup_{0<r_1<\rho} \int_0^{2\pi} |g_j(z)|^2 \, d\theta_1 < \infty.$$

Hence $g_j \in H^2(\Omega_1)$, $j = 1, 2$. Now

$$2|g_1(z)g_2(z)| \leq |g_1(z)|^2 + |g_2(z)|^2,$$

so that by integration over θ_1,

$$2\int_0^{2\pi} |g_1(r_1 e^{i\theta_1})g_2(r_1 e^{i\theta_1})| \, d\theta_1 \leq \|g_1\|_2^2 + \|g_2\|_2^2 < \infty.$$

Taking the sup on the left, we see that $g_1g_2 \in H^1(\Omega_1)$ and the assertion of the theorem follows from Theorem 3.1.

4. Transition to Bergman Spaces

In conclusion we mention that the extension of the results in Sec. 3 to Bergman spaces is immediate. It suffices to demonstrate this for the key theorem (Theorem 3.1).

By definition, a *Bergman space* B^p is the Banach space of all holomorphic L^p-functions $(p \geq 1)$ on a domain $\Omega \subset \mathbb{C}^n$. Thus the theory of these spaces extends the L^2-theory of holomorphic functions in a domain, as developed by Bergman in numerous papers and summarized in his book [2].

In connection with Bergman operators we take $\Omega = \Omega_1 \times \Omega_2$ with Ω_1 and Ω_2 as in Sec. 3. Then the norm on $B^p(\Omega_1)$ is defined by

$$\|g\|_p = \left[\int_0^\rho M_p^p(r_1, g) r_1 \, dr_1\right]^{1/p}; \tag{4.1}$$

here, M_p is given by (3.1) with $r = r_1$ and $\theta = \theta_1$. Similarly, the norm on $B^p(\Omega)$ is defined by

$$\|u\|_p = \left[\int_0^\rho \int_0^\rho M_p^p(\tilde{R}, u) r_1 \, dr_1 \, r_2 \, dr_2\right]^{1/p} \tag{4.2}$$

with M_p as in (3.2).

Space $H^p(\Omega_1)$ is a *closed* subspace of $B^p(\Omega_1)$ (see [8], 149), and this raises the question of whether the Bergman operator of the first kind can be continuously extended from $H^p(\Omega_1)$ to $B^p(\Omega_1)$. The answer is in the affirmative, as the following theorem shows. (See also Marzuq [11].)

Theorem 4.1. *The Bergman operator $\tilde{T}_1$ of the first kind, when regarded as an operator on $B^p(\Omega_1)$ into $B^p(\Omega)$, that is,*

$$\begin{aligned} &\tilde{T}_1 : B^p(\Omega_1) \to B^p(\Omega) \\ &g(z) \mapsto u(z, z^*) = \tilde{T}_1 g(z, z^*) \end{aligned} \tag{4.3}$$

is bounded. Here g is related to f in (2.6) as shown in (2.9).

PROOF. The proof is practically the same as that of Theorem 3.1. We first have

$$\begin{aligned} M_p(\tilde{R}, u) &\leq I_1 + I_2 \\ &\leq \left(\frac{1}{2\pi}\int_0^{2\pi} |g(r_1 e^{i\theta_1})|^p \, d\theta_1\right)^{1/p} + \rho A \left(\frac{1}{2\pi}\int_0^{2\pi} |g(r_1 e^{i\theta_1})|^p \, d\theta_1\right)^{1/p} \end{aligned} \tag{4.4}$$

Hence

$$M_p^p(\tilde{R}, u) \leq (1 + \rho A)^p \frac{1}{2\pi} \int_0^{2\pi} |g(r_1 e^{i\theta_1})|^p \, d\theta_1.$$

If we now integrate on both sides over r_1 and r_2 from 0 to ρ and raise the result to the power $1/p$, we obtain the assertion, namely

$$\|u\|_p \leq (1 + \rho A)\|g\|_p. \tag{4.5}$$

References

[1] Banach, S., *Théorie des opérations linéaires.* New York: Chelsea, 1932.

[2] Bergman, S., *The Kernel Function and Conformal Mapping.* 2nd ed. Providence, RI: American Mathematical Society, 1970.

[3] ——, *Integral Operators in the Theory of Linear Partial Differential Equations.* 3rd ed. New York: Springer, 1971.

[4] Bourbaki, N., *Éléments d'histoire des mathématiques.* Paris: Hermann, 1974.

[5] Dunford, N., and J.T. Schwartz, *Linear Operators.* Part II. New York: Interscience, 1963.

[6] Duren, P.L., *Theory of H^p Spaces.* New York: Academic, 1970.

[7] Erdélyi, A., (Ed.), *Higher Transcendental Functions.* 3 vols. New York: McGraw-Hill, 1953-55.

[8] Hoffman, K., *Banach Spaces of Analytic Functions.* Prentice-Hall, Englewood Cliffs, NJ, 1962.

[9] Kracht, M., and E. Kreyszig, *Methods of Complex Analysis in Partial Differential Equations with Applications.* New York: Wiley/Interscience, 1988.

[10] Kreyszig, E., *Introductory Functional Analysis with Applications.* New York: Wiley, 1989.

[11] Marzuq, M.M., Bergman spaces for the solution of linear partial differential equations. *Revista Colomb. Mat.* **18** (1984), 137-143.

[12] Wigley, N.M., A Banach algebra structure for H^p. *Canad. Math. Bull.* **18** (1975), 597-603.

[13] Yang, S.-N., *Biholomorphic mappings and H^p spaces on bounded symmetric domains.* Thesis. SUNY at Buffalo, 1976.

[14] Zygmund, A., *Trigonometric Series.* 2nd ed., reprinted. Cambridge: Cambridge University Press, 1977.

Functional Analysis, Approximation Theory
and Numerical Analysis
Ed. John M. Rassias

SOME CHARACTERIZATION PROBLEMS IN HILBERT SPACE

R. G. LAHA
Department of Mathematics and Statistics, Bowling Green State University
Bowling Green, Ohio 43403-0221 USA

ABSTRACT

In the present paper some characterization problems in Probability Theory are discussed in the general framework of a real separable Hilbert space.

Let $\mathcal{H}$ be a real separable Hilbert space with inner product $\langle\ \rangle$ and norm $\|\ \|$. Let (Ω,S,P) be a probability space and let $\mathcal{B}$ be the σ-field generated by the class of all open subsets of $\mathcal{H}$. Let X be a random variable taking values in $\mathcal{H}$, that is, X is a measurable mapping of (Ω,S) into $(\mathcal{H},\mathcal{B})$. Let μ_X be the probability distribution of the random variable X, that is μ_X is the probability measure on $\mathcal{B}$ induced by the measurable mapping X such that the relation

$$\mu_X(E) = P\{w \in \Omega:\ X(\omega) \in E\}$$

holds for all $E \in \mathcal{B}$. Then the characteristic functional $\hat{\mu}_X$ of the random variable X is the complex valued function on $\mathcal{H}$ given by the formula

$$\hat{\mu}_X(y) = \int_{\mathcal{H}} e^{i\langle x,y\rangle}\, d\mu_X(x), \quad y \in \mathcal{H}$$

A random variable X taking values in $\mathcal{H}$ is said to follow Gaussian law, if the characteristic functional $\hat{\mu}_X$ can be represented in the form

$$\hat{\mu}_X(y) = \exp\left[i\langle x_0,y\rangle - \frac{1}{2}\langle Sy,y\rangle \right], \quad y \in \mathcal{H}$$

where $x_0 \in \mathcal{H}$ is a fixed element of $\mathcal{H}$ and S is an S-operator in $\mathcal{H}$. In this connection we note that S-operator is a bounded linear positive Hermitian operator in $\mathcal{H}$ having a finite trace.

A random variable X taking values in $\mathcal{H}$ with characteristic functional $\hat{\mu}_X$ is said to be infinitely divisible, if for any positive integer $n \geq 1$, there exists a characteristic functional $\hat{\mu}_n$ such that the relation

$$\hat{\mu}_X(y) = [\hat{\mu}_n(y)]$$

holds for all $y \in \mathcal{H}$. It is well known[3] that a random variable X taking values in $\mathcal{H}$ is infinitely divisible, if and only if its characteristic functional $\hat{\mu}_X$ can be represented in the form

$$\hat{\mu}_X(y) = \exp\left[i\langle x_0, y\rangle - \frac{1}{2}\langle Sy, y\rangle + \int_{\mathcal{H}} K(x,y)\, d\upsilon(y) \right], \qquad y \in \mathcal{H}.$$

Here $x_0 \in \mathcal{H}$ is a fixed element in $\mathcal{H}$, S is an S-operator in $\mathcal{H}$ and υ is a σ-finite measure on $\mathcal{B}$ with finite mass outside every neighborhood of the origin $0 \in \mathcal{H}$ and satisfying the relation

$$\int_{\{x \in \mathcal{H}: \|x\| \leq 1\}} \|x\|^2 \, d\upsilon(x) < \infty.$$

Here the kernel K is given to the formula

$$K(x,y) = e^{i\langle x,y\rangle} - 1 - \frac{i\langle x,y\rangle}{1 + \|x\|^2}, \qquad (x, y \in \mathcal{H}).$$

Moreover the element $x_0 \in \mathcal{H}$, S-operator S and the measure υ are determined uniquely by $\hat{\mu}_X$.

Apparently it seems that the paper of Eaton and Pathak[1] is the first result on the characterization of an infinitely divisible law and Gaussian law in a real separable Hilbert space. The results of Eaton and Pathak can be summarized as follows:

Let $\hat{\mu}_X$ be the characteristic functional of a random variable X taking values in a real separable Hilbert space $\mathcal{H}$. Suppose that $\hat{\mu}_X$ satisfies the functional equation

$$\hat{\mu}_X(y) = \prod_{j=1}^{n} \hat{\mu}_X(B_j y)^{\alpha_j}, \qquad y \in \mathcal{H}$$

where $\alpha_j > 0$ and B_j is a bounded linear operator in $\mathcal{H}$ with a bounded inverse $(1 \le j \le n)$ and suppose that there exists a positive real number $\lambda_0 > 0$ $(0 < \lambda_0 < 1)$ such that $\|B_j\| \le \lambda_0$ for $1 \le j \le n$. Then the following assertions hold:

(i) The characteristic functional $\hat{\mu}_X$ is infinitely divisible.

(ii) Moreover suppose that $\sum_{j=1}^{n} \alpha_j B_j B_j^* \ge I$ where B_j^* is the adjoint of B_j and I is the identity operator in $\mathcal{H}$. Then $\hat{\mu}_X$ is the characteristic functional of a Gaussian law (possibly degenerate) in $\mathcal{H}$.

(iii) Suppose that $\sum_{j=1}^{n} \alpha_j \le 1$. Then $\hat{\mu}_X$ is the characteristic functional of a probability measure degenerate at the origin $0 \in \mathcal{H}$.

Then Rao[4] obtained the following result:

Let $\mathcal{V}$ be a real Hilbert space provided with an inner product $\langle\ \rangle$ and let f be a continuous complex valued function defined on $\mathcal{V}$ satisfying the functional equation $f(x + y) = f(x) + f(y)$ for all $x,y \in \mathcal{V}$ such that $\langle x,y \rangle = 0$. Then f is a polynomial of degree not greater than 2.

Recently the author of the present paper[2] obtained the following characterization of a Gaussian law in a real separable Hilbert space:

Let X and Y be two independently (but not necessarily identically) distributed random variables taking values in $\mathcal{H}$. Then the random variables $X + Y$ and $X - Y$ are independently distributed if and only if each of X and Y follows Gaussian law with identical S-operators.

The proof of this result depends on the solution of a functional equation in the general framework of a real separable Hilbert space as follows:

Let X be a random variable taking values in $\mathcal{H}$. Then X follows Gaussian law if and only if its characteristic functional $\hat{\mu}_X$ can be represented in the form

$$\hat{\mu}_X(y) = \exp[i\langle x_0, y\rangle - \theta(y)], \quad y \in \mathcal{H}$$

where $x_0 \in \mathcal{H}$ is a fixed element in $\mathcal{H}$ and θ is a continuous nonnegative function on $\mathcal{H}$ satisfying the functional equation

$$\theta(x + y) + \theta(x - y) = 2[\theta(x) + \theta(y)]$$

for all $x,y \in \mathcal{H}$.

Let f_1 and f_2 be two continuous complex valued functions defined on $\mathcal{H}$ satisfying the relation

$$f_1(x+y)\, f_2(x-y) = f_1(x)\, f_2(x)\, f_1(y)\, f_2(-y)$$

for all $x, y \in \mathcal{H}$. Then

$$f_1(y) = \exp[i\langle x_1, y\rangle - \theta(y)]: \ f_2(y) = \exp[i\langle x_2, y\rangle - \theta(y)]$$

for all $y \in \mathcal{H}$ where $x_1, x_2 \in \mathcal{H}$ are fixed elements in $\mathcal{H}$ and θ is a continuous nonnegative function on $\mathcal{H}$ satisfying the functional equation

$$\theta(x+y) + \theta(x-y) = 2[\theta(x) + \theta(y)]$$

for all $x, y \in \mathcal{H}$.

References

1. M. L. Eaton and P. K. Pathak, A characterization of the normal law in Hilbert space, *Sankhya*, Series A, Vol. **31**(1969), 259-268.

2. R. G. Laha, A characterization of Gaussian law in Hilbert space, *Aequationes Mathematicae*, Vol. **41**(1991), 85-93.

3. K. R. Parthasarthy, *Probability Measures on Metric Spaces*, Academic Press, New York, 1967.

4. C. R. Rao, Some problems in the characterization of multivariate normal distributions, Inaugural Linnik Memorial Lecture, *Statistical Distributions in Scientific Work*, Vol. **3**(1975), 1-13.

Functional Analysis, Approximation Theory
and Numerical Analysis
Ed. John M. Rassias

FIXED POINT PROCEDURE IN BANACH SPACES FOR CALCULATING PERIODICAL SOLUTIONS OF DUFFING TYPE EQUATIONS

SILVIO NOCILLA

Department of Aerospace Engineering, Politecnico of Torino,
Corso Duca degli Abruzzi 24–10129, Italy

Index

1. Introduction

The present study is devoted meanly to the nonlinear differential equations of Duffing type; see for instance [1], [2], [7]:

$$m\ddot{x} + c\dot{x} + Kx + K'x^3 = f_0 \sin \Omega t \tag{1.1}$$

with

$$f_0 = \begin{cases} F_0 & \text{in Ⓐ} \\ m h_0 \Omega^2 & \text{in Ⓑ} \end{cases} \tag{1.2}$$

corresponding to the mechanical models indicated in Fig. 1.

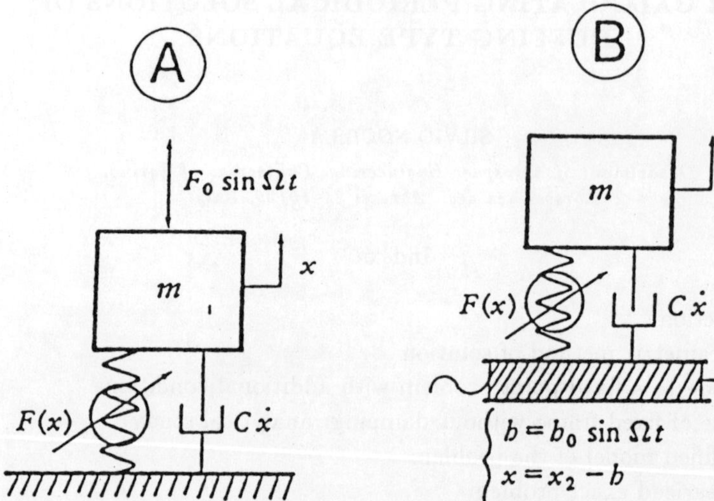

Fig. 1. Nonlinear systems with different kind of excitation: in all cases it is $F(x) = Kx + K'x^3$, $K' \gtrless 0$.

Same recent results obtained by the author and co-workers, see [3–11], are presented and discussed here. The general goal is the exact calculation of the periodical harmonical (i.e., with the same circular frequency Ω) monoscillating solutions, for arbitrary values of the parameters, in particular of the forcing circular frequency Ω. With the term "monoscillating" we indicate that the periodical solution exhibits only one oscillation in a period. In particular the response curves in amplitude and phase, as well as the wave forms, are to be calculated. The technique consists of reducing the calculation of the solution to a fixed-point problem in suitable Banach spaces and in solving that problem with iterative procedure. A general numerical procedure is indicated for both cases Ⓐ and Ⓑ, without and with damping, some analytical properties are demonstrated, in particular in the case Ⓐ without damping, i.e., with $c = 0$; and several numerical results are reported. for case Ⓐ. The application of the procedure to similar nonlinear problems is also indicated at the end of the present study.

2. The Parametric Method of Solution

The steady state harmonic monoscillating solution of Eq. (1.1) is searched in the parametric form, see [3]:

$$\begin{cases} x = x^* \sin\tau & x^* > 0 \\ \Omega t = \tau + \vartheta + \phi(\tau) & \vartheta \in [0, \pi] \\ \phi(\tau) = \int_{-\pi}^{\tau} \varphi(s) ds & \tau \in I_0 \equiv [-\pi/2, \pi/2] \end{cases} \tag{2.1}$$

$$\dot{x} = \frac{\Omega x^* \cos\tau}{1+\varphi(\tau)}, \qquad \ddot{x} = \frac{\Omega^2 x^*}{1+\varphi(\tau)} \frac{d}{d\tau}\left(\frac{\cos\tau}{1+\varphi(\tau)}\right) \tag{2.2}$$

with the additional conditions:

$$\begin{cases} \varphi(\tau+\pi) = \varphi(\tau) \\ \int_{-\pi/2}^{\pi/2} \varphi(\tau) d\tau = 0 \end{cases} \qquad \begin{cases} \varphi(\tau) > 1, \\ \varphi'(\tau) \text{ regular.} \end{cases} \tag{2.3}$$

Eq. (1.1) becomes:

$$\frac{\Omega^2 x^*}{1+\varphi(\tau)} \frac{d}{d\tau}\left(\frac{\cos\tau}{1+\varphi(\tau)}\right) + C\frac{\Omega x^* \cos\tau}{1+\varphi(\tau)}$$
$$+ K_1 x^* \sin\tau + K_3 x^{*3} \sin^3\tau = f_0 \sin(\tau + \vartheta + \phi(\tau)) \,. \tag{2.4}$$

Multiplying by $(1+\varphi(\tau))$ and integrating with respect to τ between $-\pi/2$ and τ we have:

$$\frac{\Omega^2 x^* \cos\tau}{1+\varphi(\tau)} + \int_{-\pi/2}^{\tau} [c\Omega x^* \cos s + (K_1 x^* \sin s + K_3 x^{*3} \sin^3 s)](1+\varphi(s)) ds$$
$$= f_0[\sin\vartheta - \cos(\tau + \vartheta + \phi(\tau))] \,. \tag{2.5}$$

Dividing by $K_1 x^*$ and setting:

$$\begin{cases} \xi = \Omega^2/K_1 \\ \eta = K_3 x^{*3}/K_1 \\ \gamma = c^2/(4K_1) \end{cases} \tag{2.6}$$

whence:

$$\begin{cases} c\Omega/K_1 = 2\sqrt{\gamma\xi} \\ f_0/(K_1 X^*) = \sqrt{\overline{f}_0/\eta} \\ \text{with: } \overline{f}_0 = f_0^2 K_3/K_1^3 \end{cases} \tag{2.7}$$

also taking into account the identity:

$$\sin^3 \tau = \frac{3\sin\tau}{4} - \frac{sin 3\tau}{4} . \tag{2.8}$$

Eq. (2.5) becomes:

$$\frac{\xi\cos\tau}{1+\varphi(\tau)} + \left(1+\frac{3\eta}{4}\right)\int_{-\pi/2}^{\tau} \sin s(1+\varphi(s))ds - \frac{1}{4}\eta\int_{-\pi/2}^{\tau} \sin 3s(1+\varphi(s))ds$$

$$+ 2\sqrt{\gamma\xi}(1+\sin\tau) = \sqrt{\frac{\overline{f}_0}{\eta}}[\sin\vartheta - \cos(\tau+\vartheta+\phi(\tau))] . \tag{2.9}$$

As regards the constants $(\eta, \overline{f}_0)$, see Eq. (2.6), it is clear that in the hardening case $K_3 > 0$ they are positive. At the contrary in the softening case $K_3 < 0$ they are negative. Then in the softening case we will modify Eqs. (2.6) and (2.7) setting

$$\eta = \frac{-K_3 x^{*2}}{K_1}; \qquad \overline{f}_0 = -f_0^2\frac{K_3}{K_1^3} \tag{2.10}$$

instead of the corresponding positions in Eqs. (2.6) 2nd and (2.7) 3rd. Therefore *in the softening case all formulae must be modified setting:*

$$\begin{cases} -\eta & \text{instead of } \eta \\ -\overline{f}_0 & \text{instead of } \overline{f}_0 \end{cases} .$$

Clearly, the ratio $\overline{f}_0/\eta$ remains unchanged.

3. Reduction to a Fixed-Point Problem with Additional Conditions

For $\tau = \pi/2$, also taking into account Eq. (2.3) 2nd, Eq. (2.9) gives:

$$4\sqrt{\gamma\xi} + \left(1+\frac{3\eta}{4}\right)\int_{-\pi/2}^{\pi/2} \varphi(\tau)\sin\tau d\tau - \frac{\eta}{4}\int_{-\pi/2}^{\pi/2} \varphi(\tau)\sin 3\tau d\tau = \sqrt{\frac{\overline{f}_0}{\eta}}\sin\vartheta \tag{3.1}$$

which plays the role of a "regularization condition". Replacing in Eq. (2.9) the value of $\sqrt{\gamma\xi}$ obtained by Eq. (3.1), dividing same Eq. (2.9) by $\cos\tau$, also taking into account the identity:

$$\frac{\cos 3\tau}{\cos\tau} = 2\cos 2\tau - 1$$

we obtain:

$$\frac{\xi}{1+\varphi(\tau)} - \left(1+\frac{5\eta}{6}\right) + \frac{\eta}{6}\cos 2\tau + \left(1+\frac{3\eta}{4}\right) Z - \frac{\eta}{4} V = -\sqrt{\frac{\overline{f}_0}{\eta}}(\cos\vartheta + J) \quad (3.2)$$

where following operators have been introduced:

$$\begin{cases} Z(\varphi,\tau) & = \dfrac{1}{\cos\tau}\left[\displaystyle\int_{-\pi/2}^{\tau} \varphi(s)\sin s ds - \frac{1+\sin\tau}{2}\int_{-\pi/2}^{\pi/2}\varphi(s)\sin s ds\right] \\ V(\varphi,\tau) & = \dfrac{1}{\cos\tau}\left[\displaystyle\int_{-\pi/2}^{\tau} \varphi(s)\sin 3s ds - \frac{1+\sin\tau}{2}\int_{-\pi/2}^{\pi/2}\varphi(s)\sin 3s ds\right] \\ J(\varphi,\tau,\vartheta) & = \dfrac{1}{\cos\tau}[\cos(\tau+\vartheta+\phi(\tau)) - \cos(\tau+\vartheta)] \\ & = J_0\cos\vartheta + J_1\sin\vartheta \end{cases} \quad (3.3)$$

with

$$\begin{cases} J_0(\varphi,\tau) = J(\varphi,\tau;0) = \dfrac{1}{\cos\tau}[\cos(\tau+\phi(\tau)) - \cos\tau] \\ J_1(\varphi,\tau) = J(\varphi,\tau;\pi/2) = \dfrac{1}{\cos\tau}[-\sin(\tau+\phi(\tau)) + \sin\tau] \,. \end{cases} \quad (3.4)$$

Despite the divisor $\cos\tau$ the above operators Z, V, J are regular also for $\tau = \pm\pi/2$. In fact applying the De l'Hospital rule we obtain (see also Sec. 9):

$$\begin{cases} Z(\varphi_0,\tau_0) = -\varphi_0 \\ V(\varphi_0,\tau_0) = \varphi_0 \\ J(\varphi_0,\tau_0;\vartheta) = \cos\vartheta\varphi_0 \end{cases} \quad \text{with :}\begin{cases} \tau_0 = \pm\pi/2 \\ \varphi_0 = \varphi(\tau_0) \end{cases} . \quad (3.5)$$

Furthermore, operators Z, V, J are *omogeneous in* φ, i.e.,

$$Z(0,\tau) = V(0,\tau) = J(0,\tau;\theta) \equiv 0 \,. \quad (3.6)$$

Next step consist in multiplying Eq. (3.2) by $(1+\varphi(\tau))$ rearranging the constants, i.e.,:

$$\xi - a_0 + \frac{\eta}{6}\cos 2\tau - a_0\varphi + \left(1+\frac{3\eta}{4}\right) Z - \frac{\eta}{4} V$$
$$+ \frac{\eta}{6}\cos 2\tau\varphi + \left(1+\frac{3\eta}{4}\right) Z\varphi - \frac{1}{4} V\varphi = -\sqrt{\frac{\overline{f}_0}{\eta}}(J + J\varphi) \quad (3.7)$$

with

$$a_0 = 1 + \frac{5\eta}{6} - \sqrt{\frac{\overline{f}_0}{\eta}} \cos\vartheta \tag{3.8}$$

then *imposing the zero mean value condition for* $\varphi(\tau)$, see Eq. (2.3) 2nd:

$$\begin{aligned} \xi - a_0 + \frac{\eta}{6}\overline{(\varphi\cos 2\tau)} + \left(1 + \frac{3\eta}{4}\right)\overline{(Z + Z\varphi)} \\ - \frac{\eta}{4}\overline{(V + V\varphi)} = -\sqrt{\frac{\overline{f}_0}{\eta}}\overline{(J + J\varphi)} \end{aligned} \tag{3.9}$$

where the bar $\overline{}$ indicates in general the *mean value*:

$$\overline{\psi} = \frac{1}{\pi}\int_{-\pi/2}^{\pi/2} \psi(\tau)d\tau \,. \tag{3.10}$$

By subtraction of Eq. (3.9) from Eq. (3.7), also introducing the symbol of "oscillating part" for a generical function $\psi(\tau)$:

$$\tilde{\psi}(\tau) = \psi(\tau) - \overline{\psi} \tag{3.11}$$

we obtain finally:

$$\varphi(\tau) = \tilde{T}(\varphi, \tau) \tag{3.12}$$

with

$$\begin{aligned} T(\varphi, \tau) = g_2\cos 2\tau + g_2(\varphi\cos 2\tau) + a_1(Z + Z\varphi) \\ - a_2(V + V\varphi) - a_3(J_0 + J_0\varphi) - a_5(J_1 + J_1\varphi) \end{aligned} \tag{3.13}$$

where following new constants have been introduced:

$$\left\{\begin{aligned} a_1 &= \frac{(1 + \frac{3\eta}{4})}{a_0} \\ a_2 &= \frac{(1 + \frac{\eta}{4})}{a_0} \\ g_2 &= \frac{\frac{\eta}{6}}{a_0} \end{aligned}\right. \qquad \left\{\begin{aligned} a_3 &= \frac{-\sqrt{\overline{f}_0/\eta}}{a_0}\cos\vartheta \\ a_4 &= \frac{\sqrt{\overline{f}_0/\eta}}{a_0} \\ a_5 &= \frac{-\sqrt{\overline{f}_0/\eta}}{a_0}\sin\vartheta \end{aligned}\right. \tag{3.14}$$

To the integral equation (3.12) the regularization condition (3.1) and the zero mean value condition (3.9) must be associated. Said equation can be written:

$$\begin{cases} 4\sqrt{\gamma\xi} + \left(1+\dfrac{3\eta}{4}\right)\pi(\overline{\varphi \sin \tau}) - \dfrac{\eta}{4}\pi(\overline{\varphi \sin 3\tau}) = 2\sqrt{\overline{f}_0/\eta}\sin\vartheta \\ 1+\dfrac{5\eta}{6} - \xi - \dfrac{\eta}{6}(\overline{\varphi \cos 2\tau}) - \left(1+\dfrac{3\eta}{4}\right)(\overline{Z+Z\varphi}) + \dfrac{\eta}{4}(\overline{V+V\varphi}) \\ = \sqrt{\overline{f}_0/\eta}[\cos\vartheta + (\overline{J+J\varphi})] \end{cases} \tag{3.15}$$

It is easy to verify that in the absence of damping ($\gamma = 0$) it is sufficient to consider an even function $\varphi(\tau)$ and $\sin\vartheta = 0$ (i.e., $\vartheta = 0$ or $\vartheta = \pi$). In this way, Eq. (3.15) 1st is automatically satisfied. The basic interval I_0can be reduced to the half interval $[0, \pi/2]$. On the contrary, in presence of damping ($\gamma \neq 0$), function $\varphi(\tau)$ has both the even and odd parts, and constant ϑ assumes values no more restricted to the values 0 and π.

Thus the problem of calculating the exact solutions has been reduced to the determination of the form function $\varphi(\tau)$ for $\tau \in I_0$, and of the constants (η, ϑ) satisfying the three coupled Eqs. (3.12) and (3.15). *We have a fixed point problem*, Eq. (3.12), *with additional conditions*, Eqs. (3.15).

4. The Case of Fixed Frame without Damping: Analytical Study

Let us now consider the particular case Ⓐ of fixed frame, in absence of damping, i.e., $\gamma = 0$, in which we have $\sin\vartheta = 0$, and $\varphi(\tau)$ is an *even* function of $\tau \in I_0$, as already said at the end of the previous section. In this case the constants a_0, a_1, a_2 and a_3, considered as functions of η for $\vartheta = \left\langle \begin{smallmatrix} 0 \\ \pi \end{smallmatrix} \right.$, have the behavior indicated in the Fig. 2.

It is important to point out that:

- for $\vartheta = 0$ the constants a_1, a_2, a_3 become infinite for $a_0 = 0$, i.e., for $\eta = \eta_a$, where η_a is the unique positive root of the cubic equation:

$$\eta_a \left(1+\frac{5\eta_a}{6}\right)^2 = \overline{f}_0 \qquad \text{(condition } a_0 = 0\text{)} \tag{4.1}$$

- for $\vartheta = \pi$ the constants a_1, a_2, a_3 are positive and bounded.

Another important remark in the present case is that Eq. (3.12) is splitted, for the cases $\cos\vartheta = 1$ and $\cos\vartheta = -1$, into two different equations, each of then involving function $\varphi(\tau)$ and parameter η as unknown magnitudes. Similarly

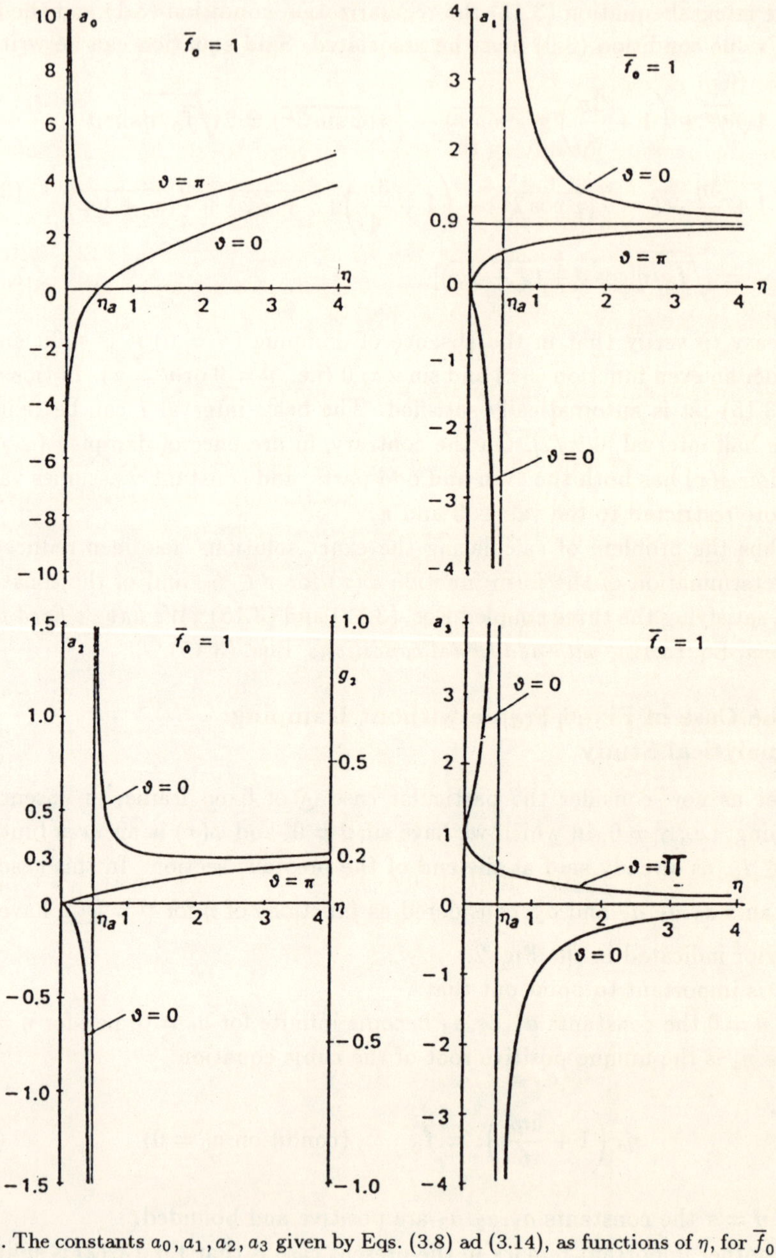

Fig. 2. The constants a_0, a_1, a_2, a_3 given by Eqs. (3.8) ad (3.14), as functions of η, for $\overline{f}_0 = 1$, and for $\vartheta = 0$ and π.

the additional condition (3.15) 2nd (the 1st one is automatically satisfied) is splitted, for the cases $\cos\vartheta = 1$ and $\cos\vartheta = -1$, into two different conditions, each of then giving the explicit value of ξ as a function of η and of suitable mean values of $\varphi(\tau)$. This particular situation suggests a different and more simplified procedure for solving the general problem of calculating the solution for all values of ξ and η, both $\in \mathbb{R}_+$. Namely we can assign an arbitrary value for $\eta \in \mathbb{R}_+$ (both in the cases $\cos\vartheta = \pm 1$), then calculate the even function $\varphi(\tau)$ for $\tau \in I_0$, only considering the fixed point problem (3.12), without additional conditions because η becomes a given constant and finally calculate parameter ξ, given by Eq. (3.15) 2nd. This way the complete relation between parameters η and ξ (with possible exclusion of exceptional intervals, as we will discuss later) is obtained, that is the general amplitude response curve, giving η as a function of ξ.

This said, let us now study the fixed point problem (3.12) with assigned η, and $\cos\vartheta = \pm 1$. For the purpose, we firstly carry out the decomposition of operator $T(\varphi, \tau)$ in a linear and in a nonlinear part as follows. Firstly, we observe that being $\sin\vartheta = 0$ following identity yields:

$$\cos(\tau + \vartheta + \phi) - \cos(\tau + \vartheta) = \cos\vartheta[\cos\tau(\cos\phi - 1) - \sin\tau\sin\phi] \qquad (\text{for } \sin\vartheta = 0) \tag{4.2}$$

Therefore from Eq. (3.3) 3rd, operator J can be written:

$$\begin{aligned} J(\varphi, \tau) &= \cos\vartheta[(\cos\phi - 1) - \tan\tau\sin\phi] \\ &= \cos\vartheta[-Q(\sigma) + S(\sigma)U(\varphi, \tau) - U(\varphi, \tau)] \end{aligned} \tag{4.3}$$

where new operators $U(\varphi, \tau), Q(\sigma)$ and $S(\sigma)$ have been introduced:

$$U(\varphi, \tau) = \tan\tau\phi(\tau) = \tan\tau \int_{-\pi/2}^{\tau} \varphi(s)ds \tag{4.4}$$

$$\begin{cases} \sigma = \phi^2(\tau) \\ Q(\sigma) = 1 - \cos\phi = \displaystyle\sum_{h=1}^{\infty}(-1)^{h-1}\frac{\sigma^h}{(2h)!} \\ S(\sigma) = 1 - \sin\phi/\phi = \displaystyle\sum_{h=1}^{\infty}(-1)^{h-1}\frac{\sigma^h}{(2h+1)!} \,. \end{cases} \tag{4.5}$$

Taking into account Eq. (4.3), Eq. (3.13) can be written:

$$T(\varphi, \tau) = g(\tau) + L(\varphi, \tau) + N(\varphi, \tau) \tag{4.6}$$

with

$$\begin{cases} g(\tau) = g_2 \cos 2\tau \\ L(\varphi, \tau) = g_2(\overline{\varphi \cos 2\tau}) + a_1 Z - a_2 V + a_3 U \\ N(\varphi, \tau) = a_1 Z\varphi - a_2 V\varphi + a_3 U\varphi + a_3(1+\varphi)(Q - SU)\ . \end{cases} \tag{4.7}$$

The different operators are given by Eqs. (3.3), (4.4), (4.5). Concluding, in this case we have the fixed point problem:

$$\varphi(\tau) = \tilde{T}(\varphi, \tau), \qquad \tilde{T} = T - \overline{T} \tag{4.8}$$

with $T(\varphi, \tau)$ given by Eq. (4.6), and where g, L, N are given by Eqs. (4.7), also depending on the assigned parameters $\vartheta = \left\langle \begin{matrix} 0 \\ \pi \end{matrix} \right.$ and $\eta \in \mathbb{R}_+$. The corresponding value of ξ is:

$$\xi = a_0 - \frac{\eta}{6}(\overline{\varphi \cos 2\tau}) - \left(1 + \frac{3\eta}{4}\right)\overline{Z} + \frac{\eta}{4}\overline{V} + \sqrt{f_0/\eta a_0}\,\overline{U} + a_0 \overline{N} \tag{4.9}$$

which, in zero approximation, reduces to (see Eq. (3.8)):

$$\xi = 1 + \frac{5\eta}{6} - \sqrt{f_0/\eta}\cos\vartheta \qquad \varphi(t) \equiv 0\ . \tag{4.10}$$

5. A Simplified Model of the Problem

The study of the above fixed point problem is somewhat difficult. In a preliminary paper [6] we considered following simplified nonlinear model of operator $T(\varphi, \tau)$ instead of (4.6):

$$T_m(\varphi, \tau) = g(\tau) + M(\varphi, \tau) \tag{5.1}$$

with

$$M(\varphi, \tau) = \lambda g(\tau)\varphi(\tau) + \frac{a(1 + \lambda\varphi(\tau))}{\cos\tau} \int_{-\pi/2}^{\tau} \sin s\varphi(s) ds \tag{5.2}$$

where a and λ are given constants. Operator M only involves operator $Z(\varphi, \tau)$ from operators contained in $T(\varphi, \tau)$. In fact we must take into account that being $\varphi(\tau)$ an even function of τ, from Eq. (3.3) 1st we have:

$$Z(\varphi, \tau) = \frac{1}{\cos\tau} \int_{-\pi/2}^{\tau} \sin s\varphi(s) ds \qquad (\text{if } \varphi(\tau) \text{ is even})\ . \tag{5.3}$$

The fixed point Eq. (4.8) is then replaced by:

$$\varphi(\tau) = g(\tau) + \tilde{M}(\varphi, \tau), \qquad \tilde{M} = M - \overline{M} . \tag{5.4}$$

This is a nonlinear integral equation also containing the mean value $\overline{M}$, having the singular nucleus $\sin s/\cos\tau$, infinite for $\tau = \pm\pi/2$. In the problem arising from the Duffing Equation (1.1) function $g(\tau)$ has the value given by Eq. (4.7), and unknown function $\varphi(\tau)$ must be even. We consider the more general situation where $g(\tau)$ is given by the Fourier expansion:

$$g(\tau) = \sum_{n=1}^{\infty} g_{2n} \cos 2n\tau \qquad (g_{2n} \text{ given constants}), \tag{5.5}$$

and we assume for $\varphi(\tau)$ a similar expansion:

$$\varphi(\tau) = \sum_{n=1}^{\infty} b_{2n} \cos 2n\tau \quad (b_{2n} \text{ unknown constants}). \tag{5.6}$$

Firstly we must state the functional class for $g(\tau)$ and $\varphi(\tau)$. Following norms have been tested:

$$\begin{cases} \|\varphi\|_a = \displaystyle\sum_{n=1}^{\infty} |b_{2n}| \\ \|\varphi\|_d = \displaystyle\sum_{n=1}^{\infty} n|b_{2n}| \end{cases}, \tag{5.7}$$

i.e., the norm of the total convergence of $\varphi(\tau)$ and $\varphi'(\tau)$ respectively, besides to the norm:

$$\|\varphi\|_c = \max_{\tau \in I_0} |\varphi(\tau)| . \tag{5.7'}$$

It results clearly:

$$\|\varphi\|_c \leq \|\varphi\|_a \leq \|\varphi\|_d . \tag{5.7''}$$

We call S_a and S_d the Banach spaces of functions $\varphi(\tau)$ with Fourier expansion (5.6) and with the norm $\|\cdot\|_a$ and $\|\cdot\|_d$ respectively. We can demonstrate that:

Preliminary Lemma. If $\varphi(\tau) \in S_a$, or S_d, then also $\tilde{Z}(\varphi, \tau) \in S_a$, or S_d respectively. In fact we have:

$$\begin{aligned} Z(\varphi,\tau) &= \frac{1}{\cos\tau} \int_{-\pi/2}^{\tau} \sum_{n=1}^{\infty} b_{2n} \cos 2ns \sin s ds \\ &= \frac{1}{2} \sum_{n=1}^{\infty} b_{2n} \left[\frac{1}{2n-1} \frac{\cos(2n-1)\tau}{\cos\tau} - \frac{1}{2n+1} \frac{\cos(2n+1)\tau}{\cos\tau} \right] \end{aligned} \tag{5.8}$$

where the term by term integration is correct in force of the considered norms for $\varphi(\tau)$. Now we introduce the trigonometric polynomials

$$C_{2n}(\tau) = \frac{\cos(2n+1)\tau}{\cos\tau} = \overline{C}_{2n} + \tilde{C}_{2n}(\tau) \qquad (n = 0, 1, 2, \dots) \tag{5.9}$$

with

$$\overline{C}_{2n} = (-1)^n; \qquad \tilde{C}_{2n}(\tau) = 2\sum_{k=1}^{n}(-1)^{n+k}\cos 2k\tau \tag{5.9'}$$

i.e., in explicit form:

$$\begin{cases} n=0 & C_0(\tau) = 1 \\ n=1 & C_2(\tau) = -1 + 2\cos 2\tau \\ n=2 & C_4(\tau) = 1 - 2\cos 2\tau + 2\cos 4\tau \\ n=3 & C_6(\tau) = -1 + 2\cos 2\tau - 2\cos 4\tau + 2\cos 6\tau \\ \dots & \dots\dots\dots\dots\dots\dots \end{cases} \tag{5.9''}$$

for which it results:

$$|\overline{C}_{2n}| = 1; \quad \|\tilde{C}_{2n}\|_a = 2n; \quad \|\tilde{C}_{2n}\|_d = 2\sum_{k=1}^{n} k = n(n+1). \tag{5.10}$$

Operator $Z(\varphi, \tau)$ can be written as follows:

$$Z(\varphi,\tau) = \frac{1}{2}\sum_{n=1}^{\infty} b_{2n}\left[\frac{C_{2n-2}(\tau)}{2n-1} - \frac{C_{2n}(\tau)}{2n+1}\right] = \overline{Z} + \tilde{Z}(\varphi,\tau)\,, \tag{5.11}$$

with:

$$\begin{aligned}\overline{Z}(\varphi,\tau) &= \frac{1}{2}\sum_{n=1}^{\infty} b_{2n}\left(\frac{\overline{C}_{2n-2}}{2n-1} - \frac{\overline{C}_{2n}}{2n+1}\right) \\ &= \frac{1}{2}\sum_{n=1}^{\infty} b_{2n}\left[\frac{(-1)^{n-1}}{2n-1} - \frac{(-1)^n}{2n+1}\right] = -\sum_{n=1}^{\infty} b_{2n}\frac{(-1)^n 2n}{4n^2-1}\,;\end{aligned} \tag{5.12}$$

$$\begin{aligned}\tilde{Z}(\varphi,\tau) &= \frac{1}{2}\sum_{n=1}^{\infty} b_{2n}\left[\frac{\tilde{C}_{2n-2}(\tau)}{2n-1} - \frac{\tilde{C}_{2n}(\tau)}{2n+1}\right] \\ &= \sum_{n=1}^{\infty} b_{2n}\left[\frac{1}{2n-1}\sum_{k=1}^{n-1}(-1)^{n+k-1}\cos 2k\tau - \frac{1}{2n+1}\sum_{k=1}^{n}(-1)^{n+k}\cos 2k\tau\right].\end{aligned} \tag{5.13}$$

It results:

$$\left\{\begin{array}{l} |\overline{Z}| \leq \sum_{n=1}^{\infty}\left(\frac{2n}{4n^2-1}\right)|b_{2n}| \leq \frac{2}{3}\|\varphi\|_a \\ |\overline{Z}| \leq \sum_{n=1}^{\infty}\left(\frac{2}{4n^2-1}\right)n|b_{2n}| \leq \frac{2}{3}\|\varphi\|_d \end{array}\right. , \tag{5.14}$$

since:

$$\max_{n\in N_+}\frac{2n}{4n^2-1} = \max_{n\in N_+}\frac{2}{4n^2-1} = \frac{2}{3} .$$

Furthermore assuming the norm $\|\cdot\|_a$ we have:

$$\begin{aligned} \|\tilde{Z}\|_a &\leq \frac{1}{2}\sum_{n=1}^{\infty}|b_{2n}|\left\{\frac{1}{2n-1}\|\tilde{C}_{2n-2}\|_a + \frac{1}{2n+1}\|\tilde{C}_{2n}\|_a\right\} \\ &= \frac{1}{2}\sum_{n=1}^{\infty}|b_{2n}|\left\{\frac{2(n-1)}{2n-1} + \frac{2n}{2n+1}\right\} . \end{aligned} \tag{5.15}$$

On account that it is

$$\frac{2(n-1)}{2n-1} + \frac{2n}{2n+1} = 2\left(1 - \frac{2n}{4n^2-1}\right) < 2 \qquad \forall n \in N_+, \tag{5.16}$$

we obtain:

$$\|\tilde{Z}\|_a \leq \sum_{n=1}^{\infty}|b_{2n}| = \|\varphi\|_a \tag{5.17}$$

In a similar way, assuming the norm $\|\cdot\|_d$ we have:

$$\begin{aligned} \|\tilde{Z}\|_d &\leq \frac{1}{2}\sum_{n=1}^{\infty}|b_{2n}|n\left(\frac{n-1}{2n-1} + \frac{n+1}{2n+1}\right) \\ &= \frac{1}{2}\sum_{n=1}^{\infty}n|b_{2n}|\left(1 - \frac{1}{4n^2-1}\right) < \frac{1}{2}\|\varphi\|_d , \end{aligned}$$

on account of same inequality (5.16), and therefore

$$\|\tilde{Z}\|_d \leq \frac{1}{2}\|\varphi\|_d . \tag{5.18}$$

Eqs. (5.13), (5.17), and (5.18) show that if $\varphi(\tau)$ has the Fourier expansion (5.6) and the bounded norm $\|\varphi\|_a$ or $\|\varphi\|_d$, then also $\tilde{Z}(\varphi, \tau)$ has a similar

Fourier expansion, and a bounded norm $||\tilde{Z}||_a$ or $||\tilde{Z}||_d$ respectively. That is, if $\varphi(\tau) \in S_a$, then $\tilde{Z}(\varphi, \tau) \in S_a$; and if $\varphi(\tau) \in S_d$, then $\tilde{Z}(\varphi, \tau) \in S_d$. Then the Lemma is true. On account of this preliminary Lemma , and other similar properties, it is demonstrate in [6] that if $g(\tau) \in S_a$, with a bounded norm $||g||_a$, and if the norm of the functions:

$$\varphi(\tau) = \sum_{n=1}^{\infty} b_{2n} \cos 2n\tau; \qquad \varphi^*(\tau) = \sum_{n=1}^{\infty} b_{2n}^* \cos 2n\tau \tag{5.19}$$

have both the same upper bound δ:

$$||\varphi||_a < \delta; \qquad ||\varphi^*||_a < \delta \ , \tag{5.20}$$

then in suitable conditions for $a, \lambda, ||g||_a$, (which are specified in the paper) there exists a constant $K < 1$ such that:

$$||\tilde{T}_m(\varphi\tau) - \tilde{T}_m(\varphi^*, \tau)||_a \leq K||\varphi - \varphi^*||_a \ . \tag{5.21}$$

In force of the basic fixed point theorems, see for instance [12] and [13], this means that in the above conditions for parameters a, λ, $||g||_a$, the integral equation:

$$\varphi(\tau) = \tilde{T}_m(\varphi, \tau), \qquad \tilde{T}_m = T_m - \overline{T}_m \ , \tag{5.22}$$

with T_m given by Eqs. (5.1), (5.2), has one and only one solution, which can be calculated by the convergent iterative procedure:

$$\varphi^{(h+1)}(\tau) = \tilde{T}_m(\varphi^{(h)}, \tau); \qquad \varphi^{(1)} = g \ . \tag{5.23}$$

It must be remarked that the obtained conditions for $a, \lambda, ||g||_a$ (not specified here for brevity sake) for the existence and unicity of the solution basically depend on the considered Banach space, i.e., on the assumed norm. If we assume the norm $||\cdot||_d$ instead of the norm $||\cdot||_a$ the equality (5.17) is replaced by the similar inequality (5.18), with a coefficient $\frac{1}{2}$ instead of 1, hence more useful. On the other hand, the regularity of $\varphi'(\tau)$ too, beside to the regularity of $\varphi(\tau)$, is required for having regularity also in $\ddot{x}(\tau)$, beside $\dot{x}(\tau)$ as pointed out after Eqs. (2.3). Thus, for a correct application to our physical problem, we are forced to assume the norm $||\cdot||_d$, paying the price of much heavier analytical developments for demonstrating the existence and unicity of the solution.

6. The Linearized Exact Problem

Taking into account the conclusions of the previous section we will study the exact fixed point problem (4.8), *in the Banach space* S_d with given $\vartheta = \left\langle \begin{matrix} 0 \\ \pi \end{matrix} \right.$

and given $\eta \in \mathbb{R}_+$. *In a first step,* we will develop in this section, *we only consider the linear part of operator* $T(\varphi, \tau)$. *In a second step,* we will develop in Sec. 7, *we will consider the complete operator* $T(\varphi, \tau)$.

Thus, we consider now the fixed point problem:

$$\begin{cases} \varphi(\tau) = \tilde{T}_l(\varphi, \tau), & \tilde{T}_l = T_l - \overline{T}_l \\ T_l(\varphi, \tau) = g(\tau) + L(\varphi, \tau) \end{cases} \tag{6.1}$$

where the linear operator L is given by Eq. (4.7) 2nd, and where we assumed the expansion (5.5) for the given function $g(\tau)$. In order to state an existence-unicity theorem, suitable inequalities for operator Z, V, U are to be taken into account. As regards operator $Z(\varphi, \tau)$ we recall inequalities (5.14) and (5.18) demonstrated in Sec. 5. Similar inequalities can be obtained for $V(\varphi, \tau)$ and $U(\varphi, \tau)$, so that we have:

$$\begin{cases} |\overline{Z}| \leq \frac{2}{3}\|\varphi\|_d; & \|\tilde{Z}\|_d \leq \frac{1}{2}\|\varphi\|_d \\ |\overline{V}| \leq \frac{3}{5}\|\varphi\|_d; & \|\tilde{V}\|_d \leq \frac{3}{5}\|\varphi\|_d \\ |\overline{U}| \leq \frac{1}{2}\|\varphi\|_d; & \|\tilde{U}\|_d \leq \frac{1}{2}\|\varphi\|_d\,. \end{cases} \tag{6.2}$$

As regards the product $(\varphi \cos 2\tau)$ contained in the first term of $L(\varphi, \tau)$, we apply the Lemma 1, which will be reported in the following Sec. 7. Taking into account that it is $\overline{\cos 2\tau} = \overline{\varphi} = 0$, we have:

$$\|\widetilde{\varphi \cos 2\tau}\|_d \leq \|\cos 2\tau\|d\|\varphi\|_d = \|\varphi\|_d \tag{6.2}$$

Setting

$$\varphi^* = \sum_{n=1}^{\infty} b_{2n}^* \cos 2n\tau \tag{6.3}$$

and furthermore:

$$\begin{cases} Z^* = Z(\varphi^*, \tau) \\ V^* = V(\varphi^*, \tau) & T_l^* = T_l(\varphi^*, \tau) \\ U^* = U(\varphi^*, \tau) \end{cases} \tag{6.4}$$

we have following consequences of inequalities (6.2) and (6.3):

$$|\overline{Z} - \overline{Z}^*| \leq \frac{2}{3}\|\varphi - \varphi^*\|_d; \quad \|\tilde{Z} - \tilde{Z}^*\|_d \leq \frac{1}{2}\|\varphi - \varphi^*\|_d \tag{6.5}$$

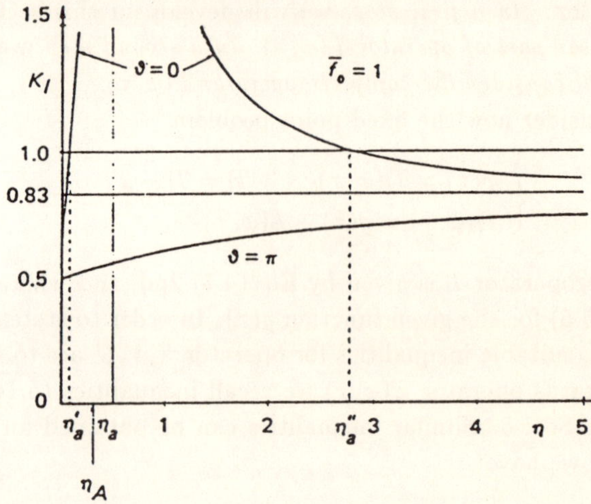

Fig. 3. The coefficient K_l of the contraction, see Eqs. (3.6), (3.7), (3.7′), related to the linearized Eq. (6.1).

and so on for operator V and U. In force of these inequalities we obtain finally:

$$\|\tilde{T}_l - \tilde{T}_l^*\|_d \le K_l \|\varphi - \varphi^*\|_d \tag{6.6}$$

with:

$$K_l = |g_2| + \frac{1}{2}|a_1| + \frac{3}{5}|a_2| + \frac{1}{2}|a_3| \tag{6.7}$$

i.e., in force of Eqs. (3.8) and (3.14):

$$K_l = \frac{1}{2} \frac{1 + \dfrac{83\eta}{60} + \sqrt{\dfrac{\bar{f}_0}{\eta}}}{\left|1 + \dfrac{5\eta}{6} - \sqrt{\dfrac{\bar{f}_0}{\eta}} \cos\vartheta\right|} \tag{6.7′}$$

The behavior of K_l as a function of $\eta \in \mathbb{R}_+$, for the two values $\vartheta = 0$ and $\vartheta = \pi$, is show in Fig. 3 for $\bar{f}_0 = 1$. We observe that:

- K_l exhibits an horizontal asymptote $K = 0.83$ for $\eta \to +\infty$.
- for $\vartheta = 0$, K_l exhibits a vertical asymptote for $\eta = \eta_a$, see Eq. (4.1), and assumes the values 0.5 for $\eta = 0$. Furthermore K_l is increasing in $]0, \eta_a[$ and decreasing in $]\eta_a, \infty[$. Therefore, since K_l is a continuous function of

η for $\eta \neq \eta_a$, there exist two values η'_a and η''_a of η, the first one at the left part of η_a, the second one at the right part, where it is $K_l = 1$. We call $I_a \equiv [\eta'_a, \eta''_a]$ the interval, containing η_a, where it is $K_l \geq 1$. Therefore, it results $K_l < 1$ for $\eta \in \mathbb{R}_+/I_a$.
- for $\vartheta = \pi$, it results always $K_l < 1$.

Thus the conclusion is that:

- in the case $\vartheta = 0$, for $\eta \in \mathbb{R}_+/I_a$, being $K_l < 1$, Eq. (6.1) has one and only one solution, whereas for $\eta \in I_a$ we cannot affirm whether or not the solution exists.
- in the case $\vartheta = \pi$, for $\eta \in \mathbb{R}_+$, Eq. (6.1) has always one and only one solution.

Therefore we have following:

1st Existence and unicity Theorem. If $g(\tau) \in S_d$, and:

$$\begin{cases} \eta \in R_+/I_a & \text{for } \vartheta = 0 \\ \eta \in R_+ & \text{for } \vartheta = \pi \end{cases} \tag{6.8}$$

there exist one and only one solution $\varphi(\tau) \in S_d$ of linearized problem (6.1). This solution, when it exists as indicated above, is calculated with following iterative procedure:

$$\varphi^{(m+1)}(\tau) = \tilde{T}_l(\varphi^{(m)}, \tau), \qquad \varphi^{(l)}(\tau) = g(\tau) \tag{6.9}$$

and reduces to the identically null solution when $g(\tau)$ vanishes, since it results also:

$$\|\varphi\|_d \leq \frac{\|g\|_d}{(1 - K_l)} . \tag{6.10}$$

7. The Exact Nonlinear Problem

Let us now consider the exact nonlinear problem, we write again for convenience:

$$\begin{cases} \varphi(\tau) = \tilde{T}(\varphi, \tau), & \tilde{T} = T - \overline{T} \\ T(\varphi, \tau) = T_l(\varphi, \tau) + N(\varphi, \tau), & T_l = g + L \end{cases} . \tag{7.1}$$

Operator T_l and N are given by Eqs. (6.1) 2nd and (4.7). We suppose again that function $g(\tau)$ is given by the Fourier expansion (5.5), with bounded norm

$||g||_d$, i.e., $g \in S_d$. We assume again for $\varphi(\tau)$ the similar expansion (5.6), and we consider again a second function $\varphi^*(\tau)$ with the expansion (6.3). Setting

$$T^* = T(\varphi^*, \tau); T_l^* = T(\varphi^*, \tau); N^* = N(\varphi^*, \tau) , \tag{7.2}$$

we have:

$$||\tilde{T} - \tilde{T}^*||_d \leq ||\tilde{T}_l - \tilde{T}_l^*||_d + ||\tilde{N} - \tilde{N}^*||_d . \tag{7.3}$$

As regards the linear operator T_l the inequality (6.6) has been already obtained in Sec. 6.

As regards the nonlinear operator N new inequalities must be taken into account. We summarize the procedure followed in [8]. Firstly, following Lemmas are demonstrated (for brevity sake, we do not report here the demonstrations, given in [8]).

Lemma 1. Let us consider two functions $\tilde{\chi}$ and $\tilde{\psi} \in S_d$:

$$\tilde{\chi} = \sum_{n=1}^{\infty} a_{2n} \cos 2n\tau; \qquad \tilde{\psi} = \sum_{n=1}^{\infty} c_{2n} \cos 2n\tau . \tag{7.4}$$

It results:

$$||\widetilde{\tilde{\chi}\tilde{\psi}}||_d \leq ||\tilde{\chi}||_d||\tilde{\psi}||_d - \frac{1}{2}\sum_{n=1}^{\infty}(n^2-1)|a_{2n}||c_{2n}| , \tag{7.5}$$

and more in particular:

$$||\widetilde{\tilde{\chi}\tilde{\psi}}||_d \leq ||\tilde{\chi}||_d||\tilde{\psi}||_d . \tag{7.6}$$

Lemma 2. Let us consider two functions χ and ψ with mean value different from zero, such that:

$$\chi = \overline{\chi} + \tilde{\chi}; \qquad \psi = \overline{\psi} + \tilde{\psi} , \tag{7.7}$$

with $\tilde{\chi}$ and $\tilde{\psi} \in S_d$. It results:

$$||\widetilde{\chi\psi}||_d \leq |\tilde{\chi}| \; ||\tilde{\psi}||_d + |\tilde{\psi}| \; ||\tilde{\chi}||_d + ||\tilde{\chi}|| \; ||\tilde{\psi}||_d . \tag{7.8}$$

Lemma 3. Let us consider two functions χ and ψ of Lemma 2. It result:

$$|\overline{\chi\psi}| \leq |\overline{\chi}| \; |\overline{\psi}| + ||\tilde{\chi}||_d \; ||\tilde{\psi}||_d . \tag{7.9}$$

Lemma 4. Let us consider two functions χ_s and $\psi_s \in S_d$:

$$\begin{cases} \chi_s = \sum_{n=1}^{\infty} a_{2n} \sin 2n\tau, & \text{with } \sum_{n=1}^{\infty} n|a_{2n}| < \infty \\ \psi_s = \sum_{n=1}^{\infty} c_{2n} \sin 2n\tau, & \text{with } \sum_{n=1}^{\infty} n|c_{2n}| < \infty \end{cases} \tag{7.10}$$

to be associated to the functions $\tilde{\chi}, \tilde{\psi} \in S_d$ given by Eqs. (7.4). It results:

$$\widetilde{\chi_s \psi_s} \in S_d, \qquad \text{and: } \|\widetilde{\chi_s \psi_s}\|_d \leq \|\tilde{\chi}\|_d \|\tilde{\psi}\|_d \,. \tag{7.11}$$

The second point is that in the present nonlinear problem, in order to apply the basic fixed point theorem, see [12], we must also introduce the upper bound for the norms $\|\varphi\|_d$ and $\|\varphi^*\|_d$, we call δ:

$$\|\varphi\|_d < \delta; \qquad \|\varphi^*\|_d < \delta \,. \tag{7.12}$$

If these inequalities are satisfied, then we can demonstrate, see [8], that following inequalities hold:

$$\begin{cases} |\overline{Q}| < Q_1(\delta) = \sum_{h=1}^{\infty} \frac{1}{(2h)!} \left(\frac{1}{4}\delta^2\right)^h \\ \|\tilde{Q}\| < Q_2(\delta) = \frac{2}{3} \sum_{h=1}^{\infty} \frac{h}{(2h)!} \left(\frac{3}{8}\delta^2\right)^h \\ |\overline{Q} - \overline{Q^*}| < \frac{1}{4}\delta \|\varphi - \varphi^*\|_d Q_3(\delta), \quad \text{with:} \\ Q_3(\delta) = \sum_{h=1}^{\infty} \frac{h}{(2h)!} \left(\frac{1}{4}\delta^2\right)^{h-1} \\ \|\tilde{Q} - \tilde{Q}^*\|_d < \frac{1}{2}\delta \|\varphi - \varphi^*\|_d Q_4(\delta), \quad \text{with:} \\ Q_4(\delta) = \sum_{h=1}^{\infty} \frac{q(h)}{(2h)!} \left(\frac{3}{8}\delta^2\right)^{h=1} \end{cases} , \tag{7.13}$$

$$\left\{\begin{aligned}
&|\overline{S}| < S_1(\delta) = \sum_{h=1}^{\infty} \frac{1}{(2h+1)!}\left(\frac{1}{4}\delta^2\right)^h \\
&\|\tilde{S}\| < S_2(\delta) = \frac{2}{3}\sum_{h=1}^{\infty} \frac{h}{(2h+1)!}\left(\frac{3}{8}\delta^2\right)^h \\
&|S - S^*| < \frac{1}{4}\delta\|\varphi - \varphi^*\|_d S_3(\delta), \quad \text{with:} \\
&S_3(\delta) = \sum_{h=1}^{\infty} \frac{h}{(2h+1)!}\left(\frac{1}{4}\delta^2\right)^{h-1} \\
&\|\tilde{S} - \tilde{S}^*\|_d < \frac{1}{2}\delta\|\varphi - \varphi^*\|_d S_4(\delta), \quad \text{with:} \\
&S_4(\delta) = \sum_{h=1}^{\infty} \frac{q(h)}{(2h+1)!}\left(\frac{3}{8}\delta^2\right)^{h=1}
\end{aligned}\right. , \tag{7.14}$$

with:

$$q(h) = \left(\frac{2}{3}\right)^{h-1} h + \sum_{k=0}^{h-1}\left\{(h-1-k)\left(\frac{2}{3}\right)^k + k\left(\frac{2}{3}\right)^{h-1-k} + \frac{2}{3}k(h-1-k)\right\}. \tag{7.14}$$

In force of these four Lemmas, and of inequalities (7.13) and (7.14), we can state the inequalities related to operator N given by Eq. (4.7) 3rd, Namely we have:

$$\|\tilde{N} - \tilde{N}^*\|_d \leq \alpha + \beta \tag{7.15}$$

with:

$$\begin{aligned}
\alpha = {} & |a_1|\,\|\widetilde{\varphi Z} - \widetilde{\varphi^* Z^*}\|_d + |a_2|\,\|\widetilde{\varphi V} - \widetilde{\varphi^* V^*}\|_d \\
& + |a_3|\,\|\widetilde{\varphi U} - \widetilde{\varphi^* U^*}\|_d
\end{aligned} \tag{7.16}$$

$$\begin{aligned}
\beta = {} & |a_3|\{\|\tilde{Q} - \widetilde{Q^*}\|_d + \|\widetilde{\varphi Q} - \widetilde{\varphi^* Q^*}\|_d \\
& + \|\widetilde{US} - \widetilde{U^* S^*}\|_d + \|\widetilde{\varphi US} - \widetilde{\varphi^* U^* S^*}\|_d\}\,.
\end{aligned} \tag{7.17}$$

As regards the term α we start from the identity:

$$\varphi Z - \varphi^* Z^* = \varphi(Z - Z^*) + Z^*(\varphi - \varphi^*) \tag{7.18}$$

and we obtain also, on account that $\overline{\varphi} = \overline{\varphi^*} = 0$:

$$\|\widetilde{\varphi Z} - \widetilde{\varphi^* Z^*}\|_d < \frac{7}{3}\delta\|\varphi - \varphi^*\|_d\,. \tag{7.19}$$

In the similar way:

$$\|\widetilde{\varphi V} - \widetilde{\varphi^* V^*}\|_d < \frac{12}{5}\delta\|\varphi - \varphi^*\|_d \tag{7.20}$$

$$\|\varphi V - \varphi^* V^*\|_d < 2\delta\|\varphi - \varphi^*\|_d \tag{7.21}$$

and therefore:

$$\alpha < \left(\frac{7}{3}|a_1| + \frac{12}{5}|a_2| + 2|a_3|\right)\delta\|\varphi - \varphi^*\|_d \,. \tag{7.22}$$

As regards the term β we start from the identities:

$$\begin{cases} \varphi Q - \varphi^* Q^* = \varphi(Q - Q^*) + Q^*(\varphi - \varphi^*) \\ US - U^*S^* = U(S - S^*) + S^*(U - U^*) \\ \varphi US - \varphi^* U^* S^* = \varphi(US - U^*S^*) + U^*S^*(\varphi - \varphi^*) \end{cases} \tag{7.23}$$

and we obtain also, on account that $\overline{\varphi} = \overline{\varphi^*} = 0$:

$$\beta < \left[\frac{1}{4}\delta + \delta^2 D(\delta)\right] |a_3| \, \|\varphi - \varphi^*\|_d \tag{7.24}$$

with:

$$D(\delta) = \frac{(Q_1 + Q_2)}{\delta^2} + \frac{1}{4}Q_3 + Q_5 + \left(\frac{1}{2} + 2\delta\right)\frac{S_1}{\delta^2} + (1 + 3\delta)\frac{S_2}{\delta^2} + \frac{1}{4}\left(\frac{1}{2} + \delta\right)S_3 + \frac{1}{2}\left(1 + \frac{3}{2}\delta\right)S_4 \,. \tag{7.25}$$

Functions $Q_1(\delta)$, $Q_2(\delta)$, etc, are given by Eqs. (7.13) and (7.14); function $Q_5(\delta)$ has the value:

$$Q_5(\delta) = \frac{1}{4} + \frac{1}{2}\sum_{h=2}^{\infty}\left(1 + \frac{1}{\delta}\right)\frac{q(h)}{(2h+1)!}\left(\frac{3}{8}\delta^2\right)^{h-1} . \tag{7.26}$$

For function $D(\delta)$ different estimates can be obtained, as for instance:

$$D(\delta) < m_0 \sum_{h=0}^{\infty}\left(\frac{\delta}{2}\right)^h = \frac{m_0}{1 - \delta/2}, \qquad \text{with } m_0 = \frac{11}{8} \,. \tag{7.27}$$

In force of inequalities (6.6), (7.22) and (7.24) from Eqs. (7.3) and (7.13), we obtain finally:

$$\|\widetilde{T} - \widetilde{T^*}\|_d < K\|\varphi - \varphi^*\|_d \,, \tag{7.28}$$

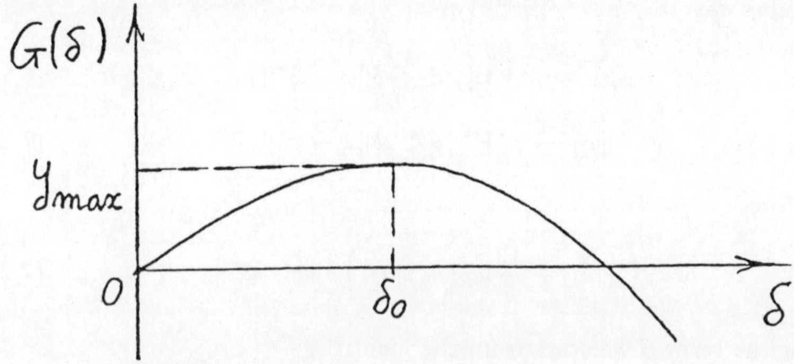

Fig. 4. Qualitative behavior of function $G(\delta)$ given by Eq. (7.33).

with

$$K = K_l + \delta H + \delta^2|a_3|D(\delta) \tag{7.29}$$

where K_l is given by Eq. (6.7) or (6.7′), $D(\delta)$ by Eq. (7.25) and H by:

$$H = \frac{7}{3}|a_1| + \frac{12}{5}|a_2| + \frac{g}{4}|a_3| \; . \tag{7.30}$$

The basic difference from the present exact, nonlinear problem, and the linearized one, studied in the previous Sec. 6, is that now the contraction coefficient K also depends on the upper bound δ for φ and φ^*, see inequalities (7.12). In force of the fixed point theorem, see [12], we know that setting:

$$y = ||g||_d \tag{7.31}$$

if the system formed by Eqs. (7.29) and:

$$y = \delta(1 - K) \; , \tag{7.32}$$

in the unknown constants (δ, K) has real solutions with $K < 1$, then there exists one and only one solution of Eq. (7.1). Thus we must show that for $y > 0$ and $\delta > 0$, from which it results automatically $K < 1$, the above system has a real solution in δ. For the purpose we eliminate K from Eqs. (7.29) and (7.32), and we obtain:

$$y = \delta(1 - K_l) - \delta^2 F(\delta) \equiv G(\delta) \; , \tag{7.33}$$

with:

$$F(\delta) = H + |a_3|\delta D(\delta) \; . \tag{7.34}$$

The right side part of Eq. (7.33) is an analytical function G of δ equal to zero for $\delta = 0$, with positive slope for $\delta = 0$ if $K_l < 1$, with the concavity always towards the down side, and going to $-\infty$ for $\delta \to +\infty$, as shown in Fig. 4.

Thus this function has positive maximum for a certain value δ_0 of δ. In order that Eq. (7.33) has real solutions in δ, it is then necessary and sufficient that it be:

$$y \leq y_{\max}, \qquad \text{with } y_{\max} = \max_{\delta \in \mathbb{R}_+}[\delta(1-K_l) - \delta^2 F(\delta)] \,. \tag{7.35}$$

There we have following:

2nd Existence and unicity Theorem. If $g(\tau) \in S_d$ with norm $y = \|g\|_d$ satisfying to the condition (7.35), and if the conditions (6.8) are satisfied in force of which it is $K_l < 1$, then there exists, both for $\vartheta = 0$ and $\vartheta = \pi$, one and only one solution $\varphi(\tau) \in S_d$ of the exact problem (7.1). This solution, when it exists as indicated above, is calculated according following iterative procedure:

$$\varphi^{(m+1)}(\tau) = \widetilde{T}(\varphi^{(m)}, \tau), \qquad \varphi^{(1)}(\tau) = g(\tau) \tag{7.36}$$

and reduce to the identically null solution when $g(\tau)$ vanishes, since it results also:

$$\|\varphi\|_d \leq \frac{\|g\|_d}{(1-K)} \,. \tag{7.37}$$

8. Some Comments on the Obtained Results

As regards to the results obtained in the previous section, we remark:

1. The conditions (6.8) in force of which it results $K_l < 1$ are basic both in the linearized and in the exact nonlinear problem. The basic point is that according to the procedure here developed the interval $I_a \equiv [\eta'_a, \eta''_a]$ for η must be excluded in the case $\vartheta = 0$. The ground of this exclusion is in the fact, pointed out in Sec. 4 , that for $\vartheta = 0$, the constant a_0 vanishes for $\eta = \eta_a$, see Eq. (4.1), and therefore constants $a_1, a_2, g_2, a_3, a_4, a_5$ become infinite for that value of η. *We express here the conjecture that such an exclusion for $\vartheta = 0$ is not due to the present procedure, but is peculiar of the problem;* that is for some values of η near to η_a the harmonical monoscillating solution does not exist. For these values of η maybe subharmonical or chaotical solutions will arise.

2. The norm $y = \|g\|_d$ reduces to $y = |g_2|$ in the specific case of Duffing Equation. In this case the condition (7.35) for $|g_2|$ is very severe. In fact some calculations here not reported, show that, in force of this condition, other intervals from η should be excluded both for $\vartheta = 0$ and $\vartheta = \pi$,

in addition to the interval I_a for $\vartheta = 0$, see remark 1. But this second kind of exclusion can be eliminated if we start from a first approximation $\varphi^{(1)}(\tau) \in S_d$ better than $g(\tau)$. Infact, if we know a good approximated solution $\varphi^{(1)}(\tau) \in S_d$, better than $g(\tau)$, setting:

$$\begin{cases} \varepsilon(\tau) = \varphi(\tau) - \varphi^{(1)}(\tau) \\ R^{(1)}(\tau) = \widetilde{T}(\varphi^{(1)}, \tau) - \varphi^{(1)}(\tau) \end{cases} , \tag{8.1}$$

and eliminating $\varphi(\tau)$ from Eq. (7.1) we have:

$$\varepsilon(\tau) = R^{(1)}(\tau) + \widetilde{E}(\varepsilon, \tau) , \tag{8.2}$$

with

$$E(\varepsilon, \tau) = T(\varphi^{(1)} + \varepsilon, \tau) - T(\varphi^{(1)}, \tau) . \tag{8.3}$$

It results from Eqs. (8.1) that ε and $R^{(1)} \in S_d$.
Introducing, as usually, a new function $\varepsilon^* \in S_d$, setting:

$$E^* = E(\varepsilon^*, \tau) , \tag{8.4}$$

and supposing:

$$\|\varepsilon\|_d < \delta, \qquad \|\varepsilon^*\|_d < \delta , \tag{8.5}$$

in force of inequality (7.28) with $\varphi = \varphi^{(1)} + \varepsilon$, and $\varphi^* = \varphi^{(1)} + \varepsilon^*$, we have

$$\|\widetilde{E} - \widetilde{E^*}\|_d = \|\widetilde{T}(\varphi^{(1)} + \varepsilon, \tau) - \widetilde{T}(\varphi^{(1)} + \varepsilon^*)\|_d < K\|\varepsilon - \varepsilon^*\|_d . \tag{8.6}$$

We can apply the same conclusions of 2nd Theorem and affirm that Eq. (7.28) has one and only one solution if in addition to the conditions (6.8) in force which it is $K_l < 1$, it results also:

$$\|R^{(1)}\|_d < y_{\max} . \tag{8.7}$$

At the light of Eq. (8.1) 2nd, this inequality indicates that the 1st approximation solution $\varphi^{(1)}(\tau)$ must be sufficiently good, in the exact signification expressed by the inequality itself. If we know such a good first approximation, the difficulty discussed in this 2nd remark is removed.
Another possibility for removing the said difficulty is to in adopt better techniques for ameliorating the extimates we have obtained here. In our opinion, this is not a easy job.

3. We remember that, as pointed out in [5], we have following exact results, in the corresponding limit cases:

A. For $\Omega \to 0$, and therefore $\xi \to 0$. Function $\varphi(\tau)$ has the value:

$$\varphi_A(\tau) = \frac{1 + 3\eta \sin^2 \tau}{\sqrt{1 + 2\eta(1 + \sin^2 \tau) + \eta^2(1 + \sin^2 \tau + \sin^4 \tau)}} - 1 \tag{8.8}$$

where $\eta = \eta_A$ is the positive solution of the cubic equation:

$$\eta_A(1 + \eta_A)^2 = \overline{f}_0 \tag{8.9}$$

B. For $\Omega \to \infty$, $x^* \to \infty$ and therefore $\xi \to \infty$, $\eta \to \infty$. The amplitude response curve has two branches approaching from opposite sides to the following oblique asymptote, independent on $\overline{f}_0$:

$$\xi = r_B \eta + s_B, \qquad \text{with } r_B = 0.71783; \quad s_B = 1.04576\,. \tag{8.10}$$

Function $\varphi(\tau)$ has the value:

$$\varphi_B(\tau) = \sqrt{\frac{2r_B}{1 + \sin^2 \tau}} - 1\,. \tag{8.11}$$

It is interesting to compare this exact value of said asymptote with the value corresponding to the zero approximation $\varphi(t) = 0$ given by Eq. (4.10), i.e.,

$$\xi = 1 + \frac{5}{6}\eta \tag{8.12}$$

C. For $\Omega \to \infty$, $x^* \to 0$, and therefore $\xi \to \infty$, $\eta \to 0$. The amplitude response curve approaches the ξ-axis as follows;

$$\eta \cong \frac{\overline{f}_0}{\xi^2} \tag{8.13}$$

and function $\varphi(\tau)$ has following limit value:

$$\varphi_c(\tau) \cong \frac{1}{3} \frac{\eta^{\frac{3}{2}}}{\sqrt{\overline{f}_0}} \cos 2\tau \tag{8.14}$$

9. Numerical Treatment of the Problem

The problem studied in Sec. 7, Eqs. (7.1) can be subjected to numerical treatment. The basic point is the numerical calculation of operators Z, V, U, contained in $T(\varphi, \tau)$, which are given explicitly by Eqs. (3.3) and (4.4), containing $\cos\tau$ as a divisor.

These operators have following basic properties. They are regular inside the interval I_0, see Eq. (2.1), but in the extremes $\tau = \pm\pi/2$ they have an indetermination form 0/0. They can be obtained by the general operator $W(v, \tau)$, for particular values of $v(\tau)$:

$$W(v, \tau) = \frac{1}{\cos\tau}\left[\int_{-\pi/2}^{\tau} v(s)ds - \frac{1}{2}(1 + \sin\tau)\int_{-\pi/2}^{\pi/2} v(s)ds\right] . \tag{9.1}$$

Despite the divisor $\cos\tau$, operator $W(v, \tau)$ is regular or $\tau \in I_0$. In fact by the De-l'hospital rule we obtain:

$$W(v, \tau_0) = \mp v(\tau_0), \qquad \tau_0 = \pm\pi/2 . \tag{9.2}$$

In addition we have:

$$W(0, \tau) \equiv 0 \qquad \forall\tau \in I_0 . \tag{9.3}$$

It is then clear that the numerical calculation of operators Z, V, U, is reduced to the calculation of operator W for different values of $v(\tau)$.

The calculation of operators $Q(\sigma)$ and $S(\sigma)$, also contained in $T(\varphi, \tau)$, given by Eqs. (4.5), do not present particular difficulty. For the numerical calculation of operator $W(v, \tau)$ see Eq. (9.1), the basic interval $I_0 \equiv [-\pi/2, \pi/2]$ is divided in three intervals:

$$I_1 \equiv [-\pi/2, \tau_1]; \qquad I_2 \equiv [\tau_1, \tau_2]; \qquad I_3 \equiv [\tau_2, \pi/2] \tag{9.4}$$

where τ_1 and τ_2 are of the order of $10^{-1}\pi/2$.

In I_2 no singularities take place, therefore standard quadrature formulae are used. In I_1 and I_3 an indetermination form 0/0 takes place at one end of each interval. We are led to calculate integrals like:

$$I(y) = \frac{1}{y}\int_0^y f(x)dx, \qquad y \in [0, y_1] . \tag{9.5}$$

The numerical calculation is carried out introducing suitable *interpolating polynomials* which assume the same values of $f(x)$ in points x_i:

$$x_i = ih; \qquad i = 0, 1, 2, \ldots, n; \quad h = y_1/n . \tag{9.6}$$

We use *Lagrange polynomials* of the 4th order, see for instance [14] Chap. 2 and [15]. Carrying out the integration we obtain:

$$I(y) = I_4(y) + E_4(y) \,, \tag{9.7}$$

with:

$$I_4(y) = f_0 + \sum_{k=1}^{4} C_4^{(k)} \left(\frac{y}{y_1}\right)^k \,. \tag{9.8}$$

Coefficients $C_4^{(k)}$ are linear combinations of $f_i = f(x_i)$, $i = 0, \dots, 4$:

$$\begin{cases} C_4^{(1)} = (-25f_0 + 48f_1 - 36f_2 + 16f_3 - 3f_4)/6 \\ C_4^{(2)} = (70f_0 - 208f_1 + 228f_2 - 112f_3 + 22f_4)/9 \\ C_4^{(3)} = (-20f_0 + 72f_1 - 96f_2 + 56f_3 - 12f_4)/3 \\ C_4^{(4)} = (32f_0 - 128f_1 + 192f_2 - 128f_3 + 32f_4)/15 \,. \end{cases} \tag{9.9}$$

The error E is neglected in the calculations.

10. Numerical Results for Case Ⓐ, without Damping

As already pointed out in Sec. 4, the forcing frequency Ω is contained only in parameter $\xi = \Omega^2/K_1$, and coefficients $a_0, a_1, a_2, g_2, a_3, a_4, a_5$ are independent on ξ.

Following procedure has been used for calculating the solution, for any prescribed value of $\overline{f}_0$:

- assigning an arbitrary value to $\eta \in \mathbb{R}_+$
- assigning to ϑ alternative values:

$$\vartheta = 0 \qquad \text{or} \qquad \vartheta = \pi \tag{10.1}$$

- calculating parameters $a_0, a_1, a_2, g_2, a_3, a_4, a_5$.
- calculating the sequences $\varphi^m(\tau)$ by the iterative procedure (7.36) till a sufficiently good convergence is obtained. Finally the value of ξ is given by Eq. (4.9).

Figure 5 shows the amplitude response curve (ξ, η), in the hardening case, for $\overline{f}_0 = 1$.

Figure 6 shows the same response curve (ξ, η), in the softening case, for $\overline{f}_0 = 1$, with the specifications given at the end of Sec. 2.

The exact amplitude response curves (ξ, η) are compared with the zero-approximation ones given by Eq. (4.10) with $\vartheta = 0$ or π. Some particular

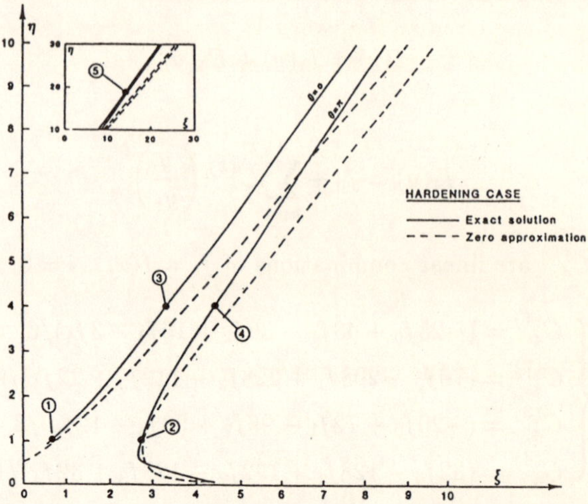

Fig. 5. Case Ⓐ :fixed frame. Amplitude response curve for $\overline{f}_0 = 1$. Hardening case.

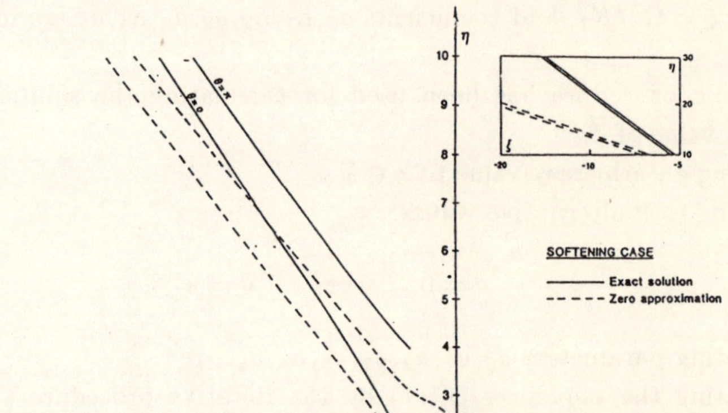

Fig. 6. As the Fig. 5. Softening case.

points of these response curves have been investigated more in detail till the calculation of the displacement $x(t)$ and of the velocity $\dot{x}(t)$, see Figs, for them also the value of the circular frequency Ω, see Eq. (2.6) 1st, and the value of the amplitude x^* given by Eq. (8) 2nd must be calculated, i.e.,

$$\Omega = \sqrt{K_1 \xi}; \qquad x^* = \sqrt{\frac{K_1 \eta}{K_3}} . \tag{10.2}$$

In the numerical calculation reported here, we have assumed:

$$K_1 = 1, \qquad K_3 = 10 . \tag{10.3}$$

The numerical values of the constants are reported in the Table.

Table
Numerical values of the constants in the tested points

Hardening **Softening**

$$\begin{cases} \overline{f}_0 = 1 & c = 0 \\ K_1 = 1 & f_0 = 1/\sqrt{10} \\ K_3 = 10 & \end{cases} \qquad \begin{cases} \overline{h}_0 = 1 & c = 0 \\ K_1 = 1 & h_0 = 1/\sqrt{10} \\ K_3 = 10 & \end{cases}$$

Case Ⓐ : fixed frame

	Hardening					Softening		
tested points	1	2	3	4	5	6	7	remarks
ξ	0.6273	2.7511	3.3420	4.4130	15.0152	2.0260	0.1905	calculated
η	1.0	1.0	4.0	4.0	20.0	0.5	1.7	given
x^*	0.3162	0.3162	0.6324	0.6324	1.4142	0.2236	0.4123	
ϑ	0	π	0	π	0	π	π	given

As regards the Figs. 5 and 6 following remarks can be made. The continue line connects several tested points, for which the present procedure is convergent. The curve is broken where our tests do not gave convergence. It can be inferred that in these cases the solution does not exist. *In the hardening case* ($K_3 > 0$) Fig. 5 we have solutions both for $\vartheta = 0$ and $\vartheta = \pi$, with some exclusions only for $\vartheta = 0$, in agreement with the discussion of point 1), of Sec. 8. *In the softening case* ($K_3 < 0$), Fig. 6, the present procedure is convergent, except for values of ξ near zero, but the corresponding calculated values of ξ are negative in the majority of the cases. From a mathematical point of view the present procedure for solving the softening case is the same as for the hardening one. On the contrary, from a physical point of view, the conclusion is that in the softening case the solution exists only for $\vartheta = \pi$, for almost every value of ξ (or Ω); the amplitude η (or x^*) of the oscillation has an upper bound, not too large.

11. Further Developments; Concluding Remarks

The analytical and numerical procedure here developed for the Duffing Equation (1.1), in the case Ⓐ with fixed frame, can be applied also to similar nonlinear problems. Firstly we recall the application to another problem, with *hysteresis* see [17] and [18], for which following mathematical model is assumed:

$$\ddot{x} + Kx + \text{ sgn } \dot{x} b(x^{*2} - x^2) = s_0 \sin \Omega t \tag{11.1}$$

where parameters (K, s_0, Ω) are known constants, and x^* is the unknown amplitude of the forced vibration. Parameter b can be either independent, or dependent on x^*. Secondarily we recall the application to problems with two degrees of freedom, as the system, studied in [16]:

$$\begin{cases} \ddot{x} + K_1 x + K_3 x^3 + a_1 y = 0 \\ \ddot{y} + h_1 y + b_1 x = 0 \end{cases} \tag{11.2}$$

and other systems corresponding to different kinds of *dynamical absorbers*, see [19], as for instance:

$$\begin{cases} \ddot{x} + \dfrac{F(x) + c\dot{x}}{m_{eq}} + \dfrac{K_1}{m_1}(x + y) = F_0 \sin \Omega t \\ \ddot{y} - \dfrac{F(x) + c\dot{x}}{m_2} = 0, \qquad F(x) = Kx + K' x^3 \,. \end{cases} \tag{11.3}$$

Of course the study of such systems with two degrees of freedom, till the exact calculation of the steady state solutions, is a more complicated job, not yet carried out by us till today. At the contrary, further numerical calculations are

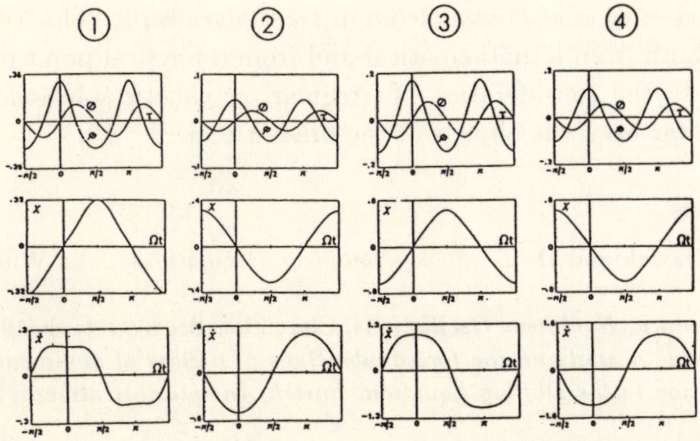

Fig. 7. Case Ⓐ : fixed frame. Some exact results for the form-function $\varphi(\tau)$ and its integral $\phi(\tau)$, and for $x(t)$ and $\dot{x}(t)$ against Ωt. Point ① , ② , ③ , ④ , of the Fig. 5.

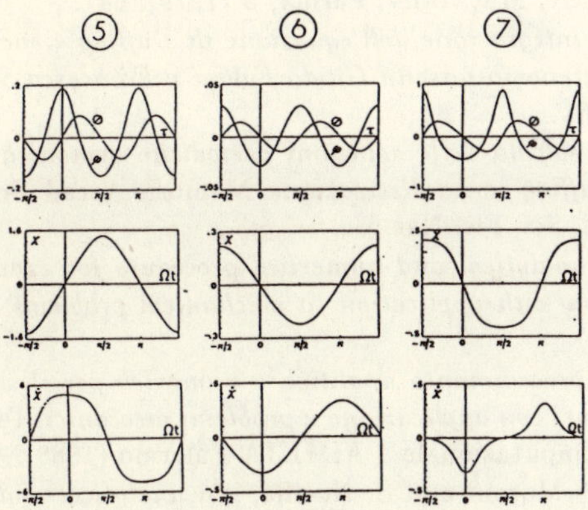

Fig. 8. As the Fig. 7. Point ⑤ of the Fig. 5 and points ⑥ and ⑦ of the Fig. 6.

now in working, see for instance [20] and [21], as regards the damping effects in the Duffing Equation (1.1), both in the cases Ⓐ of fixed, and Ⓑ of oscillating frame, and both in the hardening and in the softening cases.

We conclude with following remark. The procedure here indicated, according to which the calculation of the steady state periodical solution is reduced to the solution of a fixed-point problem in a suitable Banach space, can give question arises of *what kind of solution takes place really.* This very important question, both from a mathematical and from a physical point of view, leads, as a rule, to the consideration of irregular, or chaotic solutions. This topic however is far from the objects of the present paper.

References

1. A. H. Nayfeh and D. T. Mook, *Nonlinear Oscillations*, John-Wiley, New York, (1979).
2. P. Hagedorn, *Nonlinear Oscillations*, Clarendon Press, Oxford (1981).
3. S. Nocilla, *A study on the forced vibrations of a class of nonlinear systems with application to the Duffing Equation. Part I: Analytical treatment*, Meccanica, XI (1976), 11.
4. S. Nocilla, *Problemi insoliti di autovalori in meccanica non lineare*, Rendiconti di matematica, **10** (1977), 569.
5. S. Nocilla, *Periodical solutions of the Duffing Equation: some exact analytical results, Equadiff* **78**, Firenze (1978), 523.
6. S. Nocilla, *A proposito di una equazione integrale non lineare con nucleo singolare e valor medio*, Riv. Mat. Univ. Parma, **5** (1979), 647.
7. S. Nocilla, *Sull'integrazione dell'equazione di Duffing e sue estensioni Convegno matematica centenario nascita Guido Fubini e Francesco Severi*, Accad. Scienze Torino (1979), 81.
8. S. Nocilla, *Sul calcolo delle soluzioni periodiche armoniche monoscillanti della equazione di Duffing senza dissipazione*, Memorie Accad. Scienze Torino Serie V, **5** (1981), 23, Cl. Sci. Fis. Nat.
9. S. Nocilla, *An analytical and numerical procedure for exact calculation of non-linear oscillation with application to mechanical problems*, ICNO XI, Budapest (1987), 699.
10. S. Nocilla, *Un procedimento analitico e numerico per il calcolo esatto di oscillazioni non lineari con applicazione a problemi meccanici*, Terzo convegno italiano di meccanica computazionale - AIMETA Palermo (1988).
11. M. Di Tullio, P. Moroni and S. Nocilla, *An analytical and numerical procedure for exact calculation of nonlinear steady vibrations in system of Duffing Type*, Memoria Accad. Scienze Torino, 13–14 (1989–90), 1.
12. J. Dieudonnè, *Elements d'analyse* **I**, cap X, Gauthier Villars (1986).
13. N. Ya. Vilenkin et alii, *Functional Analysis*, Walters Noordhoff (1972).
14. S. D. Cante C. de Boor, *Elementary Numerical Analytsis*, McGrak-Hill, Tokyo (1980).
15. P. J. Davis, *Interpolation Approximation*, Blaisdell, Waltham, Mass. (1963).
16. S. Nocilla, *Steady state vibrations of a class of nonlinear system subject to non-conservative forces*, Euromech Colloquium, 190, Hamburg-Harburg (1984).
17. S. Nocilla, *Forced vibrations of a system with hysteresis, Part I: Analyical study,* ICNO X Varna (1984) 207.
18. S. Nocilla, *On exact calculation of forced vibrations of a system with hysteresis*, Atti. Sem. Mat. Fis. Univ. Moolena, XXXVIII (1990), 81.

19. S. Nocilla, *Vibrations of nonlinear dynamical absorbers*, 1st Nat. Congr. on Mechanics of the Hellenic Soc. Theor. Appl. Mech. H.S.T.A.M., Athens (1986), 554.
20. S. Nocilla, P. Moroni and M. Di Tullio, *Damping effects in nonlinear vibrating system of Duffing type*, Gauss Symposium, Guaryia, Brazil (1989).
21. M. DiTullio, P. Moroni and S. Nocilla, *Exact calculation of steady state vibrations in systems of Duffing type with damping*, Memoria Accad. Scienze Torino, 17 (1993), 1.

Functional Analysis, Approximation Theory
and Numerical Analysis
Ed. John M. Rassias

ON BANACH ALGEBRAS OF THE POTENTIAL DIFFERENTIAL AND PSEUDODIFFERENTIAL OPERATORS

IRINA E. PLESHCHINSKAYA
Kazan Chemical Institute, Kazan Karl Marx Str., 68 420015 USSR

and

NIKOLAI B. PLESHCHINSKII
Kazan State University, Kazan Lenin Str., 18 420008 USSR

Method of potential functions is widely adopted in mathematical physics. This method is based on the representing of solution of the system of the partial differential equations being under investigation through one, or more than one, auxiliary (potential) functions. These functions are a solution of some other potential system of the differential equations which is simpler or more convenient for research than the initial system. This method was used for the first time by G. Airy [1] and J. C. Maxwell [2] while researching the system of the main equations of elasticity theory. A classical example is: all components of the tensor of tension in a plane elasticity theory, without taking into account the mass forces, can be expressed through the second derivatives of Airy's function satisfying the biharmonic equation. In electrodynamics, for example, while finding an electromagnetic field harmonically dependent on time in homogeneous isotropic medium, all components of the field — (the solution of Maxwell's system of equations) — are expressed through the potential functions — the solutions of Helmholtz's equation.

Potential systems, through the solutions of which the solutions of the initial system of the partial differential equations are being expressed, can be defined arbitrarily. In the present work a set of systems being potential for the given system of the differential equations are being considered. Such approach makes easier the selection of the most suitable in one or another sense potential system.

Since it is convenient to consider many problems of the theory of the partial differential equations using the theory of pseudodifferential operators all reasonings are given for pseudodifferential operators everywhere where it is possible. Besides, one can consider the vector-valued operators or distributions as usual operators or distributions but with the set of values in the corresponding vector space. So the term 'operator' will be used both in the case of the differential operators and in the case of pseudodifferential operator, and the term 'equation' (both differential and pseudodifferential) will be used, as usual, there where it would be possible to say 'systems of equations'.

1. Let $Q \subset R^n$. In those cases when it is beyond any doubt we shall use common determination as for space of scalar functions so for the space of vector-functions or functional matrices every component of which belongs to the corresponding space of functions.

Let us consider the pseudodifferential operator

$$\begin{aligned} Su(x) &= (2\pi)^{-n} \int s(x,\xi)\tilde{u}(\xi)e^{i(x,\xi)}d\xi \\ P\varphi(x) &= (2\pi)^{-n} \int p(x,\xi)\tilde{\varphi}(\xi)e^{i(x,\xi)}d\xi \end{aligned} \tag{1}$$

with the matrix symbols $s(x,\xi)$, $p(x,\xi)$ of dimensions $l \times m$, $p \times p$ from the classes $S_0^{m_1}(Q)$, $S_0^{n_1}(Q)$ correspondingly. We shall assume for simplicity that all operators being considered in the work appear to be the proper operators.

We call operator P a *potential* for operator S if there exist operators

$$\begin{aligned} R\varphi(x) &= (2\pi)^{-n} \int r(x,\xi)\tilde{\varphi}(\xi)e^{i(x,\xi)}d\xi\ , \\ Qv(x) &= (2\pi)^{-n} \int q(x,\xi)\tilde{v}(\xi)e^{i(x,\xi)}d\xi \end{aligned} \tag{2}$$

with the matrix symbols $r(x,\xi)$, $q(x,\xi)$ of dimensions $m \times p$, $l \times p$ from the classes $S_0^{m_2}(Q)$, $S_0^{n_2}(Q)$ correspondingly such that

$$S \circ R = Q \circ P \tag{3}$$

(in the sense that $(S \circ R)\varphi(x) = (Q \circ P)\varphi(x)\ \forall\, \varphi(x) \in C_0^\infty(Q)$). Obviously, the equality $m_1 + m_2 = n_1 + n_2$ must be fulfilled.

Theorem 1.1. *Let P be a potential operator for operator S. If φ is a solution of the equation $P\varphi = 0$ then $u = R\varphi$ being a solution of the equation $Su = 0$.*

This Theorem follows directly from the above given definition.

If P is a potential operator for operator S then we say that the corresponding operator S defines the *representation* of solutions of the equation $Su = 0$. We shall say also that operator R defines a *complete* representation of solutions of the equation $Su = 0$ (or simply: R is a complete representation) if the formula $u = R\varphi$, where φ is a solution of the corresponding potential equation $Pu = 0$, gives all solutions of the equation $Su = 0$. In this case we also call the potential operator P a complete potential operator.

Theorem 1.2. *Operator P will be a potential for operator S and operator R defines the representation of solutions of the equation $Su = 0$ if and only if when there exists operator Q with a symbol $q(x,\xi)$ such that*

$$\begin{aligned} &\iint s(x,\xi)r(y,\xi_1)e^{i(x-y,\xi-\xi_1)}dyd\xi \\ &\quad = \iint q(x,\xi)p(y,\xi_1)e^{i(x-y,\xi-\xi_1)}dyd\xi \quad \forall\, x \in Q,\ \forall\, \xi_1 \in Q\,. \end{aligned} \tag{4}$$

This statement follows from the fact that the symbol is uniquely defined by the pseudodifferential operator. Besides, if two pseudodifferential operators are equal then their symbols are equal also.

Later on we shall limit ourselves to the case when operators S and P appear to be the operators of the 1st and 2nd orders correspondingly. Then the orders of operators R and Q will also be uniquely defined: the 1st and the zero order respectively.

Note that the equality (3) defines the *right factorisation* of operator R (to within the multiplier Q), i.e. the representation of operator of the 2nd order $Q \circ P$ in the form of a composition of two operators S and R of lower order. The *left factorisation* $R \circ S = Q \circ P$ means that the system of pseudodifferential equations $Su = 0$ of the 1st order can be reduced by pseudodifferenting to a system of pseudodifferential equations of the 2nd order.

Let us consider the particular case when S, R, P are differential operators with symbols

$$\begin{aligned} &s(x,\xi) = \sum_{|\alpha|\le 1} A_\alpha(x)\xi^\alpha, \quad r(x,\xi_1) = \sum_{|\beta|\le 1} h_\beta(y)\xi_1^\beta\,, \\ &p(x,\xi_1) = \sum_{|\gamma|\le 2} a_\gamma(x)\xi_1^\gamma\,, \end{aligned} \tag{5}$$

$A_\alpha(x)$, $h_\beta(y)$, $a_\gamma(x)$ being functional matrices of dimension $l \times m$, $m \times p$ and $p \times p$ correspondingly. In this case the operator Q of the zero order is equivalent to the multiplication operator by matrix $q(x)$ of dimension $l \times p$.

Theorem 1.3. *The system of partial differential equations $P\varphi = 0$ appears to be a potential one for the system $Su = 0$ and $u = R\varphi$ being a representation of solutions of the system $Su = 0$ if and only if when there exists a matrix $q(x)$ such that the conditions*

$$\begin{aligned} Sh_0 = qa_0, \quad Sh_\beta + A_\beta h_0 = qa_\beta \quad \forall\, \beta,\ |\beta| = 1 \\ A_\alpha h_\beta + A_\beta h_\alpha = qa_{\alpha+\beta} \quad \forall\, \alpha,\ |\alpha| = 1,\ \ \forall\, \beta,\ |\beta| = 1 \end{aligned} \tag{6}$$

are fulfilled.

The proof consists of the following: it is necessary to substitute the symbols (5) of the operators S, R, P into the condition (4). Then transform it by replacing the integration variable $\eta = \xi - \xi_1$ and applying the formulas $(\eta + \xi_1)^\alpha = \eta^\alpha + \xi_1^\alpha$ for $|\alpha| = 1$ and $\eta^\alpha \tilde{h}_\beta(\eta) = (\widetilde{D^\alpha h_\beta})(\eta)$.

2. Let us use $\{P\}_S$ to denote the set of all operators P being potential for operator S. As it follows from the condition (3), every operator S has infinitely many potential operators. In order to simplify the research of the set $\{P\}_S$ we introduce the separation of its elements into classes.

We call operators $P_1, P_2 \in \{P\}_S$ Q-equivalent if there exists invertible operators Q_1, Q_2 such that $Q_1P_1 = Q_2P_2$. Let us denote by $Q\{P_1\}$ a class of operators from $\{P\}_S$ being Q-equivalent to the operator $P_1 \in \{P\}_S$. Note that if P_1, P_2 are Q-equivalent then an operator Q can be found such that $P_2 = QP_1$, $P_1 = Q^{-1}P_2$.

Theorem 2.1. *$Q\{P_1\}$ is a Banach algebra.*

The proof of Theorem is based on two auxiliary statements.

Lemma 2.1. *Pseudodifferential operators Q of the zero order with symbols from the class $S_0^0(R^n)$ generate the Banach algebra $\{Q\}$.*

Proof. Operators Q with symbols from $S_0^0(R^n)$ appear to be the linear definite operators from H_s into H_s for any real s. They generate a Banach subalgebra in the Banach algebra $L(H_s, H_s)$. Ordinary operations of addition and multiplication being used for the linear continuous operators are valid in $\{Q\}$ and the norm of elements is defined in a usual sense.

Lemma 2.2. *Algebra $Q\{P_1\}$ is isometrically isomorphic to the algebra $\{Q\}$ for any $P_1 \in \{P\}_s$.*

Proof. Really, any element from the set $Q\{P_1\}$ has the form QP_1 where $Q \in \{Q\}$. Therefore it is sufficient to introduce the addition, multiplication and the norm in $Q\{P_1\}$ in such a way that operations and norm (being accepted in $\{Q\}$) would be well defined.

Note that the proper pseudodifferential operator generate a non-commutative algebra with two involutions.

Later on, as it is often accepted, one can identify the algebra $Q\{P_1\}$ with one of its elements (but not necessary with P_1 itself) which can be denoted as P_1.

In particular, let us consider a differential operator with constant coefficients. Let the initial system of the partial differential equations have the form $Su \equiv u_x + Au_y = 0$, where A is a constant matrix. As it follows from the above mentioned, if R is an arbitrary (matrix) differential operator with constant coefficients of the 1st order then any operator of the form $P = (Q)SR$ appears to be a differential operator of the 2nd order with constant coefficients, potential for the operator S. The operator Q of the zero order is written down in brackets as a sign that the question is not about a concrete potential operator but a class of Q-equivalent potential operators. To be precise, the question is about one of the representatives of this class.

In the case being considered, the operator Q of the zero order represents in itself a matrix with constant coefficients of dimension $m \times m$. It is known that the set of quadratic matrices with constant complex coefficients appears to be a complex Banach algebra with a norm

$$\|Q\| = \left(\sum_{j=1}^{m} \sum_{k=1}^{m} |a_{jk}|^2 \right)^{1/2} .$$

Example 2.1. Consider the system with constant coefficients $u_x + Au_y = 0$, where

$$A = \begin{pmatrix} 1 & 2 \\ -1 & 4 \end{pmatrix}, \qquad S = \begin{pmatrix} (\ )_x + (\ )_y & 2(\ )_y \\ -(\ )_y & (\ )_x + 4(\ )_y \end{pmatrix}$$

are a matrix of coefficients and a differential operator corresponding to the system being under consideration. Here and elsewhere we shall use some non-standard determinations of the form $(\ )_x$ for the operator of differentiating by variable x instead of commonly used $\frac{\partial}{\partial x}$.

Let

$$R_1 = \begin{pmatrix} (\,)_x & (\,)_x - (\,)_y \\ (\,)_y & 2(\,)_x \end{pmatrix}, \quad Q_1 = E\,,$$

$$R_2 = \begin{pmatrix} 2(\,)_x - 4(\,)_y & (\,)_x - 3(\,)_y \\ (\,)_x - 2(\,)_y & (\,)_x - 3(\,)_y \end{pmatrix}, \quad Q_2 = \begin{pmatrix} 1 & -1 \\ -1 & 2 \end{pmatrix}.$$

The complete potential operators

$$P_1 = \begin{pmatrix} (\,)_{xx} + (\,)_{xy} + 2(\,)_{yy} & (\,)_{xx} + 4(\,)_{xy} - (\,)_{yy} \\ 4(\,)_{yy} & 2(\,)_{xx} + 7(\,)_{xy} + (\,)_{yy} \end{pmatrix},$$

$$P_2 = \begin{pmatrix} (\,)_{xx} - 4(\,)_{yy} & 0 \\ 0 & (\,)_{xx} - 9(\,)_{yy} \end{pmatrix}$$

correspond to these representations.

We call operators P_1 and P_2 of 2nd order *S-equivalent* if they appear to be the complete potential operators for the same operator S of the 1st order. S-equivalence of two operators P_1 and P_2 enables one to set a non-trivial correspondence between solutions of two equations $P_1\varphi = 0$ and $P_2\varphi = 0$.

As if follows from the above considered example the systems of partial differential equations

$$\begin{cases} \varphi^1_{xx} + \varphi^1_{xy} + 2\varphi^1_{yy} + \varphi^2_{xx} + 4\varphi^2_{xy} - \varphi^2_{yy} = 0\,, \\ 4\varphi^1_{yy} + 2\varphi^2_{xx} + 7\varphi^2_{xy} + \varphi^2_{yy} = 0\,, \end{cases}$$

and

$$\psi^1_{xx} - 4\psi^1_{yy} = 0, \quad \psi^2_{xx} - 9\psi^2_{yy} = 0$$

appear to be S-equivalent systems.

It is interesting to single out among all elements $\{P\}_s$ the canonical one, (i.e. the simplest element in some sense). For the differential operators this can be easily done in the case when $n = 2$ (there are two variables x and y).

Let us consider the following differential operator with constant coefficients: $S = E(\,)_x + A(\,)_y$, E being a unit matrix and A being a constant matrix. Let $\lambda_1, \ldots, \lambda_n$ be the proper values of matrix A and $h^1, \ldots, h^n$ be the corresponding proper vectors. Denote by H a matrix composed of proper vectors as of columns. Let $Q = H^{-1}$ and $R = H\,\mathrm{diag}((\,)_x - \lambda_1(\,)_y, \ldots, (\,)_x - \lambda_n(\,)_y)$. Then the operator $P = QSR = \mathrm{diag}((\,)_{xx} - \lambda_1^2(\,)_{yy}, \ldots, (\,)_{xx} - \lambda_n^2(\,)_{yy})$ will have the simplest structure in comparison with all other elements of algebra, and every equation corresponding to its potential system will represent in itself the partial differential equation written down in a canonical form. In the case when matrix A has multiple proper values, it is also convenient to take as H a matrix which reduces the matrix A to a normal Jordan form. Such a matrix can be composed of proper and adjoined vectors of matrix A. In this case the matrix of the operator P will have different from, zero elements only in the stripe disposed along the main diagonal and the width of the stripe is 3 elements. In the corresponding to such a potential operator potential system of partial differential equations it will be possible to single out chains of the connected potential equations.

Now let us consider the set of classes $(Q)SR$ in the set of all operators which are potential for the operator S. We introduce the following operations for the operators R of the 1st order: the addition which is understood in the usual sense and the multiplication (for the present, for differential operators only) which can be reduced to the multiplying of the coefficients attached to derivatives of the same order, i.e. we define the product of the pseudodifferential operator with symbols

$$s^1(x,\xi) = \sum_{|\alpha|\leq 1} s^1_\alpha(x)\xi^\alpha, \quad s^2(x,\xi) = \sum_{|\alpha|\leq 1} s^2_\alpha(x)\xi^\alpha$$

as a pseudodifferential operator with symbol

$$s(x,\xi) = \sum_{|\alpha|\leq 1} s^1_\alpha(x) s^2_\alpha(x)\xi^\alpha$$

Such operations do not take us out from a class of the differential operators of the 1st order.

Theorem 2.2. *The set of $\{P\}_S$ classes of Q-equivalent operators appears to be a Banach algebra.*

In the general case we introduce a product of two pseudodifferential operators of the 1st order on the basis of the Taylor's series expansion of their symbols. Let

$$s^1(x,\xi) = \sum_{\alpha\in Z^n_+} (1/\alpha!) D^\alpha_\xi s^1(x,0)\xi^\alpha \ ,$$

$$s^2(x,\xi) = \sum_{\alpha\in Z^n_+} (1/\alpha!) D^\alpha_\xi s^2(x,0)\xi^\alpha$$

be the symbols. We define their product as a symbol

$$s(x,\xi) = \sum_{\alpha\in Z^n_+} (1/\alpha!) D^\alpha_\xi s^1(x,0) D^\alpha_\xi s^2(x,0)\xi^\alpha \ .$$

Let us call a product of two operators of the 1st order such an operator which symbol represents in itself the product of their symbols in the above mentioned sense. The statement of Theorem 2.2 remains valid in this case also.

3. Let us study a set $\{u\}_S$ of solutions of the equation $Su = 0$ by assumption $l = p = m$ (i.e. all matrices being quadratic of dimension $m \times m$).

Theorem 3.1. *If the equation $R\varphi = u$ is solvable for any u then formula $u = R\varphi$ where φ is a solution of the equation $P\varphi = 0$, $P = (Q)SR$ gives all solutions of the equation $Su = 0$, i.e. $\{u\}_S = \{u | u = R\varphi,\ (Q)SR\varphi = 0\}$.*

Proof. Obviously, $\{u | u = R\varphi,\ (Q)SR\varphi = 0\} \subset \{u\}_s$. Now let u be an arbitrary element of $\{u\}_s$ and φ be a solution of the equation $R\varphi = u_0$. Then $u_0 = R\varphi_0$, φ_0 being a solution of the equation $(Q)SR\varphi = 0$. Consequently, $u_0 \in \{u | u = R\varphi,\ (Q)SR\varphi = 0\}$.

Corollary. Any operator S has a complete potential operator P.

Really, let S be some operator of the 1st order. It is possible to pick out R in such a way that the equation $R\varphi = u$ is solvable for any u. Then $P = (Q)SR$ is a complete potential operator.

Note that the assumption that all matrices under consideration are quadratic is essential here.

So it is possible to consider the equation $P\varphi = 0$ instead of the equation $Su = 0$ but that is more complicated problem in the general case (because the order of the operator P is greater than the order of the initial operator S). But in many cases it is possible to write down the potential equation $P\varphi = 0$ as a system of independent equations, which some times coincide with each other. Then the application of the method of potential functions is quite advisable.

We call potential operator P *trivial* if the set $\{u\}_S$ contains only trivial solution of the equation $Su = 0$.

We shall say that potential operator $P \in \{P\}_S$ is *splittable* if there exists nontrivial operators $P_1, P_2 \in \{P\}_S$ such that $\{u\}_P = \{u\}_{P_1} \oplus \{u\}_{P_2}$ (the subspace of solutions of the equations $Su = 0$ associated with the operator P can be decomposed into the direct sum of the subspaces of solutions of the same equation associated with another potential operators). We call operators P_1 and P_2 the *potential suboperators* of the operator P.

Let us consider differential operators with constant coefficients as the simplest case.

Theorem 3.2. *If the operator $P \in \{P\}_S$ is splittable then there exist suboperators $P_1, P_2 \in \{P\}_S$ such that $P = P_1 + P_2$ (in the sense of the multiplication operation acceptable in the algebra $\{P\}_S$).*

Proof. If operator P is splittable then values of solutions of the equation $Su = 0$ from the classes $\{u\}_{P_1}$, $\{u\}_{P_2}$ and $\{u\}_P$ belong to linear subspaces L_1, L_2 and L of the space R^m correspondingly, by this the direct sum of the first two gives the third: $L_1 \oplus L_2 = L$. Any solution of the equation $Su = 0$ has the form $u = R\varphi$, where R is some representation and φ being a solution (vector-valued) of the corresponding potential equation. Then, for example, $u \in \{u\}_{P_1}$ if and only if when columns of matrix coefficients of the operator R belong to the subspace L_1 of the space L.

Let R be a representation of solutions of the equation $Su = 0$ generating a potential operator P. Let us decompose every column of coefficients of R into a sum of vectors from L_1 and L_2 correspondingly and construct representations $\tilde{R}_1$ and $\tilde{R}_2$ by the received addends as of columns. Then $R = \tilde{R}_1 + \tilde{R}_2$ and $P = \tilde{P}_1 + \tilde{P}_2$ where $\tilde{P}_1 = QS\tilde{R}_1$, $\tilde{P}_2 = QS\tilde{R}_2$ (the multiplier Q is just the same for $\tilde{P}_1$ and $\tilde{P}_2$; if one assumes that $Q \equiv E$ then it is possible to decompose directly the columns of the operator P into a sum of vectors from L_1 and L_2).

Example 3.1. The complete operator

$$P = \begin{pmatrix} (\,)_{xx} + (\,)_{xy} & 2(\,)_{yy} \\ -(\,)_{xy} & (\,)_{xy} + 4(\,)_{yy} \end{pmatrix} \tag{7}$$

is splittable for the operator S from the example 2.1 because $\{u\}_P = \{u\}_{P_1} \oplus \{u\}_{P_2}$ where

$$P_1 = \begin{pmatrix} (\,)_{xx} - 4(\,)_{yy} & 0 \\ 0 & 0 \end{pmatrix}, \qquad P_2 = \begin{pmatrix} 0 & 0 \\ 0 & (\,)_{xx} - 9(\,)_{yy} \end{pmatrix}.$$

and $P \neq P_1 + P_2$, but $P = \tilde{P}_1 + \tilde{P}_2$ where

$$\tilde{P}_1 = \begin{pmatrix} -(\,)_{xx} - 3(\,)_{xy} & 2(\,)_{xy} + 6(\,)_{yy} \\ -(\,)_{xx} - 3(\,)_{xy} & 2(\,)_{xy} + 6(\,)_{yy} \end{pmatrix},$$

$$\tilde{P}_2 = \begin{pmatrix} 2(\,)_{xx} + 4(\,)_{xy} & -2(\,)_{xy} - 4(\,)_{yy} \\ (\,)_{xx} + 2(\,)_{xy} & -(\,)_{xy} - (\,)_{yy} \end{pmatrix}.$$

The inverse statement also holds true.

Theorem 3.3. *If $P = P_1 + P_2$ where $P, P_1, P_2 \in \{P\}_S$ and at least one of operators P_1 and P_2 being non-trivial, then the potential operator P is splittable.*

Emphasize that the statement of Theorem concerns the representation of potential operator in the form of a sum just of potential for the initial S operators, and not of simply differential operators of the 2nd order.

Thus if a complete potential for the operator S operator is splittable then any solution of the equation $Su = 0$ can be represented as a sum of solutions of the same equation belonging to non-intersecting classes. We shall call this phenomenon a *polarisation* of solutions of the equation $Su = 0$.

4. Let us return once more to the equation $u_x + Au_y = 0$ as we considered before (the system of m equations with constant coefficients for two independent variables). Let $\lambda_1, \ldots, \lambda_m$ be different proper values of matrix A and $h^1, \ldots, h^m$ be corresponding proper vectors. As it was shown, in this case a complete potential system consists of independent potential equations of the form

$$\varphi_{xx} - \lambda_j^2 \varphi_{yy} = 0, \quad j = \overline{1, m}\ , \tag{8}$$

and a complete representation of solutions consists of the columns of the form

$$h^j((\ )_x - \lambda_j(\ )_y), \quad j = \overline{1, m}\ . \tag{9}$$

Consequently a complete potential equation is splittable into a sum of m potential operators, each of them is being generated by representation of solutions R_j of the following form: the column with the number j in matrix R_j has the form (9) and other columns are all zeros.

Now let us consider the representation of solutions of the form $R_j \times R_k$. In matrix $R_j \times R_k$ only the column with the number k is different from zero and has the form $h^k((\ )_x + \lambda_j \lambda_k(\ )_y)h^j$ (here h_j^k is a jth component of the proper vector h^k). Then the potential operator $S(R_j \times R_k)$ has different from zero the kth column only

$$h_j^k((\ )_{xx} + \lambda_j \lambda_k(\ )_{xy} + \lambda_j(\ )_{xy} + \lambda_j^2 \lambda_k(\ )_{yy})\ .$$

Obviously, there exists diagonal operator with the single element different from zero with the number k among Q-equivalent to $S(R_j \times R_k)$ potential operators. This operator generates the potential equation

$$\varphi_{xx} + \lambda_j(1 + \lambda_k)\varphi_{xy} + \lambda_j^2 \lambda_k \varphi_{yy} = 0\ . \tag{10}$$

Note that Eq. (10) can be reduced to the canonical form $\varphi_{\xi\xi} - \lambda_j^2\varphi_{\eta\eta} = 0$ by substitution of variables $\xi = x$, $\eta = (-\lambda_j x(1+\lambda_k) + 2y)/(1-\lambda_k)$. By this the particular representation of solutions of system $u = \varphi_x + \lambda_j\lambda_k\varphi_y$ turns into representation $u = \varphi_\xi - \lambda_j\varphi_\eta$ (to within the multiplier).

Now let P_j be the potential operators generated by representations of solutions R_j, $j = \overline{1,m}$. As the multiplication operation in algebra $\{P\}_S$ is induced by the multiplication operation in algebra, which is introduced into the set of representations, the above mentioned reasonings can be reduced to the following fact: in the case being considered

$$\{u\}_{P_j \times P_k} = \{u\}_{P_j}, \quad j = \overline{1,m}, \quad k = \overline{1,m}\,. \tag{11}$$

The similar statement holds true in the general case.

Lemma 4.1. *There exist elements $P_1, \dots, P_N$, $1 \le N \le m$ in algebra $\{P\}_S$ such that*

$$\{u\}_{P_j \times P_k} = \{u\}_{P_j}, \quad j = \overline{1,N}, \quad k = \overline{1,N}\,. \tag{12}$$

Proof. By corollary of Theorem 3.1 there exists an element P in algebra $\{P\}_S$ which appears to be a complete potential operator. If the operator P is non-splittable then the condition in (12) is fulfilled for $N = 1$ only. Really, the product of element P with any other element of algebra $\{P\}$ either will be the complete potential operator or a trivial potential operator.

If the operator P is splittable then by Theorem 3.2, it can be represented as a sum of non-splittable potential operators $P = P_1 + \cdots + P_N$, by this, $\{u\}_P = \{u\}_{P_1} \oplus \cdots \oplus \{u\}_{P_N}$. Since the values of the operators P_j belong to the non-intersecting linear subspaces of R^m then the condition (12) is fulfilled.

The fact is that the application of the method of potential operators appears to be the most effective just in the case when a complete potential operator is splittable into a possibly greater number of potential operators of simpler structure. So let us formulate some conditions of splittability of a complete potential operator in algebra $\{P\}_S$.

Lemma 4.2. *If there exist a non-complete, nontrivial potential operator in algebra $\{P\}_S$ then the complete potential operator is splittable.*

Really, if P is a complete operator whereas P_1 is a non-complete operator then $\{u\}_{P_1} \subset \{u\}_P$. By this values of operator P form a linear subspace L_1 in the space L of values of P. Let $L = L_1 \oplus L_2$. Decomposing the matrix of the operator R into a sum of matrices R_1 and R_2, the columns of which accept their values in L_1 and L_2 correspondingly, we receive that $\{u\}_P = \{u\}_{P_1} \oplus \{u\}_{P_2}$ where P_2 is a potential operator being generated by representation R_2.

Example 4.1. Let a complete potential operator for the system from the Example 2.1 be determined by formula (7), whereas a non-complete potential operator

$$P_1 = \begin{pmatrix} -(\,)_{xx} - 3(\,)_{xy} & 2(\,)_{xy} + 6(\,)_{yy} \\ -(\,)_{xx} - 3(\,)_{xy} & 2(\,)_{xy} + 6(\,)_{yy} \end{pmatrix}.$$

Then

$$P_2 = \begin{pmatrix} 2(\,)_{xx} + 4(\,)_{xy} & -2(\,)_{xy} - 4(\,)_{yy} \\ (\,)_{xx} + 2(\,)_{xy} & -(\,)_{xy} - 2(\,)_{yy} \end{pmatrix}$$

appears to be a non-complete potential operator also but such that $\{u\}_P = \{u\}_{P_1} \oplus \{u\}_{P_2}$ and even $P = P_1 + P_2$.

Theorem 4.1. *A complete potential for S operator is splittable if and only if when there exists at least one non-trivial right ideal in algebra* $\{P\}_S$.

Proof. Necessity. Let a complete potential operator be splittable. By virtue of Lemma 4.1 we can set in correspondence to every operator P_j, the existence which is stated in Lemma, a set of potential operators

$$\{P\}_j = \{P | P \in \{P\}_S,\ \{u\}_P = \{u\}_{P_j}\}\,.$$

Let us prove that for any j, $\{P\}_j$ is a non-trivial right ideal in algebra $\{P\}_S$.

Really by Theorem 3.2 an arbitrary element $P \in \{P\}_S$ can be represented in the form $P = \tilde{P}_1 + \cdots + \tilde{P}_N$ where, generally speaking, potential operators P_j do not coincide with operators $\tilde{P}_j$ but $\{u\}_{P_j} = \{u\}_{\tilde{P}_j}$. Let us pass on from potential operators $P, \tilde{P}_j, \ldots, \tilde{P}_N$ to representations $R, \tilde{R}_1, \ldots, \tilde{R}_N$ generating them. Obviously, $R = \tilde{R}_1 + \cdots + \tilde{R}_N$. Then

$$R_j \times R = R_j \times (\tilde{R}_1 + \cdots + \tilde{R}_N) = R_j \times \tilde{R}_1 + \cdots + R_j \times \tilde{R}_N$$

and $R_j \times \tilde{R}_k$ generate the potential operators $P_j \times \tilde{P}_k \in \{P\}_j$. Consequently, $P_j \times P \in \{P\}_j$.

Sufficiency. Let I be a non-trivial right ideal in algebra $\{P\}_S$. We shall consider the set $\{u\}_I = \bigcup_{P \in I} \{u\}_P$. If $\{u\}_I = \{u\}_S$, the ideal I would coincide with $\{P\}_S$. Consequently, $\{u\}_I \subset \{u\}_S$. Let us single out the subset of elements in I

$$\{P\}_I = \{P | P \in I,\ \{u\}_P = \{u\}_I\}\,.$$

Then any element in $\{P\}_I$ will be a non-complete potential operator and by Lemma 4.2 and the complete potential operator is splittable.

References

1. G. B. Airy, *On the strains in the interior of beams.* Philos. Trans. Roy. Soc., **153** (1863), p. 49–80.
2. J. C. Maxwell, *On reciprocal diagrams in space and their relation to Airy's function of stress.* Proc. London Math. Soc., **2** (1865/69), p. 58–60.
3. Ju. V. Egorov, *Linear Differential Equations of the Main Type.* Nauka, Moscow, Russian, 1984.
4. F. Treves, *Introduction to Pseudodifferential and Fourier Integral Operators.* Plenum Press, New York and London, 1982.
5. E. Hille and R. S. Phillips, *Functional Analysis and Semi-Groups.* A. M. S., Providence, 1957.
6. N. Dunford and J. T. Schwartz, *Linear Operators. Part 2. Spectral Theorie.* Wiley-Interscience, 1971.

Functional Analysis, Approximation Theory
and Numerical Analysis
Ed. John M. Rassias

ON RELATIVE COMPACTNESS SET OF ABSTRACT FUNCTIONS FROM SCALE OF THE BANACH SPACES

A. G. Podgaev

Institute for Applied Mathematics, USSR Academy of Sciences,
Far Eastern Branch, Khabarovsk.

Let B^t and B^t_1 for any $t \in [0,T] \subset R$ be normed linear Banach spaces; B_1 is separable; $B^t \subset B^t_1$ and embedding is continuous for every t; for $t_2 < t_1$ $B^{t_1} \subset B^{t_2}$, $B_1^{t_1} \subset B_1^{t_2}$, and the last embedding is continuous in sense that there exists bounded on $[0,T]^2$ function $\varphi_1(\cdot,\cdot)$ and a constant φ_1 that for any t_1, $t_2 \in$ $\in [0,T]$ $t_2 \le t_1$ and any $u \in B_1^{t_1}$ inequality

$$(1) \qquad \|u\|_{B_1^{t_2}} \le \varphi_1(t_1,t_2)\,\|u\|_{B_1^{t_1}} \le \varphi_1\|u\|_{B_1^{t_1}}$$

is held. Analogously there exists a constant φ_2 such that

$$\|u\|_{B_1^t} \le \varphi_2\|u\|_{B^t}.$$

Let S^t be nonlinear subset of B^t, with function $M_t : S^t \to$ $\to \bar{R}_+$, $R_+ = \{ y \in \bar{R} : y \ge 0\}$. Define $S^t_a = \{v \in S^t : M_t(v) \le a\}$, and suppose that S^t_a is relatively compact in B^t for any $a < \infty$.

Since $B^{t_1} \subset B^{t_2}$ for $t_1 > t_2$, $u(t_1) \in B^{t_2}$. Therefore $M_{t_2}[u(t_1)]$ is defined. We will assume, that function M_t such that

$$(2) \qquad M_{t_2}[u(t_1)] \le \varphi(t_1,t_2)\, M_{t_1}[u(t_1)],$$

where φ is bounded function on $[0,T]^2$.

Further, denote F_1 some subset of elements $u(t)$ such that for any $t \in [0,T]$ $u(t) \in B^t$ and

(3) $$\underset{t \in [0,T]}{\text{vrai max}} \|u(t)\|_{B^t} \le C_1, \quad \int_0^T M_t(u(t))dt \le C_2,$$

where C_1 and C_2 are common constants for all u in F_1.

Moreover, we will assume equicontinuity of norms in B^t by parameter t on subset F_1:

(4) $$\begin{cases} \text{there exist } \eta(\cdot,\cdot) \text{ such that for all elements } u \in B^{t_1} \\ \text{and all } t_2 \le t_1 \\ \left| \|u\|_{B^{t_2}} - \|u\|_{B^{t_1}} \right| \le \eta(t_1, t_2) \longrightarrow 0 \text{ as } t_2 - t_1 \longrightarrow 0. \end{cases}$$

Here η does not depend on element u in F_1.

And, at last, let either

(5) $$F = \left\{ u \in F_1 : u(t) \text{ is } B_1 \text{- measurable and} \int_0^T \| u'(t) \|^{P_1}_{B_1^t} dt \le C_3,\ P_1 > 1 \right\},$$

or F be subset of elements $u(t)$ in F_1 such that

(5') $$\sup_{\substack{t \in [0,T] \\ t+h \le T}} \|u(t+h) - u(t)\|_{B_1^t} \le H(h),$$

where $h > 0$, $H(h) \longrightarrow 0$ as $h \longrightarrow 0$ and $H(h)$ is common for all u in F_1.

In (5) we mean that $u'(t)$ is element of B_1^t, which for all $\tau < t$ is the limit in norm B_1^τ of $\dfrac{u(t+h) - u(t)}{h}$ as $h \longrightarrow 0$, $\tau \le t + h$. See Additional part for details.

In [1], C.4, § 5 compactness of embedding of reflexive Banach spaces of abstract functions was established in space like

$L_p(0,T;\ B)$. After that Yu. A. Dubinskiĭ [2] proved this result in the case when function $M(v)$ didn't define the norm, but possesses property of homogeneity: $M(\lambda\ v) = |\lambda|\ M(v)$.

In case of Orlich spaces and half-normed set similar results were proved in [3],[4].

By slightly modified assumptions in [5],[6] these statements were proved for function $M(v)$ which has arbitrary structure. All these results are essentially used for proofs of existence of solutions for nonstationary problems for partial differential equations, accordingly for linear, nonlinear degenerating equations with homogeneous nonlinearity, and, at last, with arbitrary nonlinearity. For the latter see, for example, [7],[8].

In this article we generalize results [5],[6] and give up the condition: for any $t \in [0,T]$ $u(t) \in B$, where B is a fixed (for all t) Banach space and we consider functions $u(\cdot)$ that belong to the scale ("monotone" and continuous by parameter) of Banach spaces. These results can be used for proving of existence of solution for mathematical physics of nonlinear equations in noncylindrical domains and also represent independent interest.

<u>*Theorem 1*</u>. Let embedding $B^t \subseteq B^t_1$ be compact. For any $p \geq 1$ set F is relatively compact in space of functions $u(\cdot)$ with norm $\left[\int_0^T \|u(t)\|^p_{B^t}\ d\ t\right]^{1/p}$. For corresponding limit elements the first inequality in (3) is valid.

Proof. Let $F^N = \left\{u_n\right\}_{n=1}^{\infty}$ be a countable set of elements in F. Since the first inequality in (1) is correct there is a set $E_0 \subset$ $\subset [0,T]$ with meas $E_0 = T$ such that for all $t \in E_0$

(6) $$\|u_n(t)\|_{B^t} \le C_1 < \infty .$$

Similarly, since $M_t(u_n(t)) \in L_1(0,T)$, there exist sets $E_n \subset \subset [0,T]$, with meas $E_n = T$ such that for all $t \in E_n$

(7) $$M_t(u_n(t)) \le k^t_n , \quad k^t_n < \infty , \quad n = 1,2,\ldots$$

(In other words: functions from L_1 are finite for almost all $t \in [0,T]$).

Consider $E = \bigcap_{n=0}^{\infty} E_n$. Obviously, meas $E = T$.

<u>*Lemma 1.*</u> For each $\varepsilon > 0$ there exists a constant $c(\varepsilon)$ such that for all $t \in E$ and all u, $v \in F^N$

(8) $$\|u(t) - v(t)\|_{B^t} \le \varepsilon\,[M_t(u(t)) + M_t(v(t)) + 1] + + C(\varepsilon)\,\|u(t) - v(t)\|_{B^t_1} .$$

We do not assume compactness of embedding $B^t \subset B^t_1$ in this Lemma.

Proof Of Lemma 1. First of all we note that by (6)-(7) and embedding $B^t \subset B^t_1$, the left-hand and right-hand sides of (8) are finite for all $t \in E$. Assume the contrary statement of the Lemma *1*. Then there is an $\varepsilon_o > 0$ such that for any constant $C > 0$ there exist elements u_c, $v_c \in F^N$ and number $t_c \in E$ such that

(9) $$\|u_c(t_c) - v_c(t_c)\|_{B^{t_c}} > \varepsilon_o[M_{t_c}(u_c(t_c)) + M_{t_c}(v_c(t_c)) + 1] + + C\,\|u_c(t_c) - v_c(t_c)\|_{B_1^{t_c}}$$

Let $c = c_i \longrightarrow \infty$. It follows from (6) that the left-hand is bounded by the constant $2c_1$. Then from (9) we get for elements $\omega^1_i = u_{c_i}(t_{c_i})$, $\omega^2_i = v_{c_i}(t_{c_i})$ of $B^{t_{c_i}}$ that

(10) $$\| \omega_i^1 - \omega_i^2 \|_{B_1}^{tc_i} \le 2c_1/c_i \longrightarrow 0.$$

Compactness of $[0,T]$ yields a subsequence such that $t_{c_{i(k)}} \longrightarrow \bar{t}_o \in$ $\in [0,T]$. We will assume that $t_{c_i} \longrightarrow \bar{t}_o$. Differently takes subsequence. We define $\tau_N = \min_{i \ge N} t_{c_i}$, then $\tau_N \le \tau_{N+1}$; $\tau_N \le t_{c_i}$, $i \ge N$. Therefore, from (10) and from continuity of embedding B_1^t we have

(11) $$\|\omega_i^1 - \omega_i^2\|_{B_1^{\tau_N}} \le \varphi_1(\tau_N, t_{c_i})\|\omega_i^1 - \omega_i^2\|_{B_1^{tc_i}} \le 2c_1\varphi_1/c_i \longrightarrow 0,$$
$$i \ge N,\ i \longrightarrow \infty.$$

From (9) we get

$$M_{tc_i}(\omega_i^1) \le 2c_1/\varepsilon_o,\ M_{tc_i}(\omega_i^2) \le 2c_1/\varepsilon_o.$$

Since $\tau_N \le t_{c_i}$, $i \ge N$ from (2) we get:

(12) $$M_{\tau_N}(\omega_i^1) \le \varphi\ M_{tc_i}(\omega_i^1) \le 2c_1\varphi/\varepsilon_o;\ M_{\tau_N}(\omega_i^2) \le \varphi\ M_{tc_i}(\omega_i^2) \le 2c_1\varphi/\varepsilon_o,\ i \ge N.$$

Let $N = 1$. Using the relative compactness of S_a^t in B^t (for $t = \tau_1$, $a = 2c_1\varphi/\varepsilon_o$) and the completeness of B^t, we conclude that there exists subsequence $\omega_{n_k}^1$ of sequence ω_i^1, converging to some element $\bar{u}_1$ in B^{τ_1} for $k \longrightarrow \infty$. But, sequence $\omega_{n_k^1}^1 = \omega_{n_k}^1$ satisfies (12). Therefore, from it we can select subsequence $\omega_{n_k^2}^1$ such that $\omega_{n_k^2}^1 \longrightarrow \bar{u}_2$ in B^{τ_2}. Notice, that since $\tau_1 \le \tau_2$, $\bar{u}_1 = \bar{u}_2$ in B^{τ_2}. And further the same is done for all N. If we take sequence $\omega_{n_k^k}^1 \overset{\text{def}}{=} \bar{\omega}_k^1$, it is obvious, that it converges in any B^{τ_N} to element ω_1. On analogy, for sequence $v_{c_i}(t_{c_i})$. And we may conclude that

$\omega^2_{n_k} = \bar{\omega}^2_k \longrightarrow \omega_2$ in any B^{τ_N}. Therefore, $\bar{\omega}^1_k - \bar{\omega}^2_k \longrightarrow \omega_1 - \omega_2$ in B^{τ_N}. But $\bar{\omega}^1_k$, $\bar{\omega}^2_k$ satisfy (*11*). That is $\bar{\omega}^1_k - \bar{\omega}^2_k \longrightarrow 0$ in $B_1^{\tau_N}$. We conclude that $\omega_1 = \omega_2$ and for any fixed N

$$(13) \qquad \bar{\omega}^1_k - \bar{\omega}^2_k \longrightarrow 0 \text{ in } B^{\tau_N} \text{ as } k \longrightarrow \infty.$$

Notice that $\bar{\omega}^{1,2}_k \in B^{t_{\bar{C}_k}}$, $\bar{C}_k = C_{n_k}$. Therefore from (9) we get for sufficiently large N and k:

$$(14) \qquad \|\bar{\omega}^1_k - \bar{\omega}^2_k\|_{B^{t_{\bar{C}_k}}} - \|\bar{\omega}^1_k - \bar{\omega}^2_k\|_{B^{\tau_N}} \geq \frac{\varepsilon_o}{2}.$$

This contradicts the condition (*4*) of equicontinuity of norms in B^t on elements set F_1. Actually, the left-hand side (*14*) by condition (*4*) is smaller $\eta(\tau_N, t_{\bar{C}_k})$. Since $t_{\bar{C}_k} \longrightarrow T_o$ and $\tau_N \longrightarrow T_o$, we have $\eta(\tau_N, t_{\bar{C}_k}) \longrightarrow 0$. Lemma *1* is proved.

Notice that we can prove (*8*) if in right-hand side we replace $[M(u) + M(v) + 1]$ by $[M(u) + M(v) + 1]^{1/p}$ for any fixed $p < \infty$. Using inequality $(a + b)^p \leq 2^p(a^p + b^p)$, later integrating by t we can conclude:

$$(16) \qquad \int_o^T \|u(t) - v(t)\|^p_{B^t}\, dt \leq \varepsilon \int_o^T [M_t(u(t)) + M_t(v(t)) + 1]dt + $$
$$+ c(\varepsilon) \int_o^T \|u(t) - v(t)\|^p_{B_1^t}\, dt.$$

<u>*Lemma 2.*</u> Let embedding $B^t \subseteq B_1^t$ be compact. Then F is relatively compact in space of abstract functions $u(\cdot)$ with norm

$$\operatorname*{vraimax}_{t\in[o,T]} \|u(t)\|_{B_1^t}.$$

Proof of Lemma 2. We use only first inequality in (3) and (5) or (5$'$) in this Lemma.

Let $P = \{t_o, t_1, \dots, t_n, \dots\}$ be dense countable system in E, $t_i \in E$. Now we will prove existence of P.

Let $E = \bigcup_\alpha t^\alpha$, $t^\alpha \in E$ and let $E^j_\alpha = \{t \in [0,T]: |t - t^\alpha| < 1/j\}$. Then $\bigcup_\alpha E^j_\alpha = [0,T]$. It is consequence of density set E on $[0,T]$. There is a finite number of $E^j_{\alpha^j_i}$ and $t_{\alpha^j_i}$, $i = 1, \dots, N_j$, that $\bigcup_{i=1}^{N_j} E^j_{\alpha^j_i} = [0,T]$. Consider countable system $P = \bigcup_{j=1}^{\infty} \Big[\bigcup_{i=1}^{N_j} t_{\alpha^j_i} \Big]$.

Then for all $\tau \in E$ and $\varepsilon > 0$ there is $t_{\nu(\varepsilon,\tau)} \in P$ such that $|t - t_{\nu(\varepsilon,\tau)}| < \varepsilon$. Actually, if it wasn't done, we would have $\tilde{t} \in E$ and $\tilde{\gamma} > 0$, that for all $t \in P$ $|t - \tilde{t}| \geq 1/\tilde{\gamma}$. Let integer $j > \tilde{j}$. Because $\bigcup_{i=1}^{N_j} E^j_{\alpha^j_i} = [0,T]$, then $\tilde{t}$ belong to certain $E^j_{\tilde{\alpha}} = \{\tau \in [0,T]: |\tau - t^{\tilde{\alpha}}| < 1/j\}$, $\tilde{\alpha} = \alpha^j_{i_o}$, $t^{\tilde{\alpha}} \in P$. This contradicts the last inequality. Existence of P is proved.

Let $\{u_n\}$ be sequence from F. By (6) we have that for all $t_i \in P \subset E$ $\|u_n(t_i)\|_{B^{t_i}} \leq C_1$. We may assume that $t_o = T \in E$. Compactness of embedding yields that $\{u_n(t_i)\}$ is relative compact in $B_1^{t_i}$. Diagonal process yields a subsequence u_μ such that for all i $u_\mu(t_i)$ is Cauchy sequence in $B_1^{t_i}$.

Now we prove that the sequence u_μ is convergent by norm

$$\underset{t \in [0,T]}{\mathrm{vraimax}} \; \|u(t)\|_{B_1^t} .$$

Consider case (5$'$). Let $\varepsilon > 0$ be given. There exists $\delta(\varepsilon) > 0$

such that $H(h) < \varepsilon/4$ if $h < \delta(\varepsilon)$. Consider sets $S_i = \{ t \in [0,T]: |t_i - t| < \delta(\varepsilon),\ t_i > t \}$, $i = 1,2,\dots$, $t_i \in P$, $S_o = \{ t \in [0,T]: |T - t| < \delta(\varepsilon),\ t \le T \}$, which are opened by topology of $[0,T]$.

Moreover, $\bigcup_{i=o}^{\infty} S_i = [0,T]$. It follows from density of P in E, and hence in $[0,T]$. Select finite number S_{i_ν}, $\nu = 1,2,\dots, M(\varepsilon)$. Then

$$(18) \qquad \sup_{t\in[o,T]} \|u_{\mu+k}(t) - u_\mu(t)\|_{B_1^t} = \max_{\nu\le M(\varepsilon)} \Big[\sup_{t\in S_{i_\nu}} \|u_{\mu+k}(t) - u_\mu(t)\|_{B_1^t} \Big].$$

Further,

$$(19) \qquad \sup_{t\in S_{i_\nu}} \|u_{\mu+k}(t) - u_\mu(t)\|_{B_1^t} \le \sup_{t\in S_{i_\nu}} \|u_{\mu+k}(t) - u_{\mu+k}(t_{i_\nu})\|_{B_1^t} +$$
$$+ \sup_{t\in S_{i_\nu}} \|u_{\mu+k}(t_{i_\nu}) - u_\mu(t_{i_\nu})\|_{B_1^t} + \sup_{t\in S_{i_\nu}} \|u_\mu(t_{i_\nu}) - u_\mu(t)\|_{B_1^t}.$$

Using (5′) we can estimate the first and the third members in terms of $\varepsilon/4$. Since $u_\mu(t_{i_\nu})$ converge in $B_1^{t_{i_\nu}}$, and since $i_\nu \le M(\varepsilon)$, using (1) we can make the second member not larger than $\varepsilon/2$ for sufficiently large $\mu \ge N(\varepsilon)$:

$$(20) \qquad \|u_{\mu+k}(t_{i_\nu}) - u_\mu(t_{i_\nu})\|_{B_1^t} \le \varphi_1(t_{i_\nu}, t)\|u_{\mu+k}(t_{i_\nu}) - u_\mu(t_{i_\nu})\|_{B_1^{t_{i_\nu}}} \le \varepsilon/2.$$

(18)-(20) yields that u_μ converge in norm $\operatorname{vrai}\max_{t\in[o,T]} \|u(t)\|_{B_1^t}$.

Consider case (5). See Additional part for details.

Replace S_i by $\tilde{S}_i = \{t \in [0,T]: |t-t_i| < (\varepsilon/4\ \varphi_1 c_3^{1/p_1})^{q_1}$, $t_i > t\}$, $i = 1,2,\dots$. Analogous $\tilde{S}_o$. And we reason by analogous

(5′). The first and the third members in (19) in this case are estimated in the following way. Use (0.4) and (0.5) we have

$$\sup_{t\in S_{i_\nu}} \|u_\mu(t) - u_\mu(t_{i_\nu})\|_{B_1^t} = \sup_{t\in S_{i_\nu}} \Big\|\int_t^{t_{i_\nu}} u'_\mu(\tau)d\tau\Big\|_{B_1^t} \le$$

$$\le \sup_{t\in S_{i_\nu}} \varphi_1 \int_t^{t_{i_\nu}} \|u'_\mu(\tau)\|_{B_1^\tau} d\tau \le \varphi_1 \Big(\int_0^T \|u'_\mu(\tau)\|^{p_1}_{B_1^\tau} d\tau\Big)^{1/p_1} \sup_{t\in S_{i_\nu}} |t_{i_\nu} - t|^{1/q_1} \le$$

$$\le \varphi_1 c_3^{1/p_1} \sup_{t\in S_{i_\nu}} |t_{i_\nu} - t|^{1/q_1} \le \varepsilon/4, \; 1/p_1 + 1/q_1 = 1.$$

And then by analogy with (5′).

Passing to the proof of Theorem *1*, we consider $\{u_n\} \subset F$ and take $F^N = \{u_n\}$. By Lemma *2*, there exists Cauchy subsequence with respect to norm $\operatorname{vraimax}_{t\in[0,T]} \|u(t)\|_{B_1^t}$. By (1) and by (16) we may conclude that u_μ is Cauchy sequence with respect to norm

$$\Big(\int_0^T \|u(\tau)\|^p_{B^\tau} d\tau\Big)^{1/p} \text{ for any } p < \infty .$$

Proof the last part of Theorem 1. Due to completeness space of elements $u(t)$ such that $\int_0^T \|u(\tau)\|^p_{B^\tau} d\tau < \infty$, there exists $u = \lim u_\mu$(See Add.p.,Th.4). Consider $\|u_\mu(t) - u(t)\|_{B^t}$. Due to convergence in $L_p(0,T)$ we may assume that it converges to zero for almost all $t \in [0,T]$. Therefore, $\|u_\mu(t)\|_{B^t} \longrightarrow \|u(t)\|_{B^t}$. Using (6) we conclude for enough large μ : $\|u_\mu(t)\|^p_{B^t} \le c_1^p + 1$. If we go over to the limit, we get that $\|u(t)\|^p_{B^t} \le c_1^p + 1$. The Theorem *1* is proved.

Now let S be a compact in R^n, and consider further the case, when $B^t = B$ and $B^t_1 = B_1$, $M_t = M$ for each $t \in S$. And let embedding $B \subset B_1$ be compact. Definition F with condition (5′) is similar as before.

In this case conclusion about of compactness by norm $\left[\int \|u(t)\|^p_{B_1} dt\right]^{1/p}$ (Analogy of Lemma *2*) may be obtained from Theorem *14* in [9]. Our attention was drawn to this way by Proof. Yu. Batt.

Actually, let $u(t) = 0$ for $t \bar{\in} S$ and let $G = R^n$, $E = B_1$, $K = \left\{u : \|u\|_{L_p(R^n;B_1)} \le c_1\right\}$ in that Theorem. Then for any measurable $A \subset R^n$ set $\left\{ \int_A u(t)dt, u \in F \right\}$, will be relatively compact in B_1. It is the result of inequality (5′), inequality

$$\|\int_A u(t)dt\|_B \le \int_A \|u(t)\|_B \, dt \le c ,$$

and of compactness of embedding $B \subset B_1$. Therefore,

Theorem 2. If embedding $B \subset B_1$ is compact, S is compact in R^n and F is constructed as earlier and condition (5′) is held, then F is relatively compact in $L_p(S;B)$.

This Theorem we may use for investigation of ultra parabolic equations, that is equations with many time variables.

Theorem 3. If for every $u \in F$ and every measurable set $A \subset S \subset R^n$, meas $A < \infty$, exists constant $c(A) < \infty$ such that

$$M\left[\int_A u(t)dt\right] \leq c(A) \tag{21}$$

and (5′) holds ($u(t) = \theta$ for all $t \bar{\in} S$), then F is relatively compact in $L_p(S;B)$ for any $p : 1 \leqslant p < \infty$.

Actually, (21) yields relative compactness for every fixed A of the set $\int_A u(t)dt$ in B, and, so, in B_1. Further, proof is analogous of the Theorem 1.

Additional part. *On integration of functions by parameter of the scale.*

Here we consider scale of the normed spaces B^t, $t \in [0,T]$ and assume that B^a is Banach space, $B^{t_1} \subset B^{t_2}$ for $t_1 \geq t_2$ and B^t is separable space. We write B^t instead of B^t_1 for brevity.

Definition 1. Function $u(\cdot)$, $u(t) \in B^t$ is called simple on $[a,b] \subset [0,T]$ if exists partition of $[a,b]$ by finite quantity Lebesque-measurable nonintersection sets $S_i \subset [a,b]$, $i = 1,\dots, n$ and elements $u_i \in B^{t_i}$ ($t_i = sup\,\{t : t \in S_i\}$), such that $u(t) = u_i$ for $t \in S_i$ and $u(t) = \theta$ if $t \in [a,b] \setminus \bigcup_{i=1}^{n} S_i$.

Definition 2. Integral on $[a,b]$ from simple function $u(\cdot)$ is element f of Banach space B^a defined by

$$f = \int_a^b u(\tau)d\tau = \sum_{i=1}^{n} u_i \text{ meas } S_i = \sum_{i=1}^{n} u(\bar{t}_i) \text{ meas } S_i ,$$

where $\bar{t}_i \in S_i$, and u_i are elements which correspond to this partition of $[a,b]$.

Linear combination of simple function is simple function and therefore the operator of integration of simple functions is linear.

For simple functions we have easy inequality

$$(0.1)\qquad \Big\| \int_a^b u(\tau)d\tau \Big\|_{B^a} = \Big\| \sum_{i=1}^{n} u(\bar{t}_i) \text{ meas } S_i \Big\|_{B^a} \le$$

$$\le \sum_{i=1}^{n} \|u(\bar{t}_i)\|_{B^a} \text{ meas } S_i = \int_a^b \|u(\tau)d\tau\|_{B^a} d\tau \le \varphi_2 \int_a^b \|u(\tau)d\tau\|_{B^\tau} d\tau \quad \bar{t}_i \in S_i .$$

Definition 3. Function $u(\cdot)$ is called integrable (p-integrable) on $[a,b]$, if there exists sequence of simple functions $u_n(\cdot)$ which is Cauchy sequence with respect to the norm

$$(0.2)\qquad \int_a^b \|u(\tau)\|_{B^a} d\tau , \qquad \left[\left(\int_a^b \|u(\tau)\|^p_{B^a} d\tau \right)^{1/p} \right],$$

and such that $u_n(t) \longrightarrow u(t)$ with respect to the norm in B^t for almost all $t \in [a,b]$.

Function $u(\cdot)$ that is the limit in B^t almost everywhere of simple functions is called measurable (or, more exactly, B - measurable).

Definition 4. Integral on $[a,b]$ from such function is element $f \in B^a$ defined by

$$f = \int_a^b u(\tau)d\tau = \lim_{n\to\infty} \int_a^b u_n(\tau)d\tau$$

Correctness of this definition will be established if we prove existence of the limit for $\int_a^b u_n(\tau)d\tau$ and independence of f from the choice of sequence u_n of simple functions.

Consider $f_n = \int_a^b u_n(\tau)d\tau$ and prove that f_n is Cauchy sequence in B^a. Let $\{ \tilde{S}_i^{n,s} \}$ $i = 1, \ldots , \tilde{k}(n,s)$ be partition of $[a,b]$, such that elements $\tilde{S}_i^{n,s} \subset S^n_{j(i)}$ and $\tilde{S}_i^{n,s} \subset \bar{S}^{n+s}_{l(i)}$ for some $j(i)$,

$l(i)$. There $\{S_i^n\}$, $i = 1, \ldots, k_n$, corresponds to simple function u_n and $\{\bar{S}_i^{n+s}\}$, $i = 1, \ldots, \bar{k}_{n+s}$, corresponds to simple function u_{n+s}. Then since u_n is Cauchy sequence we have

$$\|f_{n+s} - f_n\|_{B^a} = \Big\| \sum_{i=1}^{\tilde{k}(n,s)} (u_{n+s}(\tau_i^{n,s}) - u_n(\tau_i^{n,s})) \operatorname{meas} S_i^{n,s} \Big\|_{B^a} \le$$

$$\le \int_a^b \|u_{n+s}(\tau) - u_n(\tau)\|_{B^a} d\tau \longrightarrow 0, \; \tau_i^{n,s} \in S_i^{n,s}.$$

Therefore, if B^a is Banach space, then exists element $f \in B^a$ being an integral from $u(\cdot)$.

Consider $g_n = \int_a^b v_n(\tau) d\tau$ where $v(\cdot)$ is Cauchy sequence with respect to the norm $\left[\int_a^b \|v(\tau)\|^p_{B^a} d\tau \right]^{1/p}$ and $v_n(t) \longrightarrow u(t)$ for almost all $t \in [a,b]$. It is obviously that for any n $\|u_n(t) - v_n(t)\|_{B^a}$ is integrable according to Lebesque and limit is equal to zero as $n \longrightarrow \infty$ for almost all t. Moreover, $\int_a^b \|u_n(\tau) - v_n(\tau)\|_{B^a} d\tau$ is uniformly bounded by n because u_n and v_n are Cauchy sequence by this norm. Lebesque theorem yields that

$$\lim_{n\to\infty} \|f_n - g_n\|_{B^a} \le \lim_{n\to\infty} \int_a^b \|u_n(\tau) - v_n(\tau)\|_{B^a} d\tau = 0.$$

Therefore $\|f - g\|_{B^a} \le \|f - f_n\|_{B^a} + \|f_n - g_n\|_{B^a} + \|g_n - g\|_{B^a} < \varepsilon$ for sufficiently large n. Correctness of **Definition 3** is established.

We easily conclude that if $u(\cdot)$ is p - integrable, $p \ge 1$, then $u(\cdot)$ is integrable.

Let $u(\cdot)$ be arbitrary integrable on $[a,b]$ function. From inequality (*o.1*) for simple functions and because of continuation of norm, we have

$$\left\|\int_a^b u(\tau)d\tau\right\|_{B^a} = \lim_n \left\|\int_a^b u_n(\tau)d\tau\right\|_{B^a} \le \lim_n \int_a^b \|u_n(\tau)\|_{B^a}\, d\tau =$$

$$= \int_a^b \|u(\tau)\|_{B^a}\, d\tau.$$

Last equality follows from Lebesque theorem. Therefore, inequality (*0.1*) is fulfilled for every integrable function.

Definition 5. Function is called amplified *p*-integrable if Cauchy sequence from **Definition** 3 will be considered in the "norm", exactly:

$$\left[\int_a^b \|u(\tau)\|^p_{B^\tau}\, d\tau\right]^{1/p}. \tag{0.3}$$

Obviously, if for the scale of spaces B^t condition like (*1*) is fulfilled, then each amplified *p*-integrable function is *p*-integrable.

Obviously, if $u_1(\cdot)$ and $u_2(\cdot)$ are integrable functions and $u_1(t) = u_2(t)$ for almost all t, then $\int_a^b u_1(\tau)d\tau = \int_a^b u_2(\tau)d\tau$. Identification of such functions allows to introduce space L_p, *p*-integrable (and $\tilde{L}_p$, amplified *p*-integrable) functions with norm (*0.2*) ((*0.3*)).

<u>*Theorem 4*</u>. If B^a is Banach space, then L_p is Banach space. If B^t are Banach spaces for all $t \in [a,b]$, then $\tilde{L}_p$ is complete.

Proof coincides with proof of completeness of space integrable according to Bochner functions $L_p(S;X)$, see [10]. (By analogy density of step functions in L_p and in $\tilde{L}_p$ is proved).

Definition 6. Say, that function $u(\cdot)$ is B_1-differentiable at the point $t \in [0,T]$ if there exists element $\omega \in B_1^t$ such that for all $\tau < t$

$$\left\| \frac{u(t+h) - u(t)}{h} - \omega \right\|_{B_1^\tau} \longrightarrow 0, \text{ as } h \longrightarrow 0, \tau \le t + h.$$

Element ω is called derivative of u at the point t and is marked by $u'(t)$.

The following quality is proved like at the Bochner Theorem in [11], but instead of y_n it is necessary to consider $y_n(t) = x_n(t)$, if $\|x_n(t)\|_{B_1^t} \le 2\|x(t)\|_{B_1^t}$ and $y_n(t) = 0$ if conversely.

This property we use for $v(t) = u'(t)$:

<u>*Lemma 3*</u>. Measurable function $v(t)$ is p_1-integrable (amplified p_1- integrable) if and only if $\|v(t)\|_{B_1^a}^{p_1}$ $\left[\text{accordingly } \|v(t)\|_{B_1^t}^{p_1}\right]$ is Lebesque integrable.

Let $u(\cdot)$ be B_1-measurable and differentiable function at every point $t \in [0,T]$, and $u'(t)$ satisfies condition (5).

Let $t_1 < t$; v_o is some fixed functional on $B_1^{t_1}$: $\langle v_o, \cdot \rangle$: $B_1^{t_1} \longrightarrow R$. Because $B^t \subset B_1^t \subset B_1^{t_1}$, we can define function $\chi(t) = \langle v_o, u(t) \rangle$. As $u(\cdot)$ is B_1-measurable and B_1^t is separable, Pettis's Theorem [11] yields that $\chi(t)$ is Lebesque-measurable function. Moreover,

$$|\chi(t)| \le \|v_o\|_{B_1^{t_1} \to R} \cdot \|u(t)\|_{B_1^{t_1}} \le c(t_1)\, \tilde{\varphi}_1\, \varphi_2 \|u(t)\|_{B^t} \in L_p(t_1, T).$$

Further

$$\frac{\chi(t+h) - \chi(t)}{h} = \left\langle v_o, \frac{u(t+h) - u(t)}{h} \right\rangle.$$

Therefore, if h is sufficiently small, we get $t + h > t_1$ and because of B_1-differentiable of $u(\cdot)$ we conclude that

$$\frac{\chi(t+h) - \chi(t)}{h} \underset{h \to 0}{\longrightarrow} \langle v_0, u'(t)\rangle.$$

Therefore for any $t \in [0,T]$ exists $u'(t)$. On the other hand $\chi'(t)$ is measurable like limit for every t measurable functions. And inequality

$$|\chi'(t)| \le \|v_0\|_{B_1^{t_1} \to R} \cdot \|u'(t)\|_{B_1^{t_1}} \le \varphi_1 c(t_1)\|u'(t)\|_{B_1^t} \in L_{p_1}[t_1,T)$$

is held.

Then formula $\chi(t) - \chi(t_1) = \int_{t_1}^{t} \chi'(\tau)d\tau$ yields $\langle v_0, u(t) -$

$$- u(t_1)\rangle = \int_{t_1}^{t} \langle v_0, u'(\tau)\rangle d\tau = \langle v_0, \int_{t_1}^{t} u'(\tau)d\tau\rangle.$$

The last equality follows from continuity of the functional v_0 on $B_1^{t_1}$. In view of arbitrariness $v_0 \in [B_1^{t_1}]^*$ we conclude that

$$u(t) - u(t_1) = \int_{t_1}^{t} u'(\tau)d\tau \text{ in } B_1^{t_1}. \tag{0.4}$$

From inequality (0.1) we get for any pair number t, t_1 ($t_1 < t$) inequality:

$$\|u(t) - u(t_1)\|_{B_1^{t_1}} \le \int_{t_1}^{t} \|u'(\tau)\|_{B_1^{t_1}} d\tau \le \varphi_1 \int_{t_1}^{t} \|u'(\tau)\|_{B_1^\tau} d\tau. \tag{0.5}$$

This inequality is used in *Lemma 2* in the case (5), where instead of t we need take t_{i_ν}, and instead of t_1 we take t.

Bibliography

1. J.L.Lions. Quelques methodes de resolution des problemes aux limites non lineaires. Paris, 1969.
2. Yu.A.Dubinskii. Weak convergence in nonlinear elliptic and parabolic equations. Amer. Soc. Transl. (2) 67 (1968), 226-258. MR 32 # 7958.
3. C.N.Soltanov. On solvability of some nonlinear parabolic problems in the Orlich-Sobolev spaces. Partial differential equations. Transactions of Conference on Differential Equations and Numerical Mathematics. "Nauka", Novosibirsk, 1980, pp. 99 - 104 (Russian).
4. C.N.Soltanov. On solvability some nonlinear parabolic problems. Dif.Uravnenia, V.XVI, N 5, "Minsk", 1980, pp.882-887.(Russian).
5. A.G.Podgaev. Compactness of certain nonlinear sets. Soviet Math Dokl. Vol 32 (1985), N 3, p.840-842.
6. A.G.Podgaev. On conditions of compactness of sets in spaces of Bochner integrable functions. Equations nonclassical type. Institute of Mathematics, Sibirian branch Acad.Nauk SSSR, Novosibirsk, 1986, pp.111-113.(Russian).
7. A.G.Podgaev. About boundary value problems fore some quasi linear nonuniform parabolic equations with nonclassical degenerates.Sib.Math.J.,V.XXVIII,N2,1987, pp.129-139.(Russian).
8. S.G.Piatkov and A.G.Podgaev. On solvability of a boundary value problem for nonlinear parabolic equation with the change direction of time. Sib. Math. J., V.XXVIII, N 3, 1987, pp 184 - - 192. (Russian).
9. Dinculeany, N. On Kolmogorov-Tamarkin and M.Riesz Compactness Criteria in Function Space over a locally Compact Group.

Journal of Mathematical Analisis and Applications, V.89, N 1, 1982, pp. 67-85.

10. Ch.Gaevski,K. Greger and K.Zacharias. Nonlinear operator equations and operator differential equations, "Mir", M. 1978. (Russian).

11. K.Yosida. Functional Analysis, Gundleheren B. 123, Springer, Berlin, 1965.

Functional Analysis, Approximation Theory
and Numerical Analysis
Ed. John M. Rassias

ON THE EXTENDED OSTROWSKI CONSTANT

JOHN M. RASSIAS

The National University of Athens, Pedagogical Department,
4, Agamemnonos Str., Aghia Paraskevi, Attikis 15342, Greece

Abstract

A. Ostrowski (1979) established that if $f(z)$ is a polynomial of degree m and $g(z)$ a polynomial of degree n, then $M_f M_g \geq M_{fg} \geq \gamma \cdot M_f M_g$, where $M_f = \max\{|f(z)| : |z| = 1\}$, and the Ostrowski constant: $\gamma = \sin^m(\pi/8m)\sin^n(\pi/8n)$. In this paper we improve γ and extend it to $f_i, i = 1, 2, \ldots, k$, in $U_r = \{z : |z| = r\}$ by applying Jensen's formula.

Theorem. *If $f_i(z) = z^{n_i} + \ldots + f_i(0)$, $f_i(0) = 1, i = 1, 2, \ldots, k$, are polynomials of degrees n_i in $D = \{z : |z| \geq 1\}$, and if the zeros (roots) a_j^i, $j = 1, 2, \ldots, n_i$, of these polynomials are such that $|a_j^i| \geq 1$, then*

$$\prod_{i=l}^{k} M_{f_i} \geq M_{\prod_{i=1}^{k} f_i} \geqq \gamma_2 \prod_{i=1}^{k} M_{f_i}, \qquad (*)$$

where $\gamma_2 = 2^{-N}$, and $M_{f_i} = \max\{|f_i(z)| : |z| = 1\}$, $i = 1, 2, \ldots, k$, and

$$N = \sum_{i=1}^{k} n_i \quad \text{is the degree of } F = \prod_{i=1}^{k} f_i .$$

If $k = 2$ then $\gamma_1 = \gamma$: then Ostrowski constant [1], and our constant γ_2 is greater than γ_1, because

$$\sin\left(\frac{2}{k}\frac{\pi}{8n_i}\right) \geq \frac{2}{k}\frac{\pi}{8n_i} < \frac{1}{2},$$

where

$$\gamma_1 = \prod_{i=1}^{k} \sin^{n_i}\left(\frac{2}{k}\frac{\pi}{8n_i}\right),$$

and $i = 1, 2, \ldots, k$, and $k = 2, 3, \ldots$. Assume $M_{f_i}^r = \max\{|f_i(z)| : |z| = r\}$, $i = 1, 2, \ldots, k$, and

$$\gamma_2^r = (2r)^{-N} = \gamma_2 r^{-N}.$$

If f_i, $i = 1, 2, \ldots, k$, are polynomials of degrees n_i, $i = 1, 2, \ldots, k$, in U_r, then $(*)$ is extended to the following form

$$\prod_{i=1}^{k} M_{f_i}^r \geq M_F^r \geq \gamma_2^r \cdot \prod_{i=1}^{k} M_{f_i}^r. \tag{$**$}$$

Proof of Theorem. It is clear that the left hand side relation of $(*)$ holds. To prove the right hand side relation of $(*)$ we assume

$$f_i(z) = z^{n_i} + \ldots + f_i(0), \quad f_i(0) = 1,$$

are polynomials of degrees n_i in $D = \{z : |z| \geq 1\}$, $U = \partial D = \{z : |z| = 1\}$. In fact,

$$f_i(z) = \prod_{j=1}^{n_i} \left(z - a_j^i\right), \qquad i = 1, 2, \ldots, k,$$

or

$$|f_i(z)| \leq \prod_{j=1}^{ni} (1 + |a_j^i|)$$

in U, $i = 1, 2, \ldots, k$, where a_j^i, $j = 1, 2, \ldots, n_i$, are zeros of f_i in D. Therefore

$$|a_j^i| \geq 1, \text{ or } 1 + |a_j^i| \leq 2|a_j^i|.$$

Hence

$$M_{f_i} \leq 2^{n_i} \prod_{j=1}^{n_i} |a_j^i|, \qquad i = 1, 2, \ldots, k,$$

or

$$\prod_{i=1}^{k} M_{f_i} \leq 2^N \prod_{i=1}^{k} \prod_{j=1}^{n_i} |a_j^i|$$

$$= 2^N e^{\sum\limits_{i=1}^{k} \sum\limits_{j=1}^{n_i} \ln |a_j^i|} .$$

By Jensen's formula ([2], p. 128, and p. 139) we get

$$\sum_{j=1}^{n_i} \ln |a_j^i| = \frac{1}{2\pi} \int_0^{2\pi} \ln |f_i(e^{it})| dt .$$

Applying this formula we get

$$\prod_{i=1}^{k} M_{f_i} \leq 2^{\sum\limits_{i=1}^{k} n_i} e^{\frac{1}{2\pi} \int_0^{2\pi} \ln |\prod_{i=1}^{k} f_i(e^{it})| dt} ,$$

or

$$\prod_{i=1}^{k} M_{f_i} \leq \gamma_2^{-1} M_F ,$$

completing the proof of the Theorem.

Similarly, we prove (**). In fact, we employ the extended Jensen's formula ([2], p. 128, and p. 139)

$$\sum_{j=1}^{n_i} \ln |a_j^i / r| = \frac{1}{2\pi} \int_0^{2\pi} \ln |f_i(re^{it})| dt ,$$

and assume polynomials f_i, $i = 1, 2, \ldots, k$, in $D_r = \{z : |z| \geq r\}$. Thus

$$|f_i(z)| \leq r^{n_i} \prod_{j=1}^{n_i} (1 + |a_j^i / r|)$$

in $U_r = \partial D_r = \{z : |z| = r\}$, where a_j^i are zeros of f_i in D_r. The rest of the proof is omitted as analogous to the one of the above Theorem.

References

1. A. M. Ostrowski, *Notiz über Maximalwerte von Polynomen auf dem Einheitskreis*, Univ. Beograd, Publ. Elektrotehn, Fak., Sek. Mat. Fiz. 634–677, 1979, 55–56.
2. G. Polya and G. Szegö, *Problems and Theorems in Analysis*, Springer Study Edition, Transl. (in English), Vol. I, 1972, p. 128 and p. 139.

Functional Analysis, Approximation Theory
and Numerical Analysis
Ed. John M. Rassias

SOLUTION OF A STABILITY PROBLEM OF ULAM

JOHN M. RASSIAS

The National University of Athens, Pedagogical Department,
4, Agamemnonos Str., Aghia Paraskevi, Attikis 15342, Greece

In our paper [J. M. Rassias, "Solution of Problem of Ulam", J. Approx. Th. 57 (1989)] we solved the following *Ulam Problem*: "Give conditions in order for a linear mapping near an approximately linear mapping to exist" and established results involving a product of powers of norms. In this paper we state and prove a more general version of my above theorem involving a non-negative real-valued function [S. M. Ulam, "A Collection of Mathematical Problems". Interscience, New York, 1961; "Problems in Modern Mathematics", Wiley, New York, 1964; "Sets, Numbers, and Universes", M.I.T. Press, Cambridge, MA, 1974]. There has been much activity on a similar "ϵ-isometry" problem of Ulam [J. Gervirtz, Proc. Amer. Math. Soc. 89 (1983); P. Gruber, Trans. Amer. Math. Soc.. 245 (1978); J. Lindenstrauss and A. Szankowski, "Nonlinear Perturbations of Isometries", Colloquium in honor of L. Schwartz, Vol. I, Palaiseau, 1985].

Theorem 1. *Let* X *be a normed linear space and* Y *be a Banach space. Assume in addition conditions:*

(c_1) : $f : X \to Y$ *is a mapping such that* $f(t.x)$ *is continuous in* t *for each fixed* x,

(c_2) : $K : X^p \to \mathbb{R}^+ \cup \{0\}$ *a non-negative real-valued function such that*

$$R_p = R_p(x) = \sum_{i=0}^{\infty} p^{-i} K(p^i x, p^i x, \dots, p^i x) < \infty$$

is a non-negative function of x,

(c_3) :

$$\lim_{n\to\infty} p^{-n} K(p^n x_1, p^n x_2, \dots, p^n x_p) = 0 \ ,$$

(c_4) :

$$\left\| f\left(\sum_{j=1}^{p} x_j\right) - \sum_{j=1}^{p} f(x_j) \right\| \le C_2 K(x_1, x_2, \dots, x_p) \tag{1}$$

for any $x_j \in X$, C_2 *(: constant and independent of* $x_1, \dots, x_p) \ge 0$.

Then there exists a unique linear mapping $L_p : X \to Y$ *such that*

$$\|f(x) - L_p(x)\| \le C_1 R_p(x) \ , \tag{2}$$

for any $x \in X$, *where* $C_1 = C_2/p$.

If one takes $p = 2$, $x_1 = x$, $x_2 = y$, and

$$K = K(x, y) = \|x\|_1^{a+c}\|y\|_1^{b} + \|x\|_1^{a}\|y\|_1^{b+c} \tag{3}$$

such that $0 \le a + b + c < 1$, and $a, b, c :=$ constants, then there exists a unique linear mapping $L : X \to Y$ such that $L = L_2$ and

$$\|f(x) - L(x)\| \le C\|x\|_1^{a+b+c} \tag{4}$$

for any $x \in X$, where

$$C = C_2/(1 - 2^{a+b+c-1}) \ .$$

In this case we have

$$R_2 = R_2(x) = \sum_{i=0}^{\infty} (2^{-i} \times 2 \times 2^{i(a+b+c)} \|x\|_1^{a+b+c}) = \frac{2}{1 - 2^{a+b+c-1}} \|x\|_1^{a+b+c} \ .$$

If one takes $a = b = c = 0$ in (3) one obtains an additive functional L such that

$$\|f(x) - L(x)\| \le 2C_2 \tag{5}$$

for all $x \in X$. This is D. H. Hyers' result [4].

If $c = 0$, and $0 \leq a + b < 1$ in (3) we obtain our result [6], which is: *Let X be a normed linear space with norm $\|.\|_1$ and let Y be a Banach space of norm $\|.\|_2$. Assume in addition that $f : X \to Y$ is a mapping such that $f(t.x)$ is continuous in t for each fixed x. If there exist $a, b, 0 \leq a + b < 1$, and $C_2 \geq 0$ such that*

$$\|f(x+y) - [f(x) + f(y)]\|_2 \leq 2C_2 \|x\|_1^a \|y\|_1^b \tag{+}$$

For all $x, y, \in X$, then there exists a unique linear mapping $L : X \to Y$ such that

$$\|f(x) - L(x)\|_2 \leq C\|x\|_1^{a+b} \tag{++}$$

for all $x \in X$, where

$$C = \frac{C_2}{1 - 2^{a+b-1}}$$

Existence. Inequality (1) and $x_j = x$, $j = 1, 2, \ldots, p$, imply

$$\|f(px) - pf(x)\| \leq C_2 K(x, x, \ldots, x) ,$$

or

$$\|p^{-1} f(px) - f(x)\| \leq C_1 K(x, x, \ldots, x) , \tag{6}$$

where $C_1 = C_2/p$. More generally, the following lemma holds:

Lemma 1. *In the space X, for some $C_2 \geq 0$ and for any positive integer n*

$$\|f(p^n x)p^{-n} - f(x)\| \leq C_1 \sum_{i=0}^{n-1} p^{-i} K(p^i x, p^i x, \ldots, p^i x) . \tag{7}$$

To prove Lemma 1, we work by induction on n.

For $n = 1$, the result is obvious from (6). We assume then that (7) holds for $n = k$ and prove that (7) is true for $n = k + 1$. Indeed, from (7) and $n = k$ and $px = z$, we find

$$\|f(p^k z)/p^k - f(z)\| \leq C_1 \sum_{i=0}^{k-1} p^{-i} K(p^i z, p^i z, \ldots, p^i z) ,$$

or

$$\|f(p^{k+1} x)/p^k - f(px)\| \leq C_1 \sum_{i=0}^{k-1} p^{-i} K(p^{i+1} x, p^{i+1} x, \ldots, p^{i+1} x) ,$$

or

$$\|f(p^{k+1}x)/p^{k+1} - f(px)/p\| \le C_1 \sum_{i=0}^{k-1} p^{-(i+1)} K(p^{i+1}x, \ldots, p^{i+1}x) ,$$

or

$$\|f(p^{k+1}x)/p^{k+1} - f(px)/p\| \le C_1 \sum_{i=1}^{k} p^{-i} K(p^i x, p^i x, \ldots, p^i x) . \tag{8}$$

Therefore from (6) and (7) we get

$$\begin{aligned} &\|f(p^{k+1}x)/p^{k+1} - f(x)\| \\ &\le \|f(p^{k+1}x)/p^{k+1} - f(px)/p\| + \|f(px)/p^{k+1} - f(x)\| \\ &\le C_1 \sum_{i=1}^{k} p^{-i} K(p^i x, p^i x, \ldots, p^i x) + C_1 K(x, x, \ldots, x) \\ &= C_1 \sum_{i=0}^{k} p^{-i} K(p^i x, p^i x, \ldots, p^i x) , \end{aligned}$$

or (7) holds for $n = k+1$, or

$$\|f(p^{k+1}x)/p^{k+1} - f(x)\| \le C_1 \sum_{i=0}^{k} p^{-i} K(p^i x, p^i x, \ldots, p^i x) . \tag{9}$$

But (C_3) yields

$$\sum_{i=0}^{n-1} p^{-i} K(p^i x, p^i x, \ldots, p^i x) < \sum_{i=0}^{\infty} p^{-i} K(p^i x, p^i x, \ldots, p^i x) = R_p(x) . \tag{10}$$

Then Lemma 1 and inequality (10) imply

$$\|f(p^n x)/p^n - f(x)\| \le C_1 R_p(x) \tag{11}$$

for any $x \in X$, any positive integer n and some $C_1 \ge 0$.

Lemma 2. *The sequence* $\{f(p^n x)/p^n\}$ *converges.*

We first use (11) and the completeness of Y to prove that the sequence $\{f(p^n x)/p^n\}$ is a Cauchy sequence. In fact, if $i > j \ge 0$, then

$$\|f(p^i x)/p^i - f(p^j x)/p^j\| = p^{-j}\|f(p^i x)/p^{i-j} - f(p^j x)\| , \tag{12}$$

and if we set $p^j x = h$ (then $p^i x = p^{i-j} p^j x = p^{i-j} h$) in (12) and employ (11), we get

$$\begin{aligned}\|f(p^i x)p^{-i} - f(p^j x)p^{-j}\| &= p^{-j}\|f(p^{i-j}h)p^{-(i-j)} - f(h)\| \\ &\leq p^{-j} C_1 R_p(h)\,,\end{aligned}$$

or

$$\lim_{j\to\infty} \|f(p^i x)p^{-i} - f(p^j x)p^{-j}\| = 0\,, \tag{13}$$

because

$$\lim_{j\to\infty} p^{-j} R_p(p^j x) = 0\,.$$

It is obvious from (13) and the completeness of Y that the sequence $\{f(p^n x)p^{-n}\}$ converges and therefore the proof of Lemma 2 is complete.

Set

$$L_p(x) - \lim_{n\to\infty}[p^{-n} f(p^n x)]\,. \tag{14}$$

It is clear form (1), (14), and (c_3) that

$$\begin{aligned}&\|\lim_{n\to\infty} p^{-n} f(p^n \sum_{j=1}^{p} x_j) - \lim_{n\to\infty} \sum_{j=1}^{p} p^{-n} f(p^n x_j)\| \\ &\leq \lim_{n\to\infty} C_2 p^{-n} K(p^n x_1, p^n x_2, \ldots, p^n x_p) = 0\,,\end{aligned}$$

or

$$\|L_p\left(\sum_{j=1}^{p} x_j\right) - \sum_{j=1}^{p} L_p(x_j)\| = 0 \text{ for any } x_j \in X, j = 1, \ldots, p,$$

or

$$L_p\left(\sum_{j=1}^{p} x_j\right) = \sum_{j=1}^{p} L_p(x_j) \text{ for any } (x_1, x_2, \ldots x_p) \in X^p\,. \tag{15}$$

From (15) we get

$$L_p(qx) = qL_p(x) \tag{16}$$

for any $q \in Q$, where Q is the set of rationals.

Lemma 3. *Let Y^* be the space of continuous linear functionals and consider the mapping*

$$T^{(p)} : \mathbb{R} \to \mathbb{R}$$

such that

$$T^{(p)}(t) = g(L_p(tx)) \ , \tag{17}$$

where $g \in Y^*$, $t \in \mathbb{R}$, *and* $x \in X$, $x :=$ *fixed. The* $T^{(p)}$ *is continuous.*

To prove Lemma 3 we proceed as follows: Let

$$T_n^{(p)}(t) = g(p^{-n} f(p^n xt)) \tag{18}$$

such that

$$T(t) = \lim_{n\to\infty} T_n^{(p)}(t) \ , \tag{19}$$

where $x \in X$, $x :=$ fixed and $t \in \mathbb{R}$, $g \in Y^*$.

Then $T_n^{(p)}(t)$ are continuous and therefore $T^{(p)}$ is measurable as the pointwise limit of continuous mappings $T_n^{(p)}$. Moreover, $T^{(p)}$ is a homomorphism with respect to addition "+", that is,

$$T^{(p)}(t+s) = T^{(p)}(t) + T^{(p)}(s) \tag{20}$$

for any $t, s \in \mathbb{R}$. It is clear now that (20) and the measurability of $T^{(p)}$ imply that $T^{(p)}$ is a continuous mapping and thus the proof of Lemma 3 is complete ([1], p. 110–111, 116–117).

Then Lemma 3 and the fact that Y^* separates points of Y and continuity condition (c_1) yield the linearity of L_p.

If we take limits on both sides if (11) as $n \to \infty$ we obtain (2).

Uniqueness. It remains to show the uniqueness part of our theorem.

Let $M : X \to Y$ *be a linear continuous mapping, such that*

$$\|f(x) - M(x)\| \le C_1 R'(x) \ , \tag{21}$$

for any $x \in X$, *where* C_1' *is any constant:* ≥ 0. *If there exists a continuous linear mapping* $L_p : X \to Y$ *such that* (2) *holds, then*

$$L_p(x) = M(x) \tag{22}$$

for any $x \in X$.

To prove (22) we must prove the following

Lemma 4. *If* (2) *and* (21) *hold, then*

$$\|L_p(x) - M(x)\| \le C_1 R_p(x) + C_1' R'(x) \tag{23}$$

for any $x \in X$.

The required result (23) follows immediately if we use inequalities (2) and (21) the linearity of L_p and M, as well as the triangle inequality. In fact,

$$L_p(x) = p^{-j} L_p(p^j x), \qquad M(x) = p^{-j} M(p^j x) , \tag{24}$$
$$\|L_p(p^j x) - M(p^j x)\| \leq \|L_p(p^j x) - f(p^j x)\| + \|M(p^j x) - f(p^j x)\| .$$

Then if we apply (2) and (21) we obtain inequality (24) and the proof of Lemma 4 is complete.

It is clear now that (23) implies $\lim_{j\to\infty} \|L_p(x) - M(x)\| = 0$ for any $x \in X$, completing the proof of (22). Thus the *uniqueness* part of our theorem is complete, as well.

Theorem 2. *Let X be a normed linear space and let Y be a Banach space. Let N be a non-negative real-valued function on X^p such that $N(x, x, \ldots , x)$ is bounded on the unit ball of X, and $N(tx_1, tx_2, \ldots , tx_p) \leq k(t)N(x_1, x_2, \ldots, x_p)$ for all $t \geq 0$, where $k(t) < \infty$ and $\sum_{n=0}^{\infty} p^{-n}k(p^n) < \infty$. Let $f : X \to Y$ be bounded on some ball of X. Assume, furthermore, that $f(tx)$ is continuous in t for each $x \in X$. If*

$$\left\| f\left(\sum_{j=1}^{p} x_j\right) - \sum_{j=1}^{p} f(x_j) \right\| \leq N(x_1, x_2, \ldots , x_p) , \tag{25}$$

for all $x_1, x_2, \ldots , x_p \in X$, then there exists a unique linear mapping $L_p : X \to Y$ such that

$$\|f(x) - L_p(x)\| \leq PN(x, x, \ldots , x) , \tag{26}$$

for all $x \in X$, where $p = \sum_{n=0}^{\infty} p^{-(n+1)}k(p^n)$.

Proof. For $i > j \geq 0$ there holds

$$\|p^{-i} f(p^i x) - p^{-j} f(p^j x)\| \leq \left(\sum_{m=j+1}^{i} p^{-m}k(p^{m-1}) \right) N(x, \ldots , x) \tag{27}$$

This is easily proved by induction. Indeed, if $h = p^j x$, then from (25) with $x_1 = \ldots = x_p = h$ we have that

$$\begin{aligned} \|p^{-(j+1)} f(p^{j+1} x) - p^{-j} f(p^j x)\| &= p^{-(j+1)} \|f(ph) - pf(h)\| \\ &\leq p^{-(j+1)} k(p^j) N(x, \ldots , x) , \end{aligned} \tag{28}$$

so that (27) holds when $i = j+1$. If (27) is true for a given j and $i = s$, then one sees immediately that it also holds for this j and $i = s+1$ by applying (28) with j replaced by s, since

$$\begin{aligned}\|p^{-(s+1)}f(p^{s+1}) - p^{-j}f(p^j)\| &\le \|p^{-(s+1)}f(p^{s+1}) - p^{-s}f(p^s)\| \\ &\quad + \|p^{-s}f(p^s) - p^{-j}f(p^j)\| .\end{aligned}$$

From (27) and the assumption about k, it follows immediately that the sequence $\{f(p^i x)/p^i\}$ converges. From this it follows that if L_p exists, it must be unique, since

$$\begin{aligned}\|p^{-i}f(p^i x) - L_p(x)\| &= p^{-i}\|f(p^i x) - L_p(p^i x)\| \\ &\le p^{-i}PN(p^i x, \dots, p^i x) \\ &\le p^{-i}k(p^i)PN(x, \dots, x) \to 0 ,\end{aligned}$$

so that $L_p(x)$ must be the limit of the sequence $\{f(p^n x)/p^n\}$.

Thus we define $L_p(x)$ to be this limit and see at once that (20) holds by applying (27) with $j = 0$ and $i = n$.

From (25) it follows that

$$\|f(p^n(x+y)) - f(p^n x) - f(p^n y) - (p-2)f(0)\| \le k(p^n)N(x, y, 0, \dots, 0),$$

so that upon dividing by p^n and allowing $n \to \infty$ we see that

$$L_p(x+y) = L_p(x) + L_p(y) .$$

It follows from the assumption about $N(x, x, \dots, x)$ that it is bounded on any bounded subset of X. Thus by (27), for each fixed x the sequence $\{p^{-n}f(p^n xt)\}$ converges uniformly in t in any bounded subset of $\mathbb{R}$, so that $L_p(tx)$ is continuous in t. Since L_p is additive, it is therefore linear. Finally, since we have assumed that f is bounded in some open set of X, (26) implies that L_p has the same properties, so that it is continuous.

References

1. Ch. D. Aliprantis and O. Burkinshaw, *Principle of Real Analysis*, Arnold (Publ.), 1981, England.
2. J. Gervirtz, *Stability of isometries on Banach spaces*, Proc. Amer. Math. Soc. 89 (1983), 633–636.
3. P. Gruber, *Stability of isometries*, Trans. Amer. Math. Soc. 245 (1978), 263–277.
4. D. H. Hyers, *On the stability of the linear functional equation*, Proc. Nat. Acad. Sci., USA, 27 (1941), 222–224.

5. J. Lindenstrauss and A. Szankowski, *Nonlinear Perturbations of Isometries*, Colloquium in honor of Laurent Schwartz, Vol. I, Palaiseau, 1985.
6. J. M. Rassias, *Solution of a Problem of Ulam*, J. Approx. Th., Vol. 57, No. 3 (1989), 268–273.
7. S. M. Ulam, *A Collection of Mathematical Problems*, Intersci. Publ., N.Y. 1961; *Problems in Modern Mathematics*, Wiley, N.Y., 1964; *Sets, Numbers, and Universes*, M.I.T. Press, Cambridge, 1974.

Functional Analysis, Approximation Theory
and Numerical Analysis
Ed. John M. Rassias

AN INTERPRETATION OF GEGENBAUER POLYNOMIALS AND THEIR GENERALIZATION FOR THE CASE WITH MANY VARIABLES

Algimantas YANUSHAUSKAS
Institute of Mathematics, Siberian Branch of the USSR, 630090 Novosibirsk 90, USSR

Let us consider an equation

$$yu_{yy} + u_{xx} + \alpha u_y = 0 \tag{1}$$

on the half-plane $y > 0$, where α is some constant. The solution of equation (1), satisfying the condition $u(x,0) = f(x)$, where $f(x)$ is the analytical function, is formally expressed in the form of the series

$$u(x,y) = \sum_{k=0}^{\infty} \frac{(-1)^k \Gamma(\alpha)}{k!\Gamma(k+\alpha)} y^k \frac{d^{2k} f}{dx^{2k}} \tag{2}$$

that is easy to verify by substituting the series into the equation. If we put $f(x) = x^{2n}$, then from (2) we obtain

$$u_n(x,y) = x^{2n} - \frac{4(\frac{1}{2}-n)(-n)}{1\cdot \alpha} yx^{2n-2} +$$

$$+\frac{4^2(-n)(1-n)(\frac{1}{2}-n)(1+\frac{1}{2}-n)}{1\cdot 2\cdot \alpha\cdot (\alpha+1)} y^2 x^{2n-4} - \ldots =$$

$$= x^{2n} F(-n, \frac{1}{2}-n; \alpha; -\frac{4y}{x^2}),$$

where $F(\alpha,\beta;\gamma;t)$ is the Gaussian hypergeometric function [1]. If $f(x) = x^{2n+1}$, then analogously we find

$$v_n(x,y) = x^{2n+1} - \frac{4(-n)(-\frac{1}{2}-n)}{1\cdot \alpha} yx^{2n-1} +$$

$$+\frac{4^2(-n)(1-n)(-\frac{1}{2}-n)(1-\frac{1}{2}-n)}{1\cdot 2\cdot \alpha\cdot (\alpha+1)} y^2 x^{2n-3} - \ldots =$$

$$= x^{2n+1} F(-n, -\frac{1}{2}-n; \alpha; -\frac{4y}{x^2}).$$

Making use of formula [1]

$$F(\alpha,\beta;\gamma;z) = (1-z)^{-\alpha} F\left(\alpha, \gamma-\beta;\gamma;\frac{z}{1-z}\right)$$

finally we obtain

$$u_n(x,y) = (x^2+4y)^n F\left(-n, \alpha-\frac{1}{2}+n;\alpha;\frac{4y}{4y+x^2}\right),$$

$$v_n(x,y) = x(x^2+4y)^n F\big(-n, \alpha+\frac{1}{2}+n;\alpha;\frac{4y}{4y+x^2}\big).$$

These polynomials can be regarded as the analogues of homogeneous harmonic polynomials.

In the matric, determined by the quadratic form corresponding to the principal part of equation (1), the equation of the unit circle S has the shape $4y+x^2=1$. The equalities

$$u_n(x,y)\big|_S = F(-n,\alpha-\frac{1}{2}+n;\alpha;1-x^2) =$$

$$= \frac{\Gamma(\alpha-\frac{1}{2}+2n)\Gamma(\alpha)}{\Gamma(\alpha-\frac{1}{2}+n)\Gamma(n+\alpha)} x^{2n} + \ldots = Q_{2n}(x),$$

$$v_n(x,y)\big|_S = xF(-n,\alpha+\frac{1}{2}+n;\alpha;1-x^2) =$$

$$= \frac{\Gamma(\alpha+\frac{1}{2}+2n)\Gamma(\alpha)}{\Gamma(\alpha+\frac{1}{2}+n)\Gamma(n+\alpha)} x^{2n+1} + \ldots = Q_{2n+1}(x)$$

hold. The equalities

$$\frac{dQ_{2n}}{dx} = -2xF', \qquad \frac{d^2Q_n}{dx^2} = 4x^2F'' - 2F'$$

hold, whence it follows

$$F' = -\frac{1}{2x}\frac{dQ_{2n}}{dx}, \qquad F'' = \frac{1}{4x^2}\frac{d^2Q_{2n}}{dx^2} + \frac{1}{4x^3}\frac{dQ_{2n}}{dx}.$$

From the equation for the hypergeometric function

$$t(1-t)F'' + [\alpha-(\alpha+\frac{1}{2})t]F' + n(\alpha-\frac{1}{2}+n)F = 0,$$

assuming $t = 1-x^2$, we obtain

$$x^2(1-x^2)F'' + [\frac{1}{2}+(\alpha+\frac{1}{2})x^2]F' + n(\alpha-\frac{1}{2}+n)F = 0.$$

By substituting the expressions F", F' and F in terms of Q_{2n} into this equality we find

$$(1-x^2)\frac{d^2Q_{2n}}{dx^2} - 2\alpha x\frac{dQ_{2n}}{dx} + 2n(2n+2\alpha-1)Q_{2n} = 0.$$

By analogy we obtain

$$(1-x^2)\frac{d^2Q_{2n+1}}{dx^2} - 2\alpha x\frac{dQ_{2n+1}}{dx} + (2n+1)(2n+2\alpha)Q_{2n+1} = 0.$$

Consequently, the polynomials $Q_k(x)$ of the k-th degree satisfy the equation

$$(1-x^2)\frac{d^2u}{dx^2} - 2\alpha x\frac{du}{dx} + k(k+2\alpha-1)u = 0. \qquad (3)$$

Equation (3) is the same as that which is satisfied by Gegenbauer polynomials [2], therefore

$$Q_k(x) = A(k)C_k^{\alpha-\frac{1}{2}}(x),$$

where $A(k)$ is a constant. The constant $A(k)$ can be defined by comparing the terms of the highest degree in both polynomials. Obviously, the highest degree terms in $Q_k(x)$ have the form

$$Q_k(x) = \frac{\Gamma(\alpha-\frac{1}{2}+k)\Gamma(\alpha)}{\Gamma(\alpha-\frac{1}{2}+\frac{k}{2})\Gamma(\alpha+\frac{k}{2})}x^k + \dots$$

and since [2]

$$C_N^\lambda(x) = \frac{2^n\Gamma(n+\lambda)}{\Gamma(\lambda)n!}x^n + \dots,$$

by comparing these expresions we find

$$A(k) = \frac{\Gamma(\alpha)\Gamma(\alpha-\frac{1}{2})k!}{2^k\Gamma(\alpha-\frac{1}{2}+\frac{k}{2})\Gamma(\alpha+\frac{k}{2})} = \frac{\Gamma(2\alpha-1)k!}{\Gamma(2\alpha-1+k)},$$

consequently, we have

$$Q_k(x) = \frac{\Gamma(2\alpha-1)k!}{\Gamma(2\alpha-1+k)}C_k^{\alpha-\frac{1}{2}}(x), \quad \alpha > \frac{1}{2}.$$

The case $\alpha = \frac{1}{2}$ corresponds to the first genus Chebyshev polynomials snd requires special consideration. In this case we have $Q_k(x) = T_k(x)$, because with $\alpha = \frac{1}{2}$ for any solution $u(y,x)$ of equation (1) the function $u(t^2,x)$ is harmonic.

The presented approach to the construction of orthogonal polynomials admits a generalization for the case with many variables. We consider an equation

$$zu_{zz} + \Delta u + \alpha u_z = 0, \quad x = (x_1, \dots, x_n), \tag{4}$$

where Δ is the Laplace operator with respect to variables x. For $z = 0$ the bounded solution of equation (4), satisfying the condition $u(0, x) = f(x)$, can be written in the form

$$u(z, x) = \sum_{k=0}^{\infty} \frac{(-1)^k \Gamma(\alpha)}{k! \Gamma(k + \alpha)} z^k \Delta^k f. \tag{5}$$

Put $f = (x_1^2 + \dots + x_n^2)^l p_m(x)$, where $p_m(x)$ is a homogeneous harmonic polynomial of degree m. By direct calculation in this case we find

$$\Delta f = 4(-l)(1 - m - \frac{n}{2} - l)(x_1^2 + \dots + x_n^2)^{l-1} p_m(x),$$

....

$$\Delta^k f = 4^k(-l) \dots (k - 1 - l)(1 - m - \frac{n}{2} - l) \dots$$

$$(k - m - \frac{n}{2} - l)(x_1^2 + \dots + x_n^2)^{l-k} p_m(x)$$

and from (5) we obtain

$$u_l^m(z, x) = (x_1^2 + \dots + x_n^2)^l p_m(x) \times$$

$$\times F\left(-l, 1 - l - m - \frac{n}{2}; \alpha; -\frac{4z}{x_1^2 + \dots + x_n^2}\right) =$$

$$= (4z + x_1^2 + \dots + x_n^2)^l p_m(x) \times$$

$$\times F\left(-l, \alpha + \frac{n-2}{2} + l + m; \alpha; \frac{4z}{4z + x_1^2 + \dots + x_n^2}\right).$$

The unit sphere in the metric, corresponding to the principal part of equation (4) is defined by the equation $4z + x_1^2 + \dots + x_n^2 = 1$. Let us consider the trace $u_l^m(z, x)$ on the hemisphere $S : \{z \geq 0, 4z + x_1^2 + \dots + x_n^2 = 1\}$. We have

$$Q_l^m(x) = u_l^m(z, x)\big|_S = p_m(x) \times$$

$$\times F(-l, \alpha + \frac{n-2}{2} + l + m; \alpha; 1 - x_1^2 - \dots - x_n^2).$$

let us construct a partial differential equation, satisfied by the polynomials $Q_l^m(x)$.

We have

$$\frac{Q_l^m}{dx_j} = \frac{p_m}{x_j}F - 2x_j p_m F', \quad \frac{d^2 Q_l^m}{dx_j^2} = \frac{d^2 p_m}{dx_j^2}F -$$

$$-4x_j \frac{dp_m}{dx_j}F' - 2p_m F' + 4x_j^2 p_m F'',$$

$$\sum_{j=1}^{m} x_j \frac{dQ_l^m}{dx_j} = mp_m F - 2(x_2^2 + \ldots + x_n^2)p_m F',$$

$$\Delta Q_l^m = -4mp_m F' - 2np_m F' + 4(x_1^2 + \ldots + x_n^2)p_m F''.$$

However, since $Q_l^m(x) = p_m(x)F$,

$$2(x_1^2 + \ldots + x_n^2)p_m(x)F' = -\sum_{j=1}^{n} x_j \frac{dQ_l^m}{dx_j} + mQ_l^m(x),$$

$$\Delta Q_l^m = -4(m + \frac{n}{2})p_m(x)F' + 4(x_1^2 + \ldots + x_n^2)p_m(x)F''$$

whence we find

$$4(x_1^2 + \ldots + x_n^2)p_m(x)F'' = (x_1^2 + \ldots + x_n^2)\Delta Q_l^m -$$

$$-2(m + \frac{n}{2})\sum_{j=1}^{n} x_j \frac{dQ_l^m}{dx_j} + 2(m + \frac{n}{2})mQ_l^m.$$

From the equation for the hypergeometric function we get

$$(1 - \sigma)\sigma F'' + [-(m + \frac{n}{2}) + (\alpha + m + \frac{n}{2})\sigma]F' +$$

$$+l(\alpha + \frac{n-2}{2} + l + m)F = 0,$$

$$\sigma = x_1^2 + \ldots + x_n^2.$$

Substitutings expresions $p_m F'', p_m F', p_m F$ into this equation, multiplied by p_m, for $Q_l^m(x)$ we obtain the equation

$$(1 - x_1^2 - \ldots - x_n^2)\Delta Q_L^m - 2\alpha \sum_{j=1}^{n} x_j \frac{dQ_l^m}{dx_j} +$$

$$+[2l(2l + 2m + 2\alpha + n - 2) + 2m\alpha]Q_l^m = 0.$$

It follows from the expresion $Q_l^m(x)$ that

$$Q_l^m(x) = \frac{l!\Gamma(\alpha + \frac{n-2}{2} + m + 2l)\Gamma(\alpha)}{\Gamma(\alpha + \frac{n-2}{2} + m + l)\Gamma(l+\alpha)}(-1)^l \times$$

$$\times (x_1^2 + \ldots + x_n^2)^l p_m(x) + R_l^m(x), \tag{6}$$

where $R_l^m(x)$ is a polynomial of the degree not exceeding $2l + m - 2$.

Thus the polynomials $Q_l^m(x)$ yeld the solution to the following problem: to find those values of λ for which the equation

$$(1 - x_1^2 - \ldots - x_n^2)\Delta Q - 2\alpha \sum_{j=1}^{n} x_j \frac{dQ}{dx_j} + \lambda Q = 0 \tag{7}$$

has the solutions, bounded in the unit sphere, and to find these bounded solutions. It is natural to regard the polynomials $Q_l^m(x)$ as a generalization for the case with many variables of the Gegenbauer polynomials. If we represent $Q_l^m(x)$ in the form $f_l(x_1^2 + \ldots + x_n^2)p_m(x)$, then from the equation which is satisfied by $Q_l^m(x)$, for $f_l(\sigma)$ we obtain the equation

$$\sigma(1-\sigma)f'' + [m + \frac{n}{2} - (m + \frac{n}{2} + \alpha)\sigma]f' + l(l + m + \alpha + \frac{n-2}{2})f = 0. \tag{8}$$

As know [2] the Jacobian polynomials $p_l^{(a,b)}(x)$ satisfy the equation

$$(1 - x^2)u'' + [b - a - (a + b + 2)x]u' + l(l + a + b + 1)u = 0.$$

We replace the independent variable $2y = 1 + x$ and then for the function $v(y) = u(2y - 1)$ we get the equation

$$y(y-1)v'' + [b + 1 - (a + b + 2)y]v' + l(l + a + b + 1)v = 0.$$

This equation coincides with (8) for $a = \alpha - 1, \quad b = m + \frac{n-2}{2}$, consequently

$$f(\sigma) = AP_l^{(\alpha-1, m+\frac{n-2}{2})}(2\sigma - 1),$$

where A is a constant. Thus

$$Q_l^m(x) = Ap_m(x)P_l^{(\alpha-1, m+\frac{n-2}{2})}(2(x_1^2 + \ldots + x_n^2) - 1),$$

where A is a constant.

If we multiply by $(1 - x_1^2 - \ldots - x_n^2)^{\alpha-1}$, we can bring equation (7) to the self-adjoint form

$$\sum_{j=1}^{n} \frac{d}{dx_j}[(1 - x_1^2 - \ldots - x_n^2)^\alpha \frac{dQ}{dx_j}] + \lambda(1 - x_1^2 - \ldots - x_n^2)^{\alpha-1}Q = 0.$$

Hence it follows that

$$\int_{\Sigma} (1 - x_1^2 - \ldots - x_n^2)^{\alpha-1} Q_k^p(x) Q_l^m(x) dx = 0$$

if at least one of the inequalities either $k \neq l$ or $n \neq m$ is fulfilled, i.e. the polynomials $Q_l^m(x)$ comprise an orthogonal weighted system. Here Σ is a unit sphere.

For the Jacobian polynomials the representation [2]

$$P_l^{(\alpha,\beta)}(x) = \frac{(-1)^l \Gamma(l+1+\beta)}{l! \Gamma(1+\beta)} F(l+\alpha+\beta+1, -l; \beta+1; \frac{1+x}{2}) =$$

$$= \frac{\Gamma(2l+\alpha+\beta+1)2^{-l}}{\Gamma(l+\alpha+\beta+1)} (-1)^l x^l + \ldots$$

holds, consequently we have

$$P_l^{(\alpha-1, m+\frac{n-2}{2})}(2(x_1^2 + \ldots + x_n^2) - 1) =$$

$$= \frac{\Gamma(2l+\alpha+m+\frac{n-2}{2})}{\Gamma(l+\alpha+m+\frac{n-2}{2})} (-1)^l (x_1^2 + \ldots + x_n^2)^l + \ldots .$$

By comparing this formula with the highest degree terms in $Q_l^m(x)$ we find

$$Q_l^m(x) = \frac{\Gamma(\alpha) l!}{\Gamma(l+\alpha)} p_m(x) P_l^{(\alpha-1, m+\frac{n-2}{2})}(2(x_1^2 + \ldots + x_n^2) - 1).$$

REFERENCES

1. Kratzer A., Franz W. Transzendente Funktionen. Leipzig, 1960.

2. Szego G. Orthogonal Polynomials. Amer. Math. Soc. Collog. Publ. Vol. 23, Amer. Math. Soc. Providence, R. I. (1975).

1 [illegible]

Functional Analysis, Approximation Theory
and Numerical Analysis
Ed. John M. Rassias

SOLUTION OF QUASI-TRIDIAGONAL SYSTEM OF LINEAR EQUATIONS

Liu Yunkang Wu Ciqian

Computer Science Department, Lingnan (University) College,
Zhongshan University, Guangzhou, P.R.China

Abstract.

Necessary and sufficient conditions for the uniqueness, existence, and stability of solutions of a class of quasi-tridiagonal systems of linear equations, which appears in many applications, are obtained in this paper. Efficient methods for computing both exact and approximate solutions are presented.

1. Introduction.

Many problems, such as numerical solutions of differential equations[1,2] and interpolation by spline functions[3,4], especially when subject to periodic or more complicated boundary conditions, are usually reduced to the solution of quasi-tridiagonal systems of linear equations of the general form

(1.1) $$Ax=f,$$

where $x=(x_1,x_2,\cdots x_n)^T$ and $f=(f_1,f_2,\cdots f_n)^T$ are unknown and known vectors respectively, A is a quasi-tridiagonal matrix,

$$A=\begin{pmatrix} a_1 & a_2 & a_3 & \cdot & \cdot & a_{n-2} & a_{n-1} & a_n \\ a & b & c & & & & & \\ & a & b & c & & & & \\ & & \cdot & \cdot & \cdot & & & \\ & & & \cdot & \cdot & \cdot & & \\ & & & & \cdot & \cdot & \cdot & \\ & & & & & a & b & c \\ b_1 & b_2 & b_3 & \cdot & \cdot & b_{n-1} & b_{n-1} & b_n \end{pmatrix},$$

$$a\neq 0,\ c\neq 0,\ |b|>|a|+|c|.$$

The traditional double sweep method is not suited for the solution of system (1.1), and those methods, such as Gaussian elimination and iteration, have not make full use of the special structure of the coefficient matrix A, thus, they are not considered as ideal methods for this kind of systems of linear equations.

In this paper, we shall investigate the uniqueness, existence and stability of solutions of system (1.1). By analyse the inverse matrix A^{-1} of A, we arrive at some efficient methods for computing both exact and approximate solution of system (1.1). When A is a tridiagonal matrix or a circulant tridiagonal matrix, modified algorithms which need about 5n operations are presented. Numerical experiments show that these algorithms work efficiently

2. Uniqueness, existence and stability.

The uniqueness, existence and stability of solutions of system (1.1) depend mainly on its coefficient matrix A. Therefore, most part of this section is devoted to the study of matrix A.

Evidently, the two roots, say λ_1 and λ_2, of equation

(2.1) $$c\lambda^2+b\lambda+a=0$$

are distinct real ones, and $|\lambda_1|\neq|\lambda_2|$. We can assume that $|\lambda_1|<|\lambda_2|$.

Lemma. $|\lambda_1|<1, |\lambda_2|>1$.

Proof. Owing to the fact that $|b|>|a|+|c|$, we have

$$b^2-4ac>(|a|+|c|)^2-4ac\geq(|a|-|c|)^2.$$

Therefore

$$|\lambda_1|=\left|\frac{|b|-\sqrt{b^2-4ac}}{2c}\right|=\frac{2|a|}{|b|+\sqrt{b^2-4ac}}<\frac{2|a|}{|a|+|c|+|a|-|c|}=1,$$

$$|\lambda_2|=\frac{|b|+\sqrt{b^2-4ac}}{2|c|}>\frac{|a|+|c|+|c|-|a|}{2|c|}=1.$$

Theorem 2.1. Matrix A is invertible if and only if matrix

$$B=\begin{pmatrix}\sum_{i=1}^{n} a_i\lambda_1^i & \sum_{i=1}^{n} a_i\lambda_2^{i-n}\\ \sum_{i=1}^{n} b_i\lambda_1^i & \sum_{i=1}^{n} b_i\lambda_2^{i-n}\end{pmatrix}$$

is invertible, i.e.,

$$\Delta(n)=\sum_{i=1}^{n} a_i\lambda_1^i\cdot\sum_{i=1}^{n} b_i\lambda_2^{i-n}-\sum_{i=1}^{n} a_i\lambda_2^{i-n}\cdot\sum_{i=1}^{n} b_i\lambda_1^i\neq 0.$$

Proof. Matrix A is invertible if and only if the homogeneous system

$$Ax=0 \tag{2.2}$$

has trivial solution only. Suppose that $x=(x_1,x_2,\cdots x_n)^T$ is a solution of system (2.2), we have

$$\sum_{i=1}^{n} a_ix_i=0, \tag{2.3}$$

$$ax_{i-1}+bx_i+cx_{i+1}=0,\quad 1<i<n, \tag{2.4}$$

$$\sum_{i=1}^{n} b_ix_i=0. \tag{2.5}$$

Due to equation (2.4), there exist two constant α and β, such that

$$x_i=\alpha\lambda_1^i+\beta\lambda_2^{i-n},\qquad 1\leq i\leq n. \tag{2.6}$$

Thus, we know from (2.3), (2.5) and (2.6) that α and β satisfy

$$\alpha\sum_{i=1}^{n} a_i\lambda_1^i+\beta\sum_{i=1}^{n} a_i\lambda_2^{i-n}=0, \tag{2.7}$$

$$\alpha\sum_{i=1}^{n} b_i\lambda_1^i+\beta\sum_{i=1}^{n} b_i\lambda_2^{i-n}=0. \tag{2.8}$$

Consequently, from (2.6), (2.7) and (2.8), we know that if and only if the coefficient matrix B is invertible, system (2.2) has trivial solution only. Since $|B|=\Delta(n)$, theorem 2.1 is thus proved.

Corollary 2.1. System (1.1) is uniquely solvable if and only if $\Delta(n)\neq 0$.

Denote by

$$A_*=\begin{bmatrix} a_1^* & a_2^* & a_3^* & \cdot & \cdot & a_{n-2}^* & a_{n-1}^* & a_n^* \\ a & b & c & & & & & \\ & a & b & c & & & & \\ & & \cdot & \cdot & \cdot & & & \\ & & & \cdot & \cdot & \cdot & & \\ & & & & \cdot & \cdot & \cdot & \\ & & & & & a & b & c \\ b_1^* & b_2^* & b_3^* & \cdot & \cdot & b_{n-1}^* & b_{n-1}^* & b_n^* \end{bmatrix}.$$

Theorem 2.2. If matrices A and A_* are invertible, then

$$A^{-1}=(I+D)A_*^{-1},$$

where I is an n by n identity matrix,

$$D=\begin{bmatrix} \lambda_1 & \lambda_2^{1-n} \\ \lambda_1^2 & \lambda_2^{2-n} \\ \cdot & \cdot \\ \cdot & \cdot \\ \cdot & \cdot \\ \lambda_1^n & 1 \end{bmatrix} B^{-1} \begin{bmatrix} a_1^*-a_1 & a_2^*-a_2 & \cdot & \cdot & \cdot & a_n^*-a_n \\ b_1^*-b_1 & b_2^*-b_2 & \cdot & \cdot & \cdot & b_n^*-b_n \end{bmatrix}.$$

Proof. For an arbitrarily given vector $f=(f_1,f_2,\cdots f_n)^T$, suppose that $x=(x_1,x_2,\cdots x_n)^T$ is the solution of system (1.1), we have

$$\sum_{i=1}^{n} a_i(x_i-y_i)=f_1-\sum_{i=1}^{n} a_iy_i, \tag{2.9}$$

$$a(x_{i-1}-y_{i-1})+b(x_i-y_i)+c(x_{i+1}-y_{i+1})=0, \quad 1<i<n, \tag{2.10}$$

$$\sum_{i=1}^{n} b_i(x_i-y_i)=f_n-\sum_{i=1}^{n} b_iy_i \tag{2.11}$$

where $y=(y_1,y_2,\cdots y_n)^T$ is the unique solution of system

$$A_*y=f. \tag{2.12}$$

Due to equation (2.10), there exist two constants α and β, such that

$$x_i-y_i=\alpha\lambda_1^i+\beta\lambda_2^{i-n}, \quad 1\le i\le n. \tag{2.13}$$

Thus, we know from (2.9), (2.11) and (2.13) that α and β satisfy

$$\alpha\sum_{i=1}^{n} a_i\lambda_1^{i} + \beta\sum_{i=1}^{n} a_i\lambda_2^{i-n} = f_1 - \sum_{i=1}^{n} a_i y_i, \tag{2.14}$$

$$\alpha\sum_{i=1}^{n} b_i\lambda_1^{i} + \beta\sum_{i=1}^{n} b_i\lambda_2^{i-n} = f_n - \sum_{i=1}^{n} b_i y_i. \tag{2.15}$$

Noticing that

$$f_1 = \sum_{i=1}^{n} a_i^{*} y_i, \quad f_n = \sum_{i=1}^{n} b_i^{*} y_i,$$

we get from (2.14) and (2.15) that

$$\begin{pmatrix}\alpha\\ \beta\end{pmatrix} = B^{-1}\begin{pmatrix} f_1 - \sum_{i=1}^{n} a_i y_i \\ f_n - \sum_{n-1}^{n} b_i y_i \end{pmatrix} \tag{2.16}$$

$$= B^{-1}\begin{pmatrix} a_1^{*}-a_1 & a_2^{*}-a_2 & \cdots & a_n^{*}-a_n \\ b_1^{*}-b_1 & b_2^{*}-b_2 & \cdots & b_n^{*}-b_n \end{pmatrix} y.$$

Therefore, we obtain from (2.13) that

$$x_i = y_i + (\lambda_1^{i}, \lambda_2^{i-n})\begin{pmatrix}\alpha\\ \beta\end{pmatrix}, \quad 1\le i\le n, \tag{2.17}$$

Theorem 2.2 follows from

$$A^{-1}f = x = y + Dy = (I+D)y = (I+D)A_{*}^{-1}f.$$

Theorem 2.3[5]. Denote A_* by A_0 in the case of $(a_1^{*}, a_2^{*}, \cdots, a_n^{*}) = (-c\lambda_2, c, 0, \cdots, 0)$, $(b_1^{*}, b_2^{*}, \cdots, b_n^{*}) = (0, \cdots, 0, a, b)$, then A_0 is invertible and has a LU factorization $A_0 = LU$, where

$$L = \begin{pmatrix} \mu & & & & \\ a & \mu & & & \\ & a & \mu & & \\ & & \ddots & \ddots & \\ & & & a & \mu \end{pmatrix}, \quad U = \begin{pmatrix} 1 & \rho & & & \\ & 1 & \rho & & \\ & & \ddots & \ddots & \\ & & & 1 & \rho \\ & & & & 1 \end{pmatrix},$$

$\mu = -c\lambda_2 = -a/\lambda_1$, $\rho = -c\lambda_1/a = -1/\lambda_2$.

From theorem 2.2 and 2.3, we have

Corollary 2.2. The solution of system (1.1) can be formulated by

$$x = (I+D)U^{-1}L^{-1}f.$$

For the sake of numerical stability, we need to estimate the upper bound of the condition number $\varkappa_\infty(A)=\|A\|_\infty\|A^{-1}\|_\infty$. Without loss of generality, we assume that

$$\max\{\sum_{i=1}^{n}|a_i|, \sum_{i=1}^{n}|b_i|\}\le K_0,$$

where K_0 is a positive constant independent of n.

Theorem 2.4. If there exists a positive constant K_1, which is independent of n, such that

$$|\Delta(n)|\ge K_1, \tag{2.18}$$

then the condition number $\varkappa_\infty(A)$ is bounded with respect to n.

Proof. It easy to know that

$$\|A\|_\infty\le \max\{K_0, |a|+|b|+|c|\},$$

$$\|B^{-1}\|_\infty\le 2K_0/K_1.$$

From theorem 2.2 and theorem 2.3, we have

$$\|A^{-1}\|_\infty\le\|(I+D)A_0^{-1}\|_\infty\le(1+\|D\|_\infty)\|L^{-1}\|_\infty\|U^{-1}\|_\infty,$$

where

$$\|D\|_\infty\le(1+|\lambda_1|)(K_0+|c||\lambda_2|+|a|+|b|+|c|)\|B\|_\infty,$$

$$\|U^{-1}\|_\infty\le(|\mu|-|a|)^{-1}\le\frac{|\lambda_1|}{|a|(1-|\lambda_1|)},$$

$$\|L^{-1}\|_\infty\le(1-|\rho|)^{-1}\le\frac{|\lambda_2|}{|\lambda_2|-1}.$$

Summarizing the above inequalities, we know that the condition number $\varkappa_\infty(A)$ is bounded with respect to n.

Corollary 2.3. If inequality (2.18) holds, the solution of system (1.1) is stable.

3. Algorithms and numerical examples

To ensure the existence, uniqueness and stability of the solution of system (1.1), we assume that inequality (2.18) holds.

According to theorem 2.2 and theorem 2.3, we have the following algorithm for the exact solution of system (1.1):

Algorithm ES:

Step 1. Compute the solution $y=A_0^{-1}f$ of system $A_0y=f$ in the following way:

$$w_1=\mu^{-1}f_1,\quad w_i=\mu^{-1}(-aw_{i-1}+f_i),\qquad i=2,3,\cdots,n,$$

$$y_n=w_n,\quad y_i=-\rho y_{i+1}+w_i,\qquad i=n-1,n-2,\cdots,1;$$

Step 2. Compute the solution $x=A^{-1}f$ of system $Ax=f$ in the following way:

$$x_i=y_i+\alpha\lambda_1^i+\beta\lambda_2^{i-n},\quad i=1,2,\cdots,n,$$

where α and β are computed by (2.16).

To avoid redundant computations, λ_1^i and λ_2^i $(i=1,2,\cdots,n)$ should be computed recurrently in the following forms

$$\lambda_1^{i+1}=\lambda_1\cdot\lambda_1^i,\quad \lambda_2^{i+1}=\lambda_2\cdot\lambda_2^i,\quad i=1,2,\cdots,n-1.$$

Algorithm ES needs about 21n arithmetic operations. It should be noted that almost half operations are spent on the computation of the coefficients α and β, and the number of such operations are directly proportional to the number of non-zero elements in $(a_1,a_2,\dots,a_n)$ and $(b_1,b_2,\dots,b_n)$. Thus, the algorithm ES doesn't need so much operations if some elements in $(a_1,a_2,\cdots,a_n)$ and $(b_1,b_2,\cdots,b_n)$ are zero. If only a few elements are non-zero, algorithm ES needs only about 9n arithmetic operations. When $(a_1,a_2,\cdots,a_n)=(b,c,0,\cdots,0,a)$, $(b_1,b_2,\cdots,b_n)=(c,0,\cdots,0,a,b)$, Chen Mingkui[6] claim that his algorithm CTS needs about 5n arithmetic operations, that is because he didn't take into account of the left hand side terms in his algorithm which need another 4n arithmetic operations.

If n, the number of equations in system (1.1), is large, due to the fact that

$$\lim_{n\to+\infty}\lambda_1^n=\lim_{n\to+\infty}\lambda_2^{-n}=0,$$

the second step of algorithm ES can be modified to

$$x_i^*=\begin{cases} y_i+\alpha\lambda_1^i, & 1\le i<m_1,\\ y_i, & m_1\le i\le n-m_2,\\ y_i+\beta\lambda_2^{i-n}, & n-m_2<i\le n,\end{cases}$$

where $m_1+m_2\le n$. The error of the approximate solution x^* to the exact solution x satisfies

$$\|x^*-x\|_\infty\le|\alpha||\lambda_1|^{m_1}+|\beta||\lambda_2|^{-m_2}. \tag{3.1}$$

From (2.16), we know that the coefficients α and β are bounded with respect to n, Therefore, when m_1 and m_2 are comparatively large, the approximate solution x^* could be very accurate, but need less operations to compute.

In some applications, algorithm ES can be modified to approximate one which needs much less operations. Here, we consider two commonly occurred examples.

Example 1. When $(a_1,a_2,\cdots,a_n)=(b,c,0,\cdots,0)$, $(b_1,b_2,\cdots,b_n)=(0,\cdots,0,a,b)$, we have

$$\begin{aligned}\Delta(n)&=(b\lambda_1+c\lambda_1^2)(a\lambda_2^{-1}+b)-(b\lambda_2^{1-n}+c\lambda_2^{2-n})(a\lambda_1^{n-1}+b\lambda_1^n)\\ &=ac\lambda_2+0(\lambda_0^{2n}),\end{aligned}$$

where $\lambda_0=\max(|\lambda_1|,|\lambda_2^{-1}|)<1$, and

$$\begin{aligned}B^{-1}&=\frac{1}{\Delta(n)}\begin{pmatrix} a\lambda_2^{-1}+b & -b\lambda_2^{1-n}-c\lambda_2^{2-n}\\ -a\lambda_1^{n-1}-b\lambda_1^n & b\lambda_1+c\lambda_1^2\end{pmatrix}\\ &=-\frac{1}{ac\lambda_2}\begin{pmatrix} c\lambda_2 & 0\\ 0 & a\end{pmatrix}+0(|\lambda_0|^n).\end{aligned}$$

Substituting

$$\begin{pmatrix}\alpha^*\\ \beta^*\end{pmatrix}=-\frac{1}{ac\lambda_2}\begin{pmatrix} c\lambda_2 & 0\\ 0 & a\end{pmatrix}\begin{pmatrix} f_1-by_1-cy_2\\ f_n-ay_{n-1}-by_n\end{pmatrix}=\begin{pmatrix}\alpha\\ \beta\end{pmatrix}+0(\lambda_0^n)$$

for $(\alpha,\beta)^T$, the second step of algorithm ES can be modified to

$$x_i^{**}=y_i+\alpha^*\lambda_1^i+\beta^*\lambda_2^{i-n},\quad i=1,2,\cdots,n. \tag{3.2}$$

Example 2. When $(a_1,a_2,\cdots,a_n)=(b,c,0,\cdots,0,a)$, $(b_1,b_2,\cdots,b_n)=(c,0,\cdots,0,a,b)$, we have

$$\Delta(n)=(b\lambda_1+c\lambda_1^2+a\lambda_1^n)(c\lambda_2^{1-n}+a\lambda_2^{-1}+b)-(b\lambda_2^{1-n}+c\lambda_2^{2-n}+a)(c\lambda_1+a\lambda_1^{n-1}+b\lambda_1^n)$$
$$=ac(\lambda_1-\lambda_2)+0(\lambda_0^n),$$

where $\lambda_0=\max(|\lambda_1|,|\lambda_2^{-1}|)<1$, and

$$B^{-1}=\frac{1}{\Delta(n)}\begin{pmatrix} c\lambda_2^{1-n}+a\lambda_2^{-1}+b & -b\lambda_2^{1-n}-c\lambda_2^{2-n}-a \\ -c\lambda_1-a\lambda_1^{n-1}-b\lambda_1^n & b\lambda_1+c\lambda_1^2+a\lambda_1^n \end{pmatrix}$$
$$=(B^{-1})^*+0(|\lambda_0|^n),$$

where

$$(B^{-1})^*=-\frac{1}{ac(\lambda_1-\lambda_2)}\begin{pmatrix} c\lambda_2 & a \\ c\lambda_1 & a \end{pmatrix}.$$

Substituting

$$\begin{pmatrix}\alpha^* \\ \beta^*\end{pmatrix}=(B^{-1})^*\begin{pmatrix} f_1-by_1-cy_2-ay_n \\ f_n-ay_1-by_{n-1}-cy_n\end{pmatrix}=\begin{pmatrix}\alpha \\ \beta\end{pmatrix}+0(\lambda_0^n)$$

for $(\alpha,\beta)^T$, the second step of algorithm ES can also be modified to (3.2).

The modified approximate algorithm of ES in the above examples needs about 7n arithmetic operations. The error of the approximate solution x^{**} to the exact solution x satisfies the following estimate

$$\|x^{**}-x\|_\infty\le|\alpha^*-\alpha|+|\beta^*-\beta|=0(\lambda_0^n).$$

When n is large, (3.2) can be remodified to

$$x_i^{***}=\begin{cases} y_i+\alpha^*\lambda_1^i, & 1\le i<m_1, \\ y_i, & m_1\le i\le n-m_2, \\ y_i+\beta^*\lambda_2^{i-n}, & n-m_2<i\le n. \end{cases} \tag{3.3}$$

The error of x^{***} to the exact solution x satisfies the following estimates

$$\|x^{***}-x\|_\infty\le\|x^{***}-x^*\|_\infty+\|x^*-x\|_\infty$$
$$\le|\alpha^*-\alpha|+|\beta^*-\beta|+|\alpha||\lambda_1|^{m_1}+|\beta||\lambda_2|^{-m_2}.$$

If n is large and m_1, m_2 are comparatively small, the remodified approximate algorithms of ES in the above examples need about 5n arithmetic operations.

We apply the algorithms described above to solve system (1.1) in the case of a=1, b=4, c=1, and $(a_1,a_2,\cdots,a_n)=(4,1,0,\cdots,0,1)$, $(b_1,b_2,\cdots,b_n)=(1,0,\cdots,0,1,4)$, which appears in the problem of periodic spline interpolation[3]. When $f=(6,6,\cdots,6)^T$, the exact solution is $x=(1,1,\cdots,1)^T$. If we use algorithm ES, the maximum error of the computed solution to the exact solution are less than 3.65×10^{-12}. If we use the modified algorithm of ES, we have

Table 1

n	4	8	16	≥32
$\|x^{**}-x\|_\infty$	1.86×10^{-3}	9.73×10^{-6}	2.60×10^{-10}	3.65×10^{-12}

If we use the remodified algorithm of ES, let $E(n,m)=\|x^{***}-x\|_\infty$, where $m=m_1=m_2$, we have

Table 2

n	E(n,4)	E(n,8)	E(n,16)	E(n,32)
8	1.89×10^{-3}			
16	1.89×10^{-3}	9.76×10^{-6}		
32	1.89×10^{-3}	9.76×10^{-6}	2.62×10^{-10}	
≥64	1.89×10^{-3}	9.76×10^{-6}	2.62×10^{-10}	3.65×10^{-12}

From Table 1 and 2 we know that the errors are mainly come from the substitution of (3.3) for (3.2), not from the substitution of (α^*,β^*) for (α,β). General speak, when $|\lambda_1|^{m_1}$ and $|\lambda_2|^{-m_2}$ are small enough, the remodified algorithm will yield satisfactory result.

The numerical experiments are made on IBM PC using Turbo Pascal version 5.0 of Borland International, Inc.

References

[1]B.L.Buzbee,G.H.Bolub and C,W.Nielson,*On direct methods for solving Poisson's equations.*SIAM J.Numer.Anal., 7 (1970), pp 625-656.

[2]D.Fisher, G.Golub, W.Hold, C.Leiva and O.Widlund, *On Fourier- Toeplits methods for separable elliptic problems.* Math. Comp. 28 (1974), pp 349-368.

[3]J.Ahlberg, E.N.Nilson and J.L.Walsh, *The theory of splines and their applications.* Academic Press, New York,(1967).

[4]Liu Yunkang, *On quadratic and cubic splines which satisfy a general boundary conditions,* Mathematics in Practice and Theory, (1990), 3: pp 43-46.

[5]D.Kershaw, *The explicit inverse of two commonly occurring matrices,* Math. Comp., 23(1969), pp.189-191.

[6]M.A.Malcolm, J.Palmer, *A fast method for solving a class of tridiagonal linear systems.* Comm. ACM, 17(1974), pp.14-17.

[7]D.Meek, *The inverse of some matrices deviating slightly from a symmetric, tridiagonal, Toeplitz form,* SIAM J.Numer.Anal., 17(1980), pp.39-43.

[8]Gao De-ying, Shu Fong, Pai Nai-De, *The fast approximate LU method for the special tridiagonal linear simultaneous equations.* Computational Mathematics, (1981), 1: pp 10-17.

[9]Chen Mingkui, *On the solution of circulant linear systems.* SIAM.J. Numer. Anal, 24 (1987), pp 668-683.

Functional Analysis, Approximation Theory
and Numerical Analysis
Ed. John M. Rassias

The Uniqueness and Existence of Solution and Normal Boundary Condition for Thin Plate Bending Problem

Wang Zhehui Wu Ciqian
Department of Computer Science
Zhongshan University
Guang Zhou, PR China

ABSTRACT

Spline finite strip method based on equi-mesh for structure analysis has been studied by C.Q. Wu , Y.K. Cheung and S.C. Fan in 1981 [1]. We propose the generalized trapezoid element method of non-equi-mesh processing on abitrary area and its error analysis in [2]. The basic problem of structure analysis, the uniqueness and existence of solution of thin plate bending problem is discussed in this paper in order to deal with boundary condition on abitrary area uniformly . The normal boundary condition is also proposed.

Let

$$D = \begin{bmatrix} D_x & D_1 & 0 \\ D_1 & D_y & 0 \\ 0 & 0 & D_{xy} \end{bmatrix} \qquad \{\varepsilon\} = \begin{bmatrix} -W_{xx} \\ -W_{yy} \\ 2W_{xy} \end{bmatrix}$$

where D_x, D_y, D_1, D_{xy} are positive constants satisfying D_x, $D_y > D_1$ and $W(x,y)$ displacement function on the bounded area Ω.

Let $T = \partial\Omega$ be boundary of Ω and $W \in H^2(\Omega), f \in H^0(\Omega)$.

Definition (1)

$$J(W) = \iint_\Omega \left[\frac{1}{2}\{\varepsilon\}^T D\{\varepsilon\} - Wf\right]dxdy$$

$$= \iint_\Omega \left\{\frac{1}{2}[D_x W_{xx}^2 + 2D_1 W_{xx} W_{yy} + D_y W_{yy}^2 + 4D_{xy} W_{xy}^2] - Wf\right\}dxdy$$

Thin plate bending problem of structure analysis becomes the following problem.

find $W^* \in H^2(\Omega)$ such that

$$J(W^*) = \min_{W \in H^2(\Omega)} J(W) \tag{1}$$

Definition (2)

For all $u, v \in H^2(\Omega)$ $f \in H^0(\Omega)$, define

$$a(u,v) = \iint_\Omega [D_x u_{xx} v_{xx} + D_1 u_{xx} v_{yy} + D_1 u_{yy} v_{xx} + D_y u_{yy} v_{yy} + 4D_{xy} u_{xy} v_{xy}] dxdy$$

$$(f, v) = \iint_\Omega f v dxdy$$

Definition (3)

For all $u \in H^k(\Omega)$, $k = 0, 1, 2, \ldots$, define

$$\| u \|_{k,\Omega} = \left\{ \iint_\Omega \left[\sum_{0 \le \alpha+\beta \le k} \left(\frac{\partial^k u}{\partial x^\alpha \partial y^\beta} \right)^2 \right] dxdy \right\}^{1/2}$$

$$| u |_{k,\Omega} = \left\{ \iint_\Omega \left[\sum_{\alpha+\beta = k} \left(\frac{\partial^k u}{\partial x^\alpha \partial y^\beta} \right)^2 \right] dxdy \right\}^{1/2}$$

By $\| u \|_k$, $| u |_k$ denote $\| u \|_{k,\Omega}$, $| u |_{k,\Omega}$ respectively, if there is no confusion.

Lemma (1)

Let $W^* \in H^2(\Omega)$ be the solution of (1) then

$$a(W^*, v) = (f, v) \qquad \forall\ v \in H^2(\Omega) \tag{2}$$

Conversely, $W^* \in H^2(\Omega)$ satisfying (2) must be the solution of (1)

Definition (4)

Let $\Gamma = \Gamma_1 \cup \Gamma_2 \cup \Gamma_3$ with $\Gamma_i \cap \Gamma_j = \phi$ $i \neq j$ Define

$$G = \{ u; u|_{\Gamma_1} = 0,\ u_x|_{\Gamma_2} = u_y|_{\Gamma_2} = u|_{\Gamma_2} = 0 \}$$

We name Γ_1, Γ_2 and Γ_3 as simple supported boundary, clamped boundary and free boundary respectively.

Define $H_G^k = \{ u;\ u \in H^k(\Omega) \cap G \}$

especially, $E = \{ u; u = u_x = u_y = 0, (x,y) \in \Gamma \}$

$H^k_E = \{ u; u \in H^k(\Omega) \cap E \}$

Lemma (2)

Let $W^* \in H^3_G(\Omega)$ **be the solution of the problem**

$$J(W^*) = \min_{W \in H^3_G(\Omega)} J(W) \tag{3}$$

Then

$$a(W^*, v) = (f, v) \quad \forall\ v \in H^3_G(\Omega) \tag{4}$$

Conversely, $W^* \in H^3_G(\Omega)$ **satisfying (4) must be the solution of (3)**

Theorem (1)

Let $u_1, u_2 \in H^3_G(\Omega)$ **be solution of (3) Then there exist constants** $\alpha_0, \beta_0, \gamma_0$ **such that**

$$u_1 = u_2 + \alpha_0 x + \beta_0 y + \gamma_0 \qquad (x,y) \in \overline{\Omega}$$

lemma (3)

For all $u \in H^4(\Omega)$, $v \in H^2(\Omega)$, $f \in H^0(\Omega)$ **let**

$$Lu = D_x u_{xxxx} + (2D_1 + 4D_{xy}) u_{xxyy} + D_y u_{yyyy}$$

$$\Gamma_3(u) = [D_x u_{xxy} + (D_1 + 2D_{xy}) u_{xyy}] \cos\alpha + [D_y u_{yyy} + (D_1 + 2D_{xy}) u_{xxy}] \cos\beta$$

$$\Gamma_{2x}(u) = [D_x u_{xx} + D_1 u_{yy}] \cos\alpha + 2D_{xy} u_{xy} \cos\beta$$

$$\Gamma_{2y}(u) = [D_y u_{yy} + D_1 u_{xx}] \cos\beta + 2D_{xy} u_{xy} \cos\alpha$$

where $(\cos\alpha, \cos\beta)$ **is outer normal vector of** Γ**.Then**

$$a(u,v) - (f,v) = (Lu - f, v) - \int_\Gamma [v\Gamma_3(u) - v_x \Gamma_{2x}(u) - v_y \Gamma_{2y}(u)] ds$$

Theorem (2)

Let $u^* \in c^4(\overline{\Omega})$ **be the solution of boundary problem**

$$Lu = f \qquad (x,y) \in \Omega \qquad (5)$$

$$u=\Gamma_{2x}(u)=\Gamma_{2y}(u)=0 \qquad (x,y) \in \Gamma_1$$

$$u = u_x = u_y = 0 \qquad (x,y) \in \Gamma_2$$

$$\Gamma_{2x}(u)=\Gamma_{2y}(u)=\Gamma_3(u)=0 \qquad (x,y) \in \Gamma_3 \qquad (6)$$

Then u^* is also the solution of the following variational problem.

Find $u^* \in c^4(\overline{\Omega}) \cap G$ such that

$$J(u^*) = \min_{W \in H^2_G(\Omega)} J(u) \qquad (7)$$

Conversely let $u^* \in c^4(\overline{\Omega})$ be the solution of (7), then it must be the solution of (5) and (6).

Throerem (3)

Let $u \in c^4(\overline{\Omega})$ then u satisfies (5) and (6) iff $u \in H^2_G(\Omega)$ and

$$a(u,v) = (f,v) \qquad \forall\ v \in H^2_G(\Omega) \qquad (8)$$

holds.

Definition (5) Let $u^* \in H^2_G(\Omega)$ be solution of (7). Define u^* to be the generalized solution of (5), (6).

Theorem (4)

Suppose that G (cf. Definition (4)) satisfies one of the following conditions:

(i) $\Gamma_2 \neq \emptyset$ (ii) $\Gamma_1 \neq \emptyset$ and Γ_1 contains three points which are not on the same line.

If probelm (3) has solution, then it must have unique solution.

Definiion (6)

Suppose that G satisfies one of the following conditions

(i) $\Gamma_2 \neq \emptyset$ (ii) $\Gamma_1 \neq \emptyset$ there exists $A_k(x_k, y_k) \in \Gamma_1, k=1,2,3$ such that $A_1\ A_2\ A_3$ are not on the same line and $\Delta A_1\ A_2\ A_3$ contains the center of gravity of Ω.

By "normal boundary condition" we name the boundary condition of G.

Note : If the boundary condition is normal, variational problem (3) possesses solution which

has physical background. By Theorem (4) this solution W^* is unique. In this case if $W^* \in H^6(\Omega)$, we know by embedding theorem that $W^* \in C^4(\overline{\Omega})$ and determined by (5) and (6). We can study the property of W^* by (5) and (6).

Reference

[1] C.Q.Wu , Y. K. Cheung and S. C. Fan,"Spline finite stripe in structure analysis", Guangdong Scienc e Publishing House, 1986

[2] C.Q. Wu and Wang Chehui, " Generalized trapezoid element method and its error analysis for structure analysis", to be published

Functional Analysis, Approximation Theory
and Numerical Analysis
Ed. John M. Rassias

A CHARACTERIZATION OF Q-ALGEBRAS

YANNIS TSERTOS

Mathematical Institute, University of Athens
Panepistimiopolis Athens 15784 Greece

We will prove in this work the following:

(I) A t.a. (: topological algebra) (E, τ) is "Q" iff there exists $V \in W_0$ such that $r_E \leq g_V$.

(II) A l.c. (: locally convex) algebra (E, Γ) is "Q" iff there exist $M > 0$ and $p \in \Gamma$ such that $r_E \leq M \cdot p$.

(III) A l.m.c. (: locally multiplicatively convex) t.a. (E, Γ) is "Q" iff there exists $p \in \Gamma$ such that $r_E \leq p$.

Applications.

(IV) A "Q" locally convex $*$-algebra (E, Γ) with the B^*-property is normed (: there exists $p \in \Gamma$ such that $q \leq p$ for all $q \in \Gamma$).

(V) (A. Mallios) A l.c. $*$-algebra (E, Γ) with the B^*-property whose completion is a "Q"-algebra is normed.

Note 1. (III) is the generalization of the relation $r(x) \leq \|x\|$, $x \in E$ for the spectral radius r of a Banach algebra $(E, \|\cdot\|)$. All the above are contained in [7].

Definitions.

a) A *t.v.s.* (*topological vector space*) (E,τ) is in particular called a t.a. iff
a_1) E is an algebra (over $\mathbb{C}$) and
a_2) $x \mapsto ax$, $x \mapsto xa$ are both continuous for all $a \in E$ (: separately continuous multiplication in (E,τ)).

b) We put $G_E^q := \{x \in E : \exists y \in E : x+y-xy = x+y-yx = 0\}$ for the set of *quasi-invertible elements* of E (which is in fact a group under the circle-operation: $xoy := x+y-xy$, $x,y \in E$) and in case E is unital (:E has a "unit" e: $xe = ex = x$, $x \in E$) : $G_E := \{x \in E : \exists y \in E : xy = yx = e\}$ for the set of *invertible elements* of E (which is in fact a group under the multiplication of E).

c) A t.a. (E,τ) is "Q" in case $G_E^q \in \tau$ (: equivalently $G_E \in \tau$ in the unital case or equivalently $G_E^q \in W_0(\tau)$ where $W_0(\tau)$ are the neighborhoods of 0 in (E,τ)).

d) For each $x \in E$ we put $Sp_E(x) := \{\lambda \in \mathbb{C} - \{0\} : \frac{x}{\lambda} \notin G_E^q\} \cup \Phi_x$ where:

$$\Phi_x := \begin{cases} \emptyset & \text{if } E \text{ is unital and } x \in G_E \\ \{0\} & \text{if } E \text{ is unital and } x \notin G_E \\ \{0\} & \text{for a non unital } E \end{cases}$$

e) $r_E(x) := \sup\{|\lambda| : \lambda \in Sp_E(x)\}$, $x \in E$ is the spectral radius of $x \in E$ (in case $Sp_E(x) \neq \emptyset$) and the map $x \mapsto r_E(x)$ of E onto $\mathbb{R}_+$ is called the *spectral radius* of E (which is defined for all x with $Sp_E(x) \neq \emptyset$). As a consequence of the Spectral Mapping Theorem we get $r_E(x^n) = r_E(x)^n$ and $r_E(\lambda x) = |\lambda| r_E(x)$, $\lambda \in \mathbb{C}$, $x \in E$, $n \in \mathbb{N}$, ($\mathbb{N}, \mathbb{R}_+, \mathbb{C}$ above are respectively the sets of natural, positive real, complex numbers). Put

$$S(E) := \{x \in E : r_E(x) \leq 1\}$$

f) Given a balanced and absorbing $V \in W_0(\tau)$ we put $A_x(V) := \{\rho > 0 : x \in \rho V\}$, $x \in E$. It is easy to see that $[\rho, +\infty) \subseteq A_x \subseteq [0, +\infty)$, $\rho \in A_x$, $x \in E$ and we put

$$g_V(x) := \inf A_x, \quad x \in E$$

The map $g_V : E \to \mathbb{R}_+ : x \mapsto g_V(x)$ satisfies $g_V(\lambda x) = |\lambda| g_V(x)$, $\lambda \in \mathbb{C}$, $x \in E$.

g) In a l.c. algebra, $(E,\Gamma) \equiv (E,\tau_\Gamma)$ the topology τ_Γ has a base of neighborhoods of 0 spheres $S(p,\varepsilon) := \{x \in E : p(x) < \varepsilon\}$, $\varepsilon > 0$, $p \in \Gamma$, where the family Γ of (:vector space) seminorms on E is supposed to be saturated

(:$\max\{p,q\} \in \Gamma$ for all $p,q \in \Gamma$) and separating (: for $0 \neq x \in E$ there exists $p \in \Gamma : p(x) \neq 0$) (: saturability of Γ, it is not a loss of generality because $S(p,\varepsilon) \cap S(q,\varepsilon) = S(\max\{p,q\},\varepsilon)$).

h) A l.c. algebra (E,Γ) is in particular a l.m.c. algebra iff each $p \in \Gamma$ is an algebra-seminorm: $p(xy) \leq p(x)p(y)$, $x,y \in E$. (: submultiplicative seminorm p).

i) A seminorm p of a $*$-algebra $(E,*)$ (where the involution $* : E \to E$ is such that $x^{**} := (x^*)^* = x$, $(\lambda x + y)^* = \bar{\lambda} x^* + y^*$ and $(xy)^* = y^* x^*$ $x,y \in E$, $\lambda \in \mathbb{C}$) has the B^*-property iff $p(x^* x) = p(x)^2$, $x \in E$. By [6] a B^*-seminorm is submultiplicative.

j) For an algebra seminorm p on E, $\ker(p) := \{x \in E : p(x) = 0\} \equiv p^{-1}(0)$ is a (2-sided) ideal of E and the quotient space $E_p := E/\ker(p)$ equipped with the quotient-norm $\dot{p}(x_p) := \inf\{p(x+y) : p(y) = 0\} = p(x)$ (where $x_p \equiv x + \ker(p)$) it becomes a normed algebra and its completion $\hat{E}_p \equiv \widehat{(E_p, \dot{p})}$ is a Banach algebra which in particular is a C^*-algebra in case p is a B^*-seminorm.

Note 2. i) E. A. Michael [4] has the characterization "The l.c. algebra (E,Γ) is "Q" iff $S(E) \in W_0(\tau_\Gamma)$" and A. Mallios [3, Lemma II.4.2] "The t.a. (E,τ) is "Q" iff $S(E) \in W_0(\tau)$".

ii) G. Lassner [2] proved that "a "Q", complete, unital, barrelled, lmc $*$-algebra with the B^*-property is a C^*-algebra" and M. Fragoulopoulou [1, Theorem 3.3] using a much simpler technique, proved Lassner's result. At the same time, remove completeness and unit and replacing the property "Q" by a weaker condition. In the same paper (ibid., Lemma 2.1, Theorem 2.2) it was also proved that every complete "Q" l.m.c. algebra is, in fact, a C^*-algebra.

Proof. (I) (direct) Let (E,τ) be a "Q"-algebra. By $G_E^q \in \tau$ there exists a balanced, absorbing $U \in W_0(\tau)$ in such a way that for $V := \frac{1}{2}U$ we have $0 \in V \subseteq U \subseteq G_E^q$. For arbitrary $x \in E$, $\rho \in A_x(V)$ and $\lambda \in \mathbb{C} : |\lambda| \geq \rho$ we have the following: $|\lambda| \in A_x(V), x \in |\lambda| V$, $x \in \lambda V$, $\frac{x}{\lambda} \in V \subseteq U \subseteq G_E^q$, $\lambda \notin Sp_E(x)$. Thus $Sp_E(x) \subseteq \{\lambda \in \mathbb{C} : |\lambda| \leq \rho\}$, $r_E(x) \leq \rho$. Passing to the infimum we get:

$$r_E(x) \leq \inf A_x(V) \equiv g_V(x), \qquad x \in E .$$

(converse) Let $r_E \leq g_V$ for some $V \in W_0(\tau)$, and arbitrary $x \in \frac{1}{2}V$. Thus $\frac{1}{2} \in A_x(V)$, $r_E(x) \leq g_V(x) \leq \frac{1}{2} < 1$, $1 \notin Sp_E(x)$, $x = \frac{x}{1} \in G_E^q$. Thus $\frac{1}{2}V \subseteq G_E^q$ and so $G_E^q \in W_0(\tau)$. □

(II) For $v \in W_0(\tau)$ with $r_E \leq g_V$ there exists $p \in \Gamma$, $\varepsilon > 0 : \{x \in E : p(x) < \varepsilon\} \equiv S(p,\varepsilon) \subseteq V$. But $V \subseteq U$ is equivalent to $g_U \leq g_V$, $S(M \cdot p, \varepsilon) = \frac{1}{M} S(p,\varepsilon) = S(p, \frac{\varepsilon}{M})$, $A_x(\lambda V) = \frac{1}{\lambda} A_x(V)$ and $g_{\lambda V}(x) = \frac{1}{\lambda} g_V(x)$, $\lambda \in \mathbb{C}-\{0\}$, $x \in E$. Therefore $r_E \leq g_V \leq g_{S(p,\varepsilon)} = g_{\varepsilon S(p,1)} = \frac{1}{\varepsilon} g_{S(p,1)} = \frac{1}{\varepsilon} p \equiv Mp$, with $M \equiv \frac{1}{\varepsilon}$. □

(III) By (II) we have $r_E \leq M \cdot p$, for some $M > 0$, $p \in \Gamma$, for some $M > 0$, $p \in \Gamma$. Thus $r_E(x) = r_E(x^n)^{1/n}$

$$\begin{aligned}(Mp(x^n))^{1/n} &= M^{1/n} p(x^n)^{1/n} \leq M^{1/n} (p(x)^n)^{1/n} \\ &= M^{1/n} p(x) \underset{n}{\rightarrow} p(x) .\end{aligned}$$

(IV) There exists $p \in \Gamma$ such that $r_E \leq p$. By [3, p. 100] $r_E(x) = \sup_\Gamma r_{\hat{E}_p}(x_p) \leq r_{\hat{E}_p}(x_p)$, $x \in E$, $p \in \Gamma$, and by [5, Lemma 4.8.1] $\dot{q}(x_q)^2 = r_{\hat{E}_q}(x_q^* x_q)$, $q \in \Gamma$, so that for arbitrary $q \in \Gamma$ we get $q(x)^2 = \dot{q}(x_q)^2 = r_{\hat{E}_q}(x_q^* x_q) = r_{\hat{E}_q}((x^* x)_q) \leq r_E(x^* x) \leq p(x^* x) = p(x)^2$, $x \in E$, for all $x \in E$ and the proof is complete.

(V) Similarly. □

References

1. M. Fragoulopoulou, *On strong spectrally bounded lmc $*$-algebras*, Math. Nachr. **134** (1987), 161–166.
2. G. Lassner, *Über Realisierungen gewisser $*$-algebren*, Math. Nachr. **52** (1972), 161-166.
3. A. Mallios, *Topological Algebras*, Selected Topics. North-Holland, Amsterdam, (1986).
4. E. A. Michael, *Locally multiplicatively convex topological algebras*, Mem. Amer. Math. Soc. **11** (1952).
5. C. A. Rickart, *General Theory of Banach Algebras*. Van Nostrand, New York, (1960).
6. Z. Sebestyén, *Every C^*-seminorm is automatically submultiplicative*, Per. Math. Hungarica **10** (1979) 1–8.
7. Y. Tsertos, *Geometry of Topological Algebras*, Doctoral Dissertation, English summary. Univ. of Athens, Greece (1988).

Functional Analysis, Approximation Theory
and Numerical Analysis
Ed. John M. Rassias

LANDAU'S TYPE INEQUALITIES

JOHN M. RASSIAS

The National University of Athens, Pedagogical Department,
4, Agamemnonos Str., Aghia Paraskevi, Attikis 15342, Greece

Abstract

Let X be a complex Banach space, and let $t \to T(t)$ ($\|T(t)\| \leq 1, t \geq 0$) be a strongly continuous contraction semigroup (on X) with infinitesimal generator A. In this paper I prove that

$$\|Ax\|^4 \leq \frac{1024}{3}\|x\|^3\|A^4x\|, \quad \|A^2x\|^4 \leq \frac{10^4}{9}\|x\|^2\|A^4x\|^2, \quad \|A^3x\|^4 \leq 192\|x\|\,\|A^4x\|^3$$

hold for every $x \in D(A^4)$. Inequalities are established also for uniformly bounded strongly continuous semigroups, groups and cosine functions.

1. Introduction.

Edmund Landau (1913) [6] initiated the following *extremum problem*: The sharp inequality between the supremum-norms of derivatives of twice differentiable functions f such that

$$\|f'\|^2 \leq 4\|f\|\,\|f''\| \tag{+}$$

holds with norm referring to the space $C[0,\infty]$.

Then R. R. Kallman and G.-C. Rota (1970) [3] found the more general result that inequality

$$\|Ax\|^2 \leq 4\|x\|\,\|A^2x\| \tag{1}$$

holds for every $x \in D(A^2)$, and A the infinitesimal generator (i.e., the strong right derivative of T at zero) of $t \to T(t)(t \geq 0)$: a semigroup of linear contractions on a complex Banach space X

Besides Z. Ditzian (1975) [1] achieved the better inequality

$$\|Ax\|^2 \leq 2\|x\| \, \|A^2 x\| \tag{2}$$

for every $x \in D(A^2)$, where A is the infinitesimal generator of a group $t \to T(t)$ ($\|T(t)\| = 1, t \in \mathbb{R}$) of linear isometries on X.

Moreover H. Kraljević and S. Kurepa (1970) [4] established the even shaper inequality

$$\|Ax\|^2 \leq \frac{4}{3}\|x\| \, \|A^2 x\| \tag{3}$$

for every $x \in D(A^2)$, and A the infinitesimal generator (i.e., the strong right second derivative of T at zero) of $t \to T(t)(t \geq 0)$: a strongly continuous cosine function of linear contractions on X. Therefore the best *Landau's type constant* is $\frac{4}{3}$ (for cosine functions).

The above-mentioned inequality (1)–(3) were extended by H. Kraljević and J. Pečarić (1990) [5] so that new *Landau's type inequalities* hold. In particular, they proved that

$$\|Ax\|^3 \leq \frac{243}{8}\|x\|^2\|A^3 x\|, \qquad \|A^2 x\|^3 \leq 24\|x\| \, \|A^3 x\|^2 \tag{1$'$}$$

hold for every $x \in D(A^3)$, where A is the infinitesimal generator of a strongly continuous contraction semigroup on X, Besides they obtained the analogous but better inequalities

$$\|Ax\|^3 \leq \frac{9}{8}\|x\|^2\|A^3 x\|, \qquad \|A^2 x\|^3 \leq 3\|x\| \, \|A^3 x\|^2 \tag{2$'$}$$

hold for every $x \in D(A^3)$, where A is the infinitesimal generator of a strongly continuous contraction group on X. Moreover they got the set of analogous inequalities

$$\|Ax\|^3 \leq \frac{81}{40}\|x\|^2\|A^3 x\|, \qquad \|A^2 x\|^3 \leq \frac{72}{25}\|x\| \, \|A^3 x\|^2 \tag{3$'$}$$

for every $x \in D(A^3)$, where A is the infinitesimal generator of a strongly continuous cosine function on X.

In this paper, I extended above inequalities (1′)–(3′) so that other Landau's inequalities hold for every $x \in D(A^4)$, where A is infinitesimal generator of a uniformly bounded continuous semigroup (resp. group, or cosine function).

2. Semigroups

Let $t \to T(t)$ be uniformly bounded ($\|T(t)\| \leq M < \infty, t \geq 0$) strongly continuous semigroup of linear operators on X with infinitesimal generator A, such that $T(0) = I$ (:= Identity) in $B(X)$:= the Banach algebra of bounded linear operators on X, $\lim_{t \downarrow 0} T(t)x = x$, for every x, and

$$Ax = \lim_{t \downarrow 0} \frac{T(t) - I}{t} x \qquad (= T'(0)x) \tag{4}$$

for every x in a linear subspace $D(A)$ (:= Domain of A), dense in X, [2].

For every $x \in D(A)$, I have the formula

$$T(t)x = x + \int_0^t T(u)Axdu\ . \tag{5}$$

Using integration by parts, I get the formula

$$\int_0^t \left(\int_0^u TvA^2xdv \right) du = \int_0^t (t-u)TuA^2xdu\ . \tag{6}$$

Employing (6) and iterating (5), I find for every $x \in D(A^2)$ that

$$T(t)x = x + tAx + \int_0^t (t-u)TuA^2xdu\ . \tag{5'}$$

Similarly iterating (5′), I obtain for every $x \in D(A^4)$ that

$$T(t)x = x + tAx + \frac{t^2}{2}A^2x + \frac{t^3}{6}A^3x + \frac{1}{6}\int_0^t (t-u)^3TuA^4xdu\ . \tag{5''}$$

Theorem 1. *Let $t \to T(t)$ be a uniformly bounded ($\|T(t)\| \leq M < \infty, t \geq 0$) strongly continuous semigroup of linear operators on a complex Banach space X with infinitesimal generator A, such that $A^4x \neq 0$. Then the following inequalities*

$$\|Ax\| \le \left[M\frac{((ts)^2+(sr)^2+(rt)^2)+s^2(rt-sr-st)}{tsr(t-s)(s-r)}+\frac{ts+sr+rt}{tsr}\right]\|x\|$$
$$+M\frac{tsr}{24}\|A^4x\|\,, \tag{7}$$

$$\|A^2x\| \le 2\left[M\frac{(tr^2+sr^2+t^2s+t^2r)+s(rt-s^2)}{tsr(t-s)(s-r)}+\frac{t+s+r}{tsr}\right]\|x\|$$
$$+M\frac{ts+sr+rt}{12}\|A^4x\|\,, \tag{7'}$$

$$\|A^3x\| \le 6\left[M\frac{(ts+sr+rt)-s^2}{tsr(t-s)(s-r)}+\frac{1}{tsr}\right]\|x\|+M\frac{t+s+r}{4}\|A^4x\|\,, \tag{7''}$$

hold for every $x \in D(A^4)$ *and for every* $t,s,r \in \mathbb{R}^+ = (0,\infty)$, $0<t<s<r$.

Theorem 2. *Let* $t \to T(t)$ *be a uniformly bounded* ($\|T(t)\| \le M < \infty, t \ge 0$) *strongly continuous semigroup of linear operators on a complex Banach space* X *with infinitesimal generator* A, *such that* $A^4x \ne 0$. *Then the following inequalities*

$$\|Ax\|^4 \le \frac{32}{81}Mg_1(m_1,m_2)\|x\|^3\|A^4x\|\,, \tag{8}$$
$$\|A^2x\|^4 \le \frac{4}{9}M^2g_2(m_1,m_2)\|x\|^2\|A^4x\|^2\,, \tag{8'}$$
$$\|A^3x\|^4 \le \frac{8}{9}M^3g_3(m_1,m_2)\|x\|\,\|A^4x\|^3\,, \tag{8''}$$

hold for every $x \in D(A^4)$, *and for some* $m_1,m_2 \in \mathbb{R}^+$, $m_2>m_1>1$, *where*

$$g_1(m_1,m_2) = (m_1m_2)\times$$
$$\left[M\frac{(m_1^2+(m_1m_2)^2+m_2^2)+m_1^2(m_2-m_1m_2-m_1)}{m_1m_2(m_1-1)(m_2-m_1)}+\frac{m_1+m_1m_2+m_2}{m_1m_2}\right]^3,$$

$$g_2(m_1,m_2) = (m_1+m_1m_2+m_2)^2\times$$
$$\left[M\frac{(m_2^2+m_1m_2^2+m_1+m_2)+m_1(m_2-m_1^2)}{m_1m_2(m_1-1)(m_2-m_1)}+\frac{1+m_1+m_2}{m_1m_2}\right]^2,$$

$$g_3(m_1,m_2) = (1+m_1+m_2)^3\left[M\frac{(m_1+m_1m_2+m_2)-m_1^2}{m_1m_2(m_1-1)(m_2-m_1)}+\frac{1}{m_1m_2}\right].$$

Theorem 3. *Let $t \to T(t)$ be a strongly continuous contraction ($\|T(t)\| \leq 1, t \geq 0$) semigroup of linear operators on a complex Banach space X with infinitesimal generator A, such that $A^4x \neq 0$. Then the following inequalities*

$$\|Ax\|^4 \leq \frac{1024}{3}\|x\|^3\|A^4x\| , \tag{9}$$

$$\|A^2x\|^4 \leq \frac{10^4}{9}\|x\|^2\|A^4x\|^2 , \tag{9'}$$

$$\|A^3x\|^4 \leq 192\|x\| \; \|A^4x\|^3 , \tag{9''}$$

hold for every $x \in D(A^4)$.

Proof of Theorem 1. In fact, formula (5″) yield system

$$\left.\begin{aligned} 6tAx + 3t^2A^2x + t^3A^3x &= 6T(t)x - 6x - \int_0^t (t-u)^3T(u)A^4x du \\ 6sAx + 3s^2A^2x + s^3A^3x &= 6T(s)x - 6x - \int_0^s (s-u)^3T(u)A^4x du \\ 6rAx + 3r^2A^2x + r^3A^3x &= 6T(r)x - 6x - \int_0^r (r-u)^3T(u)A^4x du \end{aligned}\right\} . \tag{10}$$

The coefficient determinant D of system (10) is

$$D = 18tsr(t-s)(s-r)(r-t) . \tag{11}$$

It is clear that D is positive because of the hypothesis: $0 < t < s < r$. Therefore there is a unique solution of system (10) of the following form

$$\begin{aligned} Ax = \Big[&\frac{(sr)^2(r-s)T(t)x - (tr)^2(r-t)T(s)x + (ts)^2(s-t)T(r)x}{tsr(t-s)(s-r)(r-t)} \\ &- \frac{ts+sr+rt}{tsr}x \Big] - \int_0^r K_1(t,s,r;u)T(u)A^4x du , \end{aligned} \tag{12}$$

$$\begin{aligned} A^2x = 2\Big[&\frac{-(sr)(r^2-s^2)T(t)x + (tr)(r^2-t^2)T(s)x - (ts)(s^2-t^2)T(r)x}{tsr(t-s)(s-r)(r-t)} \\ &+ \frac{t+s+r}{tsr}x \Big] + \int_0^r K_2(t,s,r;u)T(u)A^4x du , \end{aligned} \tag{12'}$$

$$\begin{aligned} A^3x = 6 &\left[\frac{(sr)(r-s)T(t)x - (tr)(r-t)T(s)x + (ts)(s-t)T(r)x}{tsr(t-s)(s-r)(r-t)} - \frac{1}{tsr}x \right] \\ &- \int_0^r K_3(t,s,r;u)T(u)A^4x du , \end{aligned} \tag{12''}$$

where

$$K_1 = \begin{cases} \dfrac{(sr)^2(r-s)(t-u)^3 - (tr)^2(r-t)(s-u)^3 + (ts)^2(s-t)(r-u)^3}{6tsr(t-s)(s-r)(r-t)}, & 0 \le u \le t \\ \dfrac{-(tr)^2(r-t)(s-u)^3 + (ts)^2(s-t)(r-u)^3}{6tsr(t-s)(s-r)(r-t)}, & t \le u \le s, \\ \dfrac{(ts)^2(s-t)(r-u)^3}{6tsr(t-s)(s-r)(r-t)}, & s \le u \le r \end{cases}$$

$$K_2 = \begin{cases} \dfrac{(sr)(r^2-s^2)(t-u)^3 - (tr)(r^2-t^2)(s-u)^3 + (ts)(s^2-t^2)(r-u)^3}{3tsr(t-s)(s-r)(r-t)}, & 0 \le u \le t \\ \dfrac{-(tr)(r^2-t^2)(s-u)^3 + (ts)(s^2-t^2)(r-u)^3}{3tsr(t-s)(s-r)(r-t)}, & t \le u \le s, \\ \dfrac{(ts)(s^2-t^2)(r-u)^3}{3tsr(t-s)(s-r)(r-t)}, & s \le u \le r \end{cases}$$

$$K_3 = \begin{cases} \dfrac{(sr)(r-s)(t-u)^3 - (tr)(r-t)(s-u)^3 + (ts)(s-t)(r-u)^3}{tsr(t-s)(s-r)(r-t)}, & 0 \le u \le t \\ \dfrac{-(tr)(r-t)(s-u)^3 + (ts)(s-t)(r-u)^3}{tsr(t-s)(s-r)(r-t)}, & t \le u \le s, \\ \dfrac{(ts)(s-t)(r-u)^3}{tsr(t-s)(s-r)(r-t)}, & s \le u \le r. \end{cases}$$

It is obvious that $K_i = K_i(t,s,r;u) \ge 0$, $i = 1,2,3$, for every $u \in [0,r]$ $(0 < t < s < r)$, and that the following equalities

$$\int_0^r K_1 du = \frac{tsr}{24}, \qquad \int_0^r K_2 du = \frac{ts+sr+rt}{12}, \qquad \int_0^r K_3 du = \frac{t+s+r}{4}, \tag{13}$$

hold. Note that (12)–(12″) hold because the identities

$$\left.\begin{aligned}&(sr)^2(r-s)-(tr)^2(r-t)+(ts)^2(s-t)\\&=(t-s)(s-r)(r-t)(ts+sr+rt)\ ,\\&-(sr)(r^2-s^2)+(tr)(r^2-t^2)-(ts)(s^2-t^2)\\&=-(t-s)(s-r)(r-t)(t+s+r)\ ,\\&(sr)(r-s)-(tr)(r-t)+(ts)(s-t)\\&=(t-s)(s-r)(r-t)\end{aligned}\right\}$$

hold.

Therefore from formulas (12)–(12″), (13), and triangle inequality, I get inequalities (7)–(7″). This completes the proof of Theorem 1.

Proof of Theorem 2. Setting

$$s=m_1t,\quad r=m_2t,\quad m_2>m_1>1,\quad t>0 \tag{15}$$

in (7)–(7″), I obtain the following inequalities

$$\|Ax\|\le a_1\frac{1}{t}+b_1t^3,\quad \|A^2x\|\le a_2\frac{1}{t^2}+b_2t^2,\quad \|A^3x\|\le a_3\frac{1}{t^3}+b_3t, \tag{16}$$

where

$$\begin{aligned}a_1&=\left[M\frac{(m_1^2+(m_1m_2)^2+m_2^2)+m_1^2(m_2-m_1m_2-m_1)}{m_1m_2(m_1-1)(m_2-m_1)}\right.\\&\qquad\left.+\frac{m_1+m_1m_2+m_2}{m_1m_2}\right]\|x\|\ ,\\b_1&=M\frac{m_1m_2}{24}\|A^4x\|\ ,\\a_2&=2\left[M\frac{(m_2^2+m_1m_2^2+m_1+m_2)+m_1(m_2-m_1^2)}{m_1m_2(m_1-1)(m_2-m_1)}+\frac{1+m_1+m_2}{m_1m_2}\right]\|x\|\ ,\\b_2&=M\frac{m_1+m_1m_2+m_2}{12}\|A^4x\|\ ,\\a_3&=6\left[M\frac{(m_1+m_1m_2+m_2)-m_1^2}{m_1m_2(m_1-1)(m_2-m_1)}+\frac{1}{m_1m_2}\right]\|x\|\ ,\\b_3&=M\frac{1+m_1+m_2}{4}\|A^4x\|\ .\end{aligned}$$

Minimizing the right-hand side functions of t of (16), I get the sharper inequalities.

$$\|Ax\|^4 \leq \frac{256}{27}a_1^3b_1, \qquad \|A^2x\|^4 \leq 16a_2^2b_2^2, \qquad \|A^3x\|^4 \leq \frac{256}{27}a_3b_3^3 . \tag{17}$$

But

$$a_1^3b_1 = \frac{M}{24}g_1(m_1,m_2)\|x\|^3\|A^4x\|, a_2^2b_2^2 = \frac{M^2}{36}g_2(m_1,m_2)\|x\|^2\|A^4x\|^2,$$

and

$$a_3b_3^3 = \frac{3M^3}{32}g_3(m_1,m_2)\|x\| \; \|A^4x\|^3 .$$

Therefore from (15)–(17), I obtain inequalities (8)–(8″). This completes the proof of Theorem 2.

Proof of Theorem 3. Taking $M = 1$, I have

$$g_1(m_1,m_2) = 8g_1^+(m_1,m_2), \qquad g_2(m_1,m_2) = 4g_2^+(m_1,m_2),$$
$$g_3(m_1,m_2) = 2g_3^+(m_1,m_2),$$

where

$$g_1^+(m_1,m_2) = (m_1m_2)\left[\frac{m_1(1+m_2+m_2^2-m_1-m_1m_2)}{m_2(m_1-1)(m_2-m_1)}\right]^3,$$

$$g_2^+(m_1,m_2) = (m_1+m_1m_2+m_2)^2\left[\frac{1+m_2+m_2^2-m_1^2}{m_2(m_1-1)(m_2-m_1)}\right]^2,$$

$$g_3^+(m_1,m_2) = (1+m_1+m_2)^3\left[\frac{1+m_2-m_1}{m_2(m_1-1)(m_2-m_1)}\right].$$

Hence inequalities (8)–(8″) are written, as follows:

$$\|Ax\|^4 \leq \frac{256}{81}g_1^+(m_1,m_2)\|x\|^3\|A^4x\| , \tag{18}$$

$$\|A^2x\|^4 \leq \frac{16}{9}g_2^+(m_1,m_2)\|x\|^2\|A^4x\|^2 , \tag{18′}$$

$$\|A^3x\|^4 \leq \frac{16}{9}g_3^+(m_1,m_2)\|x\| \; \|A^4x\|^3 , \tag{18″}$$

for some $m_1, m_2 \in \mathbb{R}^+$: $m_2 > m_1 > 1$.

All functions $g_i^+ = g_i^+(m_1, m_2)$, $i = 1, 2, 3$, attain their minimum at the same m_1, m_2: $m_1 = 2 + \sqrt{2}$, $m_2 = 3 + 2\sqrt{2}$, so that

$$\min g_1^+(m_1, m_2) = 108 = \min g_3^+(m_1, m_2), \qquad \min g_2^+(m_1, m_2) = 625 \ . \quad (19)$$

Therefore inequalities (18)–(18″) and minima (19) yield the even sharper inequalities (9)–(9″). This completes the proof of Theorem 3.

3. Groups

Let $t \to T(t)$ be a uniformly bounded ($\|T(t)\| \leq M < \infty$, $t \in \mathbb{R} = (-\infty, \infty)$) strongly continuous group of linear operators on X with infinitesimal generator A. It is clear that analogous inequalities (as those in the aforementioned Theorems 1–3) hold for every $t, s, r \in \mathbb{R}^- = (-\infty, 0)$, $t < s < r < 0$.

Case I: $s < 0 < t < r$.

Denote

$$s = m_1 t, \qquad r = m_2 t, \qquad m_1 < 0, \quad m_2 > 1, \quad t > 0 \, , \qquad (15')$$

and

$$x_1 = 6tAx, \qquad x_2 = 3t^2A^2x, \qquad x_3 = t^3A^3x \, , \qquad (20)$$

as well as

$$a = 6T(t)x - 6x - \int_0^t (t-u)^3 T(u) A^4 x du,$$

$$b = 6T(m_1 t)x - 6x - \int_0^{m_1 t} (m_1 t - u)^3 T(u) A^4 x du$$

$$\left(= 6T(m_1 t)x - 6x - m_1^4 \int_0^t (t-u)^3 T(m_1 u) A^4 x du \right),$$

$$c = 6T(m_2 t)x - 6x - \int_0^{m_2 t} (m_2 t - u)^3 T(u) A^4 x du \, .$$

Then system (10) takes the following form:

$$x_1 + x_2 + x_3 = a, \qquad m_1 x_1 + m_1^2 x_2 + m_1^3 x_3 = b, \qquad m_2 x_1 + m_2^2 x_2 + m_2^3 x_3 = c \, . \qquad (10')$$

Solving system (10′), I find the unique solution

$$x_1 = \frac{(m_1m_2)^2(m_2-m_1)a - m_2^2(m_2-1)b + m_1^2(m_1-1)c}{m_1m_2(m_1-1)(m_2-1)(m_2-m_1)} \tag{21}$$

$$x_2 = \frac{-(m_1m_2)(m_2^2-m_1^2)a + m_2(m_2^2-1)b - m_1(m_1^2-1)c}{m_1m_2(m_1-1)(m_2-1)(m_2-m_1)} \tag{21′}$$

$$x_3 = \frac{(m_1m_2)(m_2-m_1)a - m_2(m_2-1)b + m_1(m_1-1)c}{m_1m_2(m_1-1)(m_2-1)(m_2-m_1)} \tag{21″}$$

Theorem 4. *Let $t \to T(t)$ be a uniformly bounded ($\|T(t)\| \le M < \infty, t \in \mathbb{R}$) strongly continuous group of linear operators on complex Banach space X with infinitesimal generator A, such that $A^4x \neq 0$. Then*

$$\|Ax\| \le \left[M\frac{(m_1m_2)^2+m_1^2+m_2^2-m_1+m_1m_2-m_2}{m_1m_2(m_1-1)(m_2-1)} + \frac{m_1+m_1m_2+m_2}{m_1m_2}\right] \times \|x\|\frac{1}{t} + M\frac{(m_1m_2)(1+m_1-m_1m_2+m_2)}{24(m_1-1)(m_2-1)}\|A^4x\|t^3 \,, \tag{22}$$

$$\|A^2x\| \le 2(M+1)\left(-\frac{1+m_1+m_2}{m_1m_2}\right)\|x\|\frac{1}{t^2} + \frac{M}{12}(-(m_1+m_1m_2+m_2))\|A^4x\|t^2 \,, \tag{22′}$$

$$\|A^3x\| \le 6\left[M\frac{m_1+m_1m_2+m_2-m_2^2}{m_1m_2(m_2-1)(m_2-m_1)} - \frac{1}{m_1m_2}\right]\|x\|\frac{1}{t^3} + M\frac{-m_1^2m_2-m_1m_2+m_1m_2^2+(m_1m_2)^2+m_2^2+m_2^4}{4m_2(m_2-1)(m_2-m_1)}\|A^4x\|t \,, \tag{22″}$$

hold for every $x \in D(A^4)$, for every $t \in \mathbb{R}^+$, and for some $m_1 \in \mathbb{R}^-$, $m_2 \in \mathbb{R}^+$,

$$-(m_2+1) < m_1 < -\frac{m_2}{m_2+1}, \qquad m_2 > 1 \,.$$

Theorem 5. *Let $t \to T(t)$ be a contraction ($\|T(t)\| \le 1, t \in \mathbb{R}$) strongly continuous group of linear operators on complex Banach space x with infinitesimal generator A, such that $A^4x \neq 0$. Then the following inequalities*

$$\|Ax\|^4 \leq \frac{256}{81} f_1(m_1, m_2) \|x\|^3 \|A^4x\| \,, \tag{23}$$

$$\|A^2x\|^4 \leq \frac{16}{9} f_2(m_1, m_2) \|x\|^2 \|A^4x\|^2 \,, \tag{23'}$$

$$\|A^3x\|^4 \leq \frac{16}{9} f_3(m_1, m_2) \|x\| \, \|A^4x\|^3 \,, \tag{23''}$$

hold for every $x \in D(A^4)$, and for some $m_1 \in \mathbb{R}^-$, $m_2 \in \mathbb{R}^+$,

$$-(m_2+1) < m_1 < -\frac{m_2}{m_2+1}, \qquad m_2 > 1 \,,$$

where

$$f_1(m_1, m_2) = \left(\frac{m_1 m_2}{(m_1-1)(m_2-1)} \right)^4 (1 + m_1 - m_1 m_2 + m_2) \,,$$

$$f_2(m_1, m_2) = \left(\frac{1 + m_1 + m_2}{m_1 m_2} \right)^2 (m_1 + m_1 m_2 + m_2)^2 \,,$$

$$f_3(m_1, m_2) = \frac{(1 + m_1 - m_2)(-m_1^2 m_2 - m_1 m_2 + m_1 m_2^2 + (m_1 m_2)^2 + m_2^2 + m_2^4)^3}{(m_1 m_2^3)((m_2-1)(m_2-m_1))^4} \,.$$

Theorem 6. *Let $t \to T(t)$ be a strongly continuous contraction ($\|T(t)\| \leq 1, t \in \mathbb{R}$) group of linear operators on a complex Banach space X with infinitesimal generator A, such that $A^4x \neq 0$. Then the following inequalities*

$$\|Ax\|^4 \leq 10 \left(\frac{5}{6} \right)^4 \|x\|^3 \|A^4x\| \,, \tag{24}$$

$$\|A^2x\|^4 \leq \frac{16}{9} \|x\|^2 \|A^4x\|^2 \,, \tag{24'}$$

$$\|A^3x\|^4 \leq 10 \frac{5^4 13^3}{2^5 3^6} \|x\| \, \|A^4x\|^3 \,, \tag{24''}$$

hold for every $x \in D(A^4)$.

Theorem 7. *Let $t \to T(t)$ be a strongly continuous contraction ($\|T(t)\| \leq 1, t \in \mathbb{R}$) group of linear operators on a complex Banach space X with infinitesimal generator A, such that $A^4x \neq 0$. Then the following inequalities*

$$\|Ax\|^4 \le \frac{32}{81}\frac{m_{20}^5}{(m_{20}-1)^4}\|x\|^3\|A^4x\| \,, \tag{25}$$

$$\|A^2x\|^4 \le \frac{16}{9}\|x\|^2\|A^4x\|^2 \,, \tag{25'}$$

$$\|A^3x\|^4 \le \frac{16}{9}\frac{m_{20}^4(1+m_{20}^2)^3}{(m_{20}^2-1)^4}\|x\|\,\|A^4x\|^3 \,, \tag{25''}$$

hold for every $x \in D(A^4)$, *where*

$$m_{20} = \sqrt{\frac{7+\sqrt{57}}{2}} \,.$$

Proof of Theorem 4. In fact, from (20) and (21)–(21″), I get

$$Ax = \frac{x_1}{6t} =$$

$$\left(\frac{(m_1m_2)^2(m_2-m_1)T(t)x - m_2^2(m_2-1)T(m_1t)x + m_1^2(m_1-1)T(m_2t)x}{m_1m_2(m_1-1)(m_2-1)(m_2-m_1)}\right.$$

$$\left. - \frac{(m_1m_2)^2(m_2-m_1) - m_2^2(m_2-1) + m_1^2(m_1-1)}{m_1m_2(m_1-1)(m_2-1)(m_2-m_1)}x\right)\frac{1}{t}$$

$$- \left[\int_0^t (t-u)^3\left(\frac{(m_1m_2)^2(m_2-m_1)T(u) - m_2^2(m_2-1)m_1^4T(m_1u)}{6m_1m_2(m_1-1)(m_2-1)(m_2-m_1)}\right.\right.$$

$$\left.\left. + \frac{m_1^2(m_1-1)m_2^4T(m_2u)}{6m_1m_2(m_1-1)(m_2-1)(m_2-m_1)}\right)A^4x\,du\right]\frac{1}{t} \,, \tag{26}$$

$$A^2x = \frac{x_2}{3t^2} =$$

$$2\left(\frac{-m_1m_2(m_2^2-m_1^2)T(t)x + m_2(m_2^2-1)T(m_1t)x - m_1(m_1^2-1)T(m_2t)x}{m_1m_2(m_1-1)(m_2-1)(m_2-m_1)}\right.$$

$$\left. - \frac{-m_1m_2(m_2^2-m_1^2) + m_2(m_2^2-1) - m_1(m_1^2-1)}{m_1m_2(m_1-1)(m_2-1)(m_2-m_1)}x\right)\frac{1}{t^2}$$

$$- \left[\int_0^t (t-u)^3\left(\frac{-m_1m_2(m_2^2-m_1^2)T(u) + m_2(m_2^2-1)m_1^4T(m_1u)}{3m_1m_2(m_1-1)(m_2-1)(m_2-m_1)}\right.\right.$$

$$\left.\left. - \frac{m_1(m_1^2-1)m_2^4T(m_2u)}{3m_1m_2(m_1-1)(m_2-1)(m_2-m_1)}\right)A^4x\,du\right]\frac{1}{t^2}, \tag{26'}$$

$$A^3x = \frac{x_3}{t^3} =$$

$$6\left(\frac{m_1m_2(m_2-m_1)T(t)x - m_2(m_2-1)T(m_1t)x + m_1(m_1-1)T(m_2t)x}{m_1m_2(m_1-1)(m_2-1)(m_2-m_1)}\right.$$

$$\left. - \frac{m_1m_2(m_2-m_1) - m_2(m_2-1) + m_1(m_1-1)}{m_1m_2(m_1-1)(m_2-1)(m_2-m_1)}x\right)\frac{1}{t^3}$$

$$- \left[\int_0^t (t-u)^3\left(\frac{m_1m_2(m_2-m_1)T(u) - m_2(m_2-1)m_1^4T(m_1u)}{m_1m_2(m_1-1)(m_2-1)(m_2-m_1)}\right.\right.$$

$$\left.\left. + \frac{m_1(m_1-1)m_2^4T(m_2u)}{m_1m_2(m_1-1)(m_2-1)(m_2-m_1)}\right)A^4x\,du\right]\frac{1}{t^3}. \tag{26''}$$

But it is clear that the following identities:

$$(m_1m_2)^2(m_2-m_1) - m_2^2(m_2-1) + m_1^2(m_1-1)$$
$$= (m_1-1)(m_2-1)(m_2-m_1)(m_1+m_1m_2+m_2),$$

$$-m_1m_2(m_2^2-m_1^2) + m_2(m_2^2-1) - m_1(m_1^2-1)$$
$$= -(m_1-1)(m_2-1)(m_2-m_1)(1+m_1+m_2),$$

$$- m_1m_2(m_2^2 - m_1^2) + m_2(m_2^2 - 1)m_1^4 - m_1(m_1^2 - 1)m_2^4$$
$$= -m_1m_2(m_1 - 1)(m_2 - 1)(m_2 - m_1)(m_1 + m_1m_2 + m_2)$$

hold. Applying these identities and formulas (26)–(26″), I obtain inequalities (22)–(22″). This completes the proof of Theorem 4.

Proof of Theorem 5. In fact, from inequalities (22)–(22″) I get

$$\|Ax\| \le a_1^+\frac{1}{t} + b_1^+t^3, \quad \|A^2x\| \le a_2^+\frac{1}{t^2} + b_2^+t^2, \quad \|A^3x\| \le a_3^+\frac{1}{t^3} + b_3^+t, \tag{27}$$

where

$$a_1^+ = 2\frac{m_1m_2}{(m_1-1)(m_2-1)}\|x\| ,$$
$$b_1^+ = \frac{(m_1m_2)(1+m_1-m_1m_2+m_2)}{24(m_1-1)(m_2-1)}\|A^4x\| ,$$
$$a_2^+ = (-4)\frac{1+m_1+m_2}{m_1m_2}\|x\| ,$$
$$b_2^+ = -\frac{1}{12}(m_1+m_1m_2+m_2)\|A^4x\| ,$$
$$a_3^+ = 12\frac{1+m_1-m_2}{m_1(m_2-1)(m_2-m_1)}\|x\| ,$$
$$b_3^+ = \frac{-m_1^2m_2 - m_1m_2 + m_1m_2^2 + (m_1m_2)^2 + m_2^2 + m_2^4}{4m_2(m_2-1)(m_2-m_1)}\|A^4x\| ,$$

and new identities:

$$m_1m_2(m_1 - m_2) + m_2(m_2 - 1) + m_1(m_1 - 1)$$
$$= (m_1 - 1)(m_1 + m_1m_2 + m_2 - m_2^2),$$
$$m_1m_2(m_1 - m_2) + m_2(m_2 - 1)m_1^4 + m_1(m_1 - 1)m_2^4$$
$$= m_1(m_1 - 1)(-m_1^2m_2 - m_1m_2 + m_1m_2^2 + (m_1m_2)^2 + m_2^2 + m_2^4) .$$

Minimizing the right-hand side of (27), I find

$$\|Ax\|^4 \le \frac{256}{27}(a_1^+)^3(b_1^+), \qquad \|A^2x\|^4 \le 16(a_2^+)^2(b_2^+)^2,$$
$$\|A^3x\|^4 \le \frac{256}{27}(a_3^+)(b_3^+)^3, \tag{28}$$

where

$$(a_1^+)^3(b_1^+) = \frac{8}{24} f_1(m_1, m_2),$$

$$(a_2^+)^2(b_2^+)^2 = \frac{1}{9} f_2(m_1, m_2),$$

$$(a_3^+)(b_3^+)^3 = \frac{3}{16} f_3(m_1, m_2).$$

Therefore inequalities (28) yield inequalities (23)–(23″). This completes the proof of Theorem 5.

Proof of Theorem 6. Setting $m_1 = -1$, I get $\min f_1(-1, m_2) = 10(\frac{5}{8})^4$ at $m_2 = 5$. Hence $f_2(-1, 5) = 1$, $f_3(-1, 5) = 5^4(13)^3/2^9 3^4$. Therefore from formulas (23)–(23″) and $m_1 = -1$, $m_2 = 5$, I get inequalities (24)–(24″). This completes the proof of Theorem 6.

Proof of Theorem 7. In fact, $\min f_3(-1, m_2) = f_3(-1, m_{20}) = \frac{m_{20}^4(1+m_{20}^2)^3}{(m_{20}^2-1)^4}$, where $m_{20} = \sqrt{\frac{7+\sqrt{57}}{2}}$ (= root (> 1) of equation: $m_2^4 - 7m_2^2 - 2 = 0$). Besides I have

$$f_1(-1, m_{20}) = \frac{1}{8} \frac{m_{20}^5}{(m_{20} - 1)^4}.$$

Therefore from formulas (24)–(24″), and $m_1 = -1$, $m_2 = m_{20}$, I obtain inequalities (25)–(25″). This completes the proof of Theorem 7.

Case II: $r < s < 0 < t$.

Denote $s = m_1 t$, $r = m_2 t$, $m_2 = m < -1$, $m_1 = -1$, $t > 0$. From (21)–(21″), I have

$$x_1 = \frac{m^2(m+1)a - m^2(m-1)b - 2c}{2m(m^2-1)},$$

$$x_2 = \frac{a+b}{2},$$

$$x_3 = \frac{-m(m+1)a - m(m-1)b + 2c}{2m(m^2-1)}. \tag{29}$$

Therefore from (29) and (20), I obtain

$$Ax = \left(\frac{-m^2(m+1)T(t)x + m^2(m-1)T(-t)x + 2T(mt)x}{2m(1-m^2)} - \frac{1}{m}x\right)\frac{1}{t}, \tag{30}$$

$$+ \left(\int_0^t (t-u)^3 \frac{m^2(m+1)T(u) - m^2(m-1)T(-u) - 2m^4T(mu)}{12m(1-m^2)} A^4x\,du\right)\frac{1}{t},$$

$$A^2x = 2\left(\frac{T(t)x + T(-t)x}{2} - x\right)\frac{1}{t^2} - \left(\int_0^t (t-u)^3 \frac{T(u)+T(-u)}{6} A^4x\,du\right)\frac{1}{t^2}, \tag{30'}$$

$$A^3x = 6\left(\frac{m(m+1)T(t)x + m(m-1)T(-t)x - 2T(mt)x}{2m(1-m^2)} + \frac{1}{m}x\right)\frac{1}{t^3} \tag{30''}$$

$$+ \left(\int_0^t (t-u)^3 \frac{-m(m+1)T(u) - m(m-1)T(-u) + 2m^4T(mu)}{2m(1-m^2)} A^4x\,du\right)\frac{1}{t^3}.$$

Thus from formulas (30)–(30″) I get

$$\|Ax\| \le \frac{m}{m+1}\|x\|\frac{1}{t} + \left(-\frac{m^2}{24(m+1)}\right)\|A^4x\|t^3, \tag{31}$$

$$\|A^2x\| \le 4\|x\|\frac{1}{t^2} + \frac{1}{12}\|A^4x\|t^2, \tag{31'}$$

$$\|A^3x\| \le 12\frac{m}{1-m^2}\|x\|\frac{1}{t^3} + \frac{1}{4}\frac{m(1+m^2)}{1-m^2}\|A^4x\|t. \tag{31''}$$

Minimizing the right-hand side of inequalities (31)–(31″), I find

$$\|Ax\|^4 \le \frac{32}{81}h_1(m)\|x\|^3\|A^4x\|, \tag{32}$$

$$\|A^2x\|^4 \le \frac{16}{9}\|x\|^2\|A^4x\|^2, \tag{32'}$$

$$\|A^3x\|^4 \le \frac{16}{9}h_3(m)\|x\|\,\|A^4x\|^3, \tag{32''}$$

where

$$h_1(m) = -\frac{m^5}{(m+1)^4}, \quad h_3(m) = \frac{m^4(1+m^2)^3}{(1-m^2)^4}, \quad m < -1.$$

First, minimizing $h_1(m)$, $m < -1$, I get $m = -5$. Then inequalities (32)–(32″) (with $m = -5$) are the same as the inequalities (24)–(24″). Finally, minimizing $h_3(m)$, $m < -1$, I obtain $m = m_0 = -\sqrt{(7+\sqrt{57})/2}$. Then inequalities (32)–(32″) (with $m = m_0$) are the same as the inequalities (25)–(25″).

4. Cosine functions

Let $t \to T(t)$ $(t \geq 0)$ be a uniformly bounded $(\|T(t)\| \leq M < \infty, t \geq 0)$ strongly continuous cosine function with infinitesimal operator A, such that $T(0) = I$ (:=identity) in $B(X)$, $\lim_{t \downarrow 0} T(t)x = x$, $\forall x$, and A is defined as the strong second derivatives of T at zero:

$$Ax = T''(0)x \tag{33}$$

for every x in a linear subspace $D(A)$, which is dense in X, [5].

For every $x \in D(A)$, I have the formula

$$T(t)x = x + \int_0^t (t-u)T(u)Axdu \ . \tag{34}$$

Using integration by parts, I get from (34) the formula

$$\int_0^t (t-u)\left(\int_0^u (u-v)f(v)dv\right) du = \frac{1}{6}\int_0^t (t-v)^3 f(v)dv \ , \tag{35}$$

where $f(v) = TvA^2x$. Note the *Leibniz formula*:

$$\frac{d}{du}\left(\int_0^u (u-v)^n f(v)dv\right) = n\left(\int_0^u (u-v)^{n-1} f(v)dv\right) \ . \tag{36}$$

Employing (35)–(36) and iterating (34), I find for every $x \in D(A^2)$ that

$$T(t)x = x + \frac{t^2}{2!}Ax + \frac{1}{3!}\int_0^t (t-u)^3 T(u)A^2xdx \ . \tag{34'}$$

Similarly iterating (34′) I obtain for every $x \in D(A^4)$ that

$$T(t)x = x + \frac{t^2}{2!}Ax + \frac{t^4}{4!}A^2x + \frac{t^6}{6!}A^3x + \frac{1}{7!}\int_0^t (t-u)^7 T(u)A^4xdu \ . \tag{34''}$$

Theorem 8. *Let $t \to T(t)$ be a uniformly bounded $(\|T(t)\| \leq M < \infty, t \geq 0)$ strongly continuous cosine function on a complex Banach space X with infinitesimal generator A, such that $A^4x \neq 0$. Then the following inequalities*

$$\|Ax\| \le 2\left[M\frac{(ts)^2+(sr)^2+(rt)^2+s^2(rt-sr-st)}{tsr(t-s)(s-r)}+\frac{ts+sr+rt}{tsr}\right]\|x\|$$

$$+M\frac{tsr}{20160}\|A^4x\|\,, \tag{37}$$

$$\|A^2x\| \le 24\left[M\frac{(ts^2+sr^2+t^2s+t^2r)+s(rt-s^2)}{tsr(t-s)(s-r)}+\frac{t+s+r}{tsr}\right]\|x\|$$

$$+M\frac{ts+sr+rt}{1680}\|A^4x\|\,, \tag{37'}$$

$$\|A^3x\| \le 720\left[M\frac{(ts+sr+rt)-s^2}{tsr(t-s)(s-r)}+\frac{1}{tsr}\right]\|x\|$$

$$+M\frac{t+s+r}{56}\|A^4x\|\,, \tag{37''}$$

hold for every $x \in D(A^4)$, *and for every* $t, s, r \in \mathbb{R}^+$, $0 < t < s < r$.

Theorem 9. *Let* $t \to T(t)$ *be a uniformly bounded* ($\|T(t)\| \le M < \infty, t \ge 0$) *strongly continuous cosine function on a complex Banach space* X *with infinitesimal generator* A, *such that* $A^4x \ne 0$. *Then the following inequalities*

$$\|Ax\|^4 \le \frac{32}{8505}Mg_1(m_1,m_2)\|x\|^3\|A^4x\|, \tag{38}$$

$$\|A^2x\|^4 \le \frac{4}{1225}M^2g_2(m_1,m_2)\|x\|^2\|A^4x\|^2\,, \tag{38'}$$

$$\|A^3x\|^4 \le \frac{40}{1029}\,M^3g_3(m_1,m_2)\|x\|\,\|A^4x\|^3\,, \tag{38''}$$

hold for every $x \in D(A^4)$, *and for some* $m_1, m_2 \in \mathbb{R}^+$, $m_2 > m_1 > 1$, *where* $g_i = g_i(m_1,m_2)$ *are the same as those* g_i, $i = 1,2,3$, *in Theorem 2.*

Theorem 10. *Let* $t \to T(t)$ *be a strongly continuous contractions* ($\|T(t)\| \le 1, t \ge 0$) *cosine function on a complex Banach space* X *with infinitesimal generator* A, *such that* $A^4x \ne 0$. *Then the following inequalities*

$$\|Ax\|^4 \le \frac{1024}{315}\|x\|^3\|A^4x\| , \tag{39}$$

$$\|A^2x\|^4 \le \frac{400}{49}\|x\|^2\|A^4x\|^2 , \tag{39'}$$

$$\|A^3x\|^4 \le \frac{2880}{343}\|x\|\,\|A^4x\|^3 , \tag{39''}$$

hold for every $x \in D(A^4)$.

Proof of Theorem 8. In fact, setting $t(> 0)$ instead of t^2 in (34″), I get

$$T(\sqrt{t})x = x + \frac{t}{2}Ax + \frac{t^2}{24}A^2x + \frac{t^3}{720}A^3x + \frac{1}{5040}\int_0^{\sqrt{t}} (\sqrt{t} - u)^7 T(u)A^4x du . \tag{34'''}$$

Formulas (34‴) yields

$$\left.\begin{aligned} &2520tAx + 210t^2A^2x + 7t^3A^3x \\ &\qquad = 5040T(\sqrt{t})x - 5040x - \int_0^{\sqrt{t}} (\sqrt{t} - u)^7 T(u)A^4x du \\ &2520sAx + 210s^2A^2x + 7s^3A^3x \\ &\qquad = 5040T(\sqrt{s})x - 5040x - \int_0^{\sqrt{s}} (\sqrt{s} - u)^7 T(u)A^4x du \\ &2520rAx + 210r^2A^2x + 7r^3A^3x \\ &\qquad = 5040T(\sqrt{r})x - 5040x - \int_0^{\sqrt{r}} (\sqrt{r} - u)^7 T(u)A^4x du \end{aligned}\right\} . \tag{10'}$$

The coefficient determinant D^+ of system (10′) is

$$D^+ = 3704400tsr(t-s)(s-r)(r-t)(= D/205800) . \tag{11'}$$

It is clear that $D^+ > 0$ because $0 < t < s < r$. Therefore there is a unique solution of system (10′) of the form

$$Ax = 2\left(\frac{(sr)^2(r-s)T(\sqrt{t})x-(tr)^2(r-t)T(\sqrt{s})x+(ts)^2(s-t)T(\sqrt{r})x}{tsr(t-s)(s-r)(r-t)}\right.$$

$$\left.-\frac{ts+sr+rt}{tsr}x\right)-\int_0^{\sqrt{r}}K_1^+(t,s,r;u)T(u)A^4x\,du\ , \tag{40}$$

$$A^2x = 24\left(\frac{-sr(r^2-s^2)T(\sqrt{t})x+tr(r^2-t^2)T(\sqrt{s})x-ts(s^2-t^2)T(\sqrt{r})x}{tsr(t-s)(s-r)(r-t)}\right.$$

$$\left.+\frac{t+s+r}{tsr}x\right)+\int_0^{\sqrt{r}}K_2^+(t,s,r;u)T(u)A^4x\,du\ , \tag{40'}$$

$$A^3x = 720\left(\frac{sr(r-s)T(\sqrt{t})x-tr(r-t)T(\sqrt{s})x+ts(s-t)T(\sqrt{r})x}{tsr(t-s)(s-r)(r-t)}\right.$$

$$\left.-\frac{1}{tsr}x\right)-\int_0^{\sqrt{r}}K_3^+(t,s,r;u)T(u)A^4x\,du\ , \tag{40''}$$

where

$$K_1^+ = \begin{cases} \dfrac{(sr)^2(r-s)(\sqrt{t}-u)^7-(tr)^2(r-t)(\sqrt{s}-u)^7+(ts)^2(s-t)(\sqrt{r}-u)^7}{2520tsr(t-s)(s-r)(r-t)}, \\ \qquad\qquad\qquad\qquad\qquad\qquad 0 \le u \le \sqrt{t}, \\ \dfrac{-(tr)^2(r-t)(\sqrt{s}-u)^7+(ts)^2(s-t)(\sqrt{r}-u)^7}{2520tsr(t-s)(s-r)(r-t)}, \quad \sqrt{t} \le u \le \sqrt{s}, \\ \dfrac{(ts)^2(s-t)(\sqrt{r}-u)^7}{2520tsr(t-s)(s-r)(r-t)}, \quad \sqrt{s} \le u \le \sqrt{r}, \end{cases}$$

$$K_2^+ = \begin{cases} \dfrac{sr(r^2-s^2)(\sqrt{t}-u)^7-(tr)(r^2-t^2)(\sqrt{s}-u)^7+ts(s^2-t^2)(\sqrt{r}-u)^7}{210tsr(t-s)(s-r)(r-t)}, \\ \qquad\qquad\qquad\qquad\qquad\qquad 0 \le u \le \sqrt{t}, \\ \dfrac{-tr(r^2-t^2)(\sqrt{s}-u)^7+ts(s^2-t^2)(\sqrt{r}-u)^7}{210tsr(t-s)(s-r)(r-t)}, \quad \sqrt{t} \le u \le \sqrt{s}, \\ \dfrac{ts(s^2-t^2)(\sqrt{r}-u)^7}{210tsr(t-s)(s-r)(r-t)}, \quad \sqrt{s} \le u \le \sqrt{r}, \end{cases}$$

$$K_3^+ = \begin{cases} \dfrac{sr(r-s)(\sqrt{t}-u)^7 - tr(r-t)(\sqrt{s}-u)^7 + ts(s-t)(\sqrt{r}-u)^7}{7tsr(t-s)(s-r)(r-t)}, & \\ & 0 \le u \le \sqrt{t}, \\ \dfrac{-tr(r-t)(\sqrt{s}-u)^7 + ts(s-t)(\sqrt{r}-u)^7}{7tsr(t-s)(s-r)(r-t)}, & \sqrt{t} \le u \le \sqrt{s}, \\ \dfrac{ts(s-t)(\sqrt{r}-u)^7}{7tsr(t-s)(s-r)(r-t)}, & \sqrt{s} \le u \le \sqrt{r}. \end{cases}$$

Note that $K_i^+ \ge 0$, $i = 1,2,3$, for every $u \in [0, \sqrt{r}]$ $(0 < t < s < r)$. Besides

$$\int_0^{\sqrt{r}} K_1^+ du = \frac{tsr}{20160}, \qquad \int_0^{\sqrt{r}} K_2^+ du = \frac{ts + sr + rt}{1680},$$
$$\int_0^{\sqrt{r}} K_3^+ du = \frac{t + s + r}{56}. \tag{13'}$$

From (40)–(40″), triangle inequality, (13′), (15) and similarly as in the previous section "on semigroups", I get inequalities (37)–(37″). This completes the proof of Theorem 8.

Proof of Theorem 9. From (15) and (37)–(37″) and similar calculations as in section "on semigroups", I find inequalities (38)–(38″). This completes the proof of Theorem 9.

Proof of Theorem 10. Setting $M = 1$, using (38)–(38″), and minimizing $g_i(m_1, m_2)$, $i = 1,2,3$, as in section "on semigroups", I obtain inequalities (39)–(39″). This completes the proof of Theorem 10.

References

1. Z. Ditzian, *Some remarks on inequalities of Landau and Kolmogorov*, Aequat. Math. 12 (1975), 145–151.
2. E. Hille and R. S. Phillips, *Functional Analysis and Semigroups*, Amer. Math. Soc. Colloq. Publ. Vol. 31, AMS, Providence, RI, 1957.
3. R. R. Kallman and G.-C. Rota, *On the inequality* $\|f'\|^2 \le 4\|f\|\|f''\|$, In *Inequalities II* (ed. O. Shisha), Acad. Press, New York, London, 1970, 187–192.
4. Kraljević and S. Kurepa, *Semigroups on Banach Spaces*, Glas. Mat. 5 (1970), 109–117.
5. H. Kraljević and J. Pečarić, *Some Landau's type inequalities for infinitesimal generators*, Aequat. Math. 40 (1990), 147–153.
6. E. Landau, *Einige Ungleichungen für zweimal differentzierbare Funktionen*, Proc. London Math. Soc., (2) **13** (1913), 43–49.

Functional Analysis, Approximation Theory
and Numerical Analysis
Ed. John M. Rassias

GENERALIZED LANDAU'S TYPE INEQUALITIES

JOHN M. RASSIAS

The National University of Athens, Pedagogical Department,
4, Agamemnonos Str., Aghia Paraskevi, Attikis 15342, Greece

Abstract

Let X be a complex Banach space, and let $t \to T(t)$ ($\|T(t)\| \leq 1, t \geq 0$) be a strongly continuous contraction semigroup (on X) with infinitesimal generator A. In this paper, I prove that

$$\|Ax\|^5 \leq \frac{5^9}{3\ 2^7}\|x\|^4\|A^5x\|, \qquad \|A^2x\|^5 \leq \frac{2\ 5^8}{3^2}\|x\|^3\|A^5x\|^2 \ ,$$
$$\|A^3x\|^5 \leq \frac{3^2\ 5^2\ 7^5}{2^6}\|x\|^2\|A^5x\|^3, \qquad \|A^4x\|^5 \leq 3\ 5\ 2^7\|x\|\ \|A^5x\|^4 \ ,$$

hold for every $x \in D(A^5)$. Inequalities are established also for uniformly bounded strongly continuous semigroups, groups and cosine functions.

1. Introduction

Edmund Landau (1913) [6] initiated the following *extremum problem*: The sharp inequality between the supremum-norms of derivatives of twice differentiable functions f such that

$$\|f'\|^2 \leq 4\|f\|\ \|f''\| \qquad (+)$$

holds with norm referring to the space $C[0, \infty]$.

Then R. R. Kallman and G. C. Rota (1970) [3] found the more general result that inequality

$$\|Ax\|^2 \leq 4\|x\|\,\|A^2x\| \tag{1}$$

holds for every $x \in D(A^2)$, and A the infinitesimal generator (i.e., the strong right derivative of T at zero) of $t \to T(t)(t \geq 0)$: a semigroup of linear contractions on a complex Banach space X.

Besides Z. Ditzian (1975) [1] achieved the better inequality

$$\|Ax\|^2 \leq 2\|x\|\,\|A^2x\| \tag{2}$$

for every $x \in D(A^2)$, where A is the infinitesimal generator of a group $t \to T(t)$ $(\|T(t)\| = 1, t \in \mathbb{R})$ of linear isometries on X.

Moreover H. Kraljević and S. Kurepa (1970) [4] established the even sharper inequality

$$\|Ax\|^2 \leq \frac{4}{3}\|x\|\,\|A^2x\| \tag{3}$$

for every $x \in D(A^2)$, and A the infinitesimal generator (i.e., the strong right second derivative of T at zero) of $t \to T(t)(t \geq 0)$: a strongly continuous cosine function of linear contractions on X. Therefore the best *Landau's type constant* is $\frac{4}{3}$ (for cosine functions).

The above-mentioned inequalities (1)–(3) were extended by H. Kraljević and J. Pečarić (1990) [5] so that new *Landau's type inequalities* hold. In particular, they proved that

$$\|Ax\|^3 \leq \frac{243}{8}\|x\|^2\|A^3x\|, \qquad \|A^2x\|^3 \leq 24\|x\|\,\|A^3x\|^2 \tag{1'}$$

hold for every $x \in D(A^3)$, where A is the infinitesimal generator of a strongly continuous contraction semigroup on X. Besides they obtained the analogous but better inequalities

$$\|Ax\|^3 \leq \frac{9}{8}\|x\|^2\|A^3x\|, \qquad \|A^2x\|^3 \leq 3\|x\|\,\|A^3x\|^2 \tag{2'}$$

hold for every $x \in D(A^3)$, where A is the infinitesimal generator of a strongly continuous contraction group on X. Moreover they got the set of analogous inequalities

$$\|Ax\|^3 \leq \frac{81}{40}\|x\|^2\|A^3x\|, \qquad \|A^2x\|^3 \leq \frac{72}{25}\|x\|\,\|A^3x\|^2 \tag{3'}$$

for every $x \in D(A^3)$, where A is the infinitesimal generator of a strongly continuous cosine function on X.

The above *Landau-Kraljević-Pečarić inequalities* (1′)–(3′) have been extended further by the author of this paper [7], for every $x \in D(A^4)$, where A is the infinitesimal generator of a uniformly bounded continuous semigroup (resp. group, or cosine function). In this paper I extend even further my results [7], for every $x \in D(A^5)$, where A is the infinitesimal generator of a uniformly bounded continuous semigroup (resp. group, or cosine function).

2. Semigroups

Let $t \to T(t)$ be a uniformly bounded ($\|T(t)\| \leq M < \infty, t \geq 0$) strongly continuous semigroup of linear operators on X with infinitesimal generator A, such that $T(0) = I$ (:= Identity) in $B(X)$:= the Banach algebra of bounded linear operators on X, $\lim_{t \downarrow 0} T(t)x = x$, for every x, and

$$Ax = \lim_{t \downarrow 0} \frac{T(t) - I}{t} x \qquad (= T'(0)x) \tag{4}$$

for every x in a linear subspace $D(A)$ (:= Domain of A), dense in X, [2].

For every $x \in D(A)$, I have the formula

$$T(t)x = x + \int_0^t T(u)Ax\,du\,. \tag{5}$$

Using integration by parts, I get the formula

$$\int_0^t \left(\int_0^u TvA^2x\,dv \right) du = \int_0^t (t-u)TuA^2x\,du\,. \tag{6}$$

Employing (6) and iterating (5), I find for every $x \in D(A^2)$ that

$$T(t)x = x + tAx + \int_0^t (t-u)TuA^2x\,du\,. \tag{5′}$$

Similarly iterating (5′); I obtain for every $x \in D(A^5)$ that

$$T(t)x = x + tAx + \frac{t^2}{2}A^2x + \frac{t^3}{6}A^3x + \frac{t^4}{24}A^4x + \frac{1}{24}\int_0^t (t-u)^4 T(u)A^5x\,du\,. \tag{5″}$$

Theorem 1. *Let $t \to T(t)$ be a uniformly bounded ($\|T(t)\| \leq M < \infty, t \geq 0$) strongly continuous semigroup of linear operators on a complex Banach space*

X with infinitesimal generator A, such that $A^5x \neq 0$. *Then the following inequalities*

$$\|Ax\| \le \Big\{ M[(srp)^2(s-r)(r-p)(p-s) + (trp)^2(t-r)(r-p)(p-t) + (tsp)^2(t-s)(s-p)(p-t) + (tsr)^2(t-s)(s-r)(r-t)]/ tsrp(t-s)(t-r)(t-p)(s-r)(s-p)(r-p) + [(tsr+tsp+trp+srp)/tsrp] \Big\} \|x\| + M\frac{tsrp}{120}\|A^5x\| , \tag{7}$$

$$\|A^2x\| \le 2\Big\{ M[srp(s-r)(r-p)(p-s)(sr+rp+ps) + trp(t-r)(r-p)(p-t)(tr+rp+pt) + tsp(t-s)(s-p)(p-t)(ts+sp+pt) + tsr(t-s)(s-r)(r-t)(ts+sr+rt)]/ tsrp(t-s)(t-r)(t-p)(s-r)(s-p)(r-p) + [(ts+tr+tp+sr+sp+rp)/tsrp] \Big\} \|x\| + M\frac{tsr+tsp+trp+srp}{60}\|A^5x\| , \tag{7'}$$

$$\|A^3x\| \le 6\Big\{ M[srp(s-r)(r-p)(p-s)(s+r+p) + trp(t-r)(r-p)(p-t)(t+r+p) + tsp(t-s)(s-p)(p-t)(t+s+p) + tsr(t-s)(s-r)(r-t)(t+s+r)]/ tsrp(t-s)(t-r)(t-p)(s-r)(s-p)(r-p) + \frac{t+s+r+p}{tsrp} \Big\} \|x\| + M\frac{ts+tr+tp+sr+sp+rp}{20}\|A^5x\| , \tag{7''}$$

$$\|A^4x\| \le 24\Big\{ M[srp(s-r)(r-p)(p-s) + trp(t-r)(r-p)(p-t) + tsp(t-s)(s-p)(p-t) + tsr(t-s)(s-r)(r-t)]/ tsrp(t-s)(t-r)(t-p)(s-r)(s-p)(r-p) + \frac{1}{tsrp} \Big\} \|x\| + M\frac{t+s+r+p}{5}\|A^5x\| , \tag{7'''}$$

hold for every $x \in D(A^5)$, *and for every* $t,s,r,p \in \mathbb{R}^+ = (0,\infty)$, $0 < t < s < r < p$.

Theorem 2. *Let $t \to T(t)$ be a uniformly bounded ($\|T(t)\| \leq M < \infty, t \geq 0$) strongly continuous semigroup of linear operators on a complex Banach space X with infinitesimal generator A, such that $A^5 x \neq 0$. Then the following inequalities*

$$\|Ax\|^5 \leq \frac{625}{6144} M g_1(m_1, m_2, m_3) \|x\|^4 \|A^5 x\| , \tag{8}$$

$$\|A^2 x\|^5 \leq \frac{125}{1944} M^2 g_2(m_1, m_2, m_3) \|x\|^3 \|A^5 x\|^2 , \tag{8'}$$

$$\|A^3 x\|^5 \leq \frac{225}{1728} M^3 g_3(m_1, m_2, m_3) \|x\|^2 \|A^5 x\|^3 , \tag{8''}$$

$$\|A^4 x\|^5 \leq \frac{15}{32} M^4 g_4(m_1, m_2, m_3) \|x\| \, \|A^5 x\|^4 , \tag{8'''}$$

hold for every $x \in D(A^5)$, and for some $m_1, m_2, m_3 \in \mathbb{R}^+$, $m_3 > m_2 > m_1 > 1$, where

$$\begin{aligned}
g_1(m_1, m_2, m_3) &= m_1 m_2 m_3 \\
&\Big\{ M[(m_1 m_2 m_3)^2 (m_1 - m_2)(m_2 - m_3)(m_3 - m_1) \\
&+ (m_2 m_3)^2 (1 - m_2)(m_2 - m_3)(m_3 - 1) \\
&+ (m_1 m_3)^2 (1 - m_1)(m_1 - m_3)(m_3 - 1) \\
&+ (m_1 m_2)^2 (1 - m_1)(m_1 - m_2)(m_2 - 1)] / \\
&m_1 m_2 m_3 (1 - m_1)(1 - m_2)(1 - m_3)(m_1 - m_2)(m_1 - m_3)(m_2 - m_3) \\
&+ \frac{m_1 m_2 + m_1 m_3 + m_2 m_3 + m_1 m_2 m_3}{m_1 m_2 m_3} \Big\}^4 ,
\end{aligned}$$

$$\begin{aligned}
g_2(m_1, m_2, m_3) &= (m_1 m_2 + m_1 m_3 + m_2 m_3 + m_1 m_2 m_3)^2 \\
&\Big\{ M[(m_1 m_2 m_3)(m_1 - m_2)(m_2 - m_3)(m_3 - m_1)(m_1 m_2 + m_2 m_3 + m_3 m_1) \\
&+ m_2 m_3 (1 - m_2)(m_2 - m_3)(m_3 - 1)(m_2 + m_2 m_3 + m_3) \\
&+ m_1 m_3 (1 - m_1)(m_1 - m_3)(m_3 - 1)(m_1 + m_1 m_3 + m_3) \\
&+ m_1 m_2 (1 - m_1)(m_1 - m_2)(m_2 - 1)(m_1 + m_1 m_2 + m_2)] / \\
&+ m_1 m_2 m_3 (1 - m_1)(1 - m_2)(1 - m_3)(m_1 - m_2)(m_1 - m_3)(m_2 - m_3) \\
&+ \frac{m_1 + m_2 + m_3 + m_1 m_2 + m_1 m_3 + m_2 m_3}{m_1 m_2 m_3} \Big\}^3 ,
\end{aligned}$$

$$g_3(m_1, m_2, m_3) = (m_1 + m_2 + m_3 + m_1m_2 + m_1m_3 + m_2m_3)^3$$
$$\Big\{ M[(m_1m_2m_3)(m_1 - m_2)(m_2 - m_3)(m_3 - m_1)(m_1 + m_2 + m_3)$$
$$+ m_2m_3(1 - m_2)(m_2 - m_3)(m_3 - 1)(1 + m_2 + m_3)$$
$$+ m_1m_3(1 - m_1)(m_1 - m_3)(m_3 - 1)(1 + m_1 + m_3)$$
$$+ m_1m_2(1 - m_1)(m_1 - m_2)(m_2 - 1)(1 + m_1 + m_2)]/$$
$$m_1m_2m_3(1 - m_1)(1 - m_2)(1 - m_3)(m_1 - m_2)(m_1 - m_3)(m_2 - m_3)$$
$$+ \frac{1 + m_1 + m_2 + m_3}{m_1m_2m_3} \Big\}^2 ,$$

$$g_4(m_1, m_2, m_3) = (1 + m_1 + m_2 + m_3)^4$$
$$\Big\{ M[(m_1m_2m_3)(m_1 - m_2)(m_2 - m_3)(m_3 - m_1)$$
$$+ m_2m_3(1 - m_2)(m_2 - m_3)(m_3 - 1)$$
$$+ m_1m_3(1 - m_1)(m_1 - m_3)(m_3 - 1)$$
$$+ m_1m_2(1 - m_1)(m_1 - m_2)(m_2 - 1)]/$$
$$m_1m_2m_3(1 - m_1)(1 - m_2)(1 - m_3)(m_1 - m_2)(m_1 - m_3)(m_2 - m_3)$$
$$+ \frac{1}{m_1m_2m_3} \Big\} .$$

Theorem 3. *Let $t \to T(t)$ be a strongly continuous contraction ($\|T(t)\| \leq 1, t \geq 0$) semigroup of linear operators on a complex Banach space X with infinitesimal generator A, such that $A^5x \neq 0$. Then the following inequalities*

$$\|Ax\|^5 \leq \frac{5^9}{3\ 2^7} \|x\|^4 \|A^5x\| , \tag{9}$$

$$\|A^2x\|^5 \leq \frac{2\ 5^8}{3^2} \|x\|^3 \|A^5x\|^2 , \tag{9'}$$

$$\|A^3x\|^5 \leq \frac{3^2\ 5^2\ 7^5}{2^6} \|x\|^2 \|A^5x\|^3 , \tag{9''}$$

$$\|A^4x\|^5 \leq 3\ 5\ 2^7 \|x\|\ \|A^5x\|^4 , \tag{9'''}$$

hold for every $x \in D(A^5)$.

Proof of Theorem 1. In fact, formula (5″) yields system

$$\left.\begin{aligned}
&24tAx + 12t^2A^2x + 4t^3A^3x + t^4A^4x \\
&\qquad = 24T(t)x - 24x - \int_0^t (t-u)^4 T(u)A^5 x\,du \\
&24sAx + 12s^2A^2x + 4s^3A^3x + s^4A^4x \\
&\qquad = 24T(s)x - 24x - \int_0^s (s-u)^4 T(u)A^5 x\,du \\
&24rAx + 12r^2A^2x + 4r^3A^3x + r^4A^4x \\
&\qquad = 24T(r)x - 24x - \int_0^r (r-u)^4 T(u)A^5 x\,du \\
&24pAx + 12p^2A^2x + 4p^3A^3x + p^4A^4x \\
&\qquad = 24T(p)x - 24x - \int_0^p (p-u)^4 T(u)A^5 x\,du
\end{aligned}\right\} . \tag{10}$$

The coefficient determinant D of system (10) is

$$D = 1152tsrp(t-s)(t-r)(t-p)(s-r)(s-p)(r-p) . \tag{11}$$

It is clear that D is positive because of the hypothesis: $0 < t < s < r < p$. Therefore there is a unique solution of system (10) of the following form:

$$\begin{aligned}
Ax = {} & [(srp)^2(s-r)(r-p)(p-s)T(t)x \\
& - (trp)^2(t-r)(r-p)(p-t)T(s)x \\
& + (tsp)^2(t-s)(s-p)(p-t)T(r)x \\
& - (tsr)^2(t-s)(s-r)(r-t)T(p)x]/ \\
& tsrp(t-s)(t-r)(t-p)(s-r)(s-p)(r-p) \\
& - \frac{tsr + tsp + trp + srp}{tsrp}x - \int_0^p K_1(t,s,r,p;u)T(u)A^5 x\,du ,
\end{aligned} \tag{12}$$

$$\begin{aligned}
A^2x = {} & -2[srp(s-r)(r-p)(p-s)(sr+rp+ps)T(t)x \\
& - trp(t-r)(r-p)(p-t)(tr+rp+pt)T(s)x \\
& + tsp(t-s)(s-p)(p-t)(ts+sp+pt)T(r)x \\
& - tsr(t-s)(s-r)(r-t)(ts+sr+rt)T(p)x]/ \\
& tsrp(t-s)(t-r)(t-p)(s-r)(s-p)(r-p) \\
& + 2\frac{ts+tr+tp+sr+sp+rp}{tsrp}x + \int_0^p K_2(t,s,r,p;u)T(u)A^5 x\,du ,
\end{aligned} \tag{12'}$$

$$
\begin{aligned}
A^3x = 6[&srp(s-r)(r-p)(p-s)(s+r+p)T(t)x \\
&- trp(t-r)(r-p)(p-t)(t+r+p)T(s)x \\
&+ tsp(t-s)(s-p)(p-t)(t+s+p)T(r)x \\
&- tsr(t-s)(s-r)(r-t)(t+s+r)T(p)x]/ \\
&tsrp(t-s)(t-r)(t-p)(s-r)(s-p)(r-p) \\
&- 6\frac{t+s+r+p}{tsrp}x - \int_0^p K_3(t,s,r,p;u)T(u)A^5x\,du\ ,
\end{aligned}
\tag{12''}
$$

$$
\begin{aligned}
A^4x = -24[&srp(s-r)(r-p)(p-s)T(t)x \\
&- trp(t-r)(r-p)(p-t)T(s)x \\
&+ tsp(t-s)(s-p)(p-t)T(r)x \\
&- tsr(t-s)(s-r)(r-t)T(p)x]/ \\
&tsrp(t-s)(t-r)(t-p)(s-r)(s-p)(r-p) \\
&+ 24\frac{1}{tsrp}x + \int_0^p K_4(t,s,r,p;u)T(u)A^5x\,du\ ,
\end{aligned}
\tag{12'''}
$$

where

$$
K_i(t,s,r,p;u) = -\hat{K}_i(t,s,r,p;u)/D, \qquad i = 1,2,3,4\ ,
$$

and

$$
\hat{K}_1 = \begin{cases}
48[-(srp)^2(s-r)(r-p)(p-s)(t-u)^4 \\
\quad +(trp)^2(t-r)(r-p)(p-t)(s-u)^4 \\
\quad -(tsp)^2(t-s)(s-p)(p-t)(r-u)^4 \\
\quad +(tsr)^2(t-s)(s-r)(r-t)(p-u)^4], & 0 \le u \le t \\
48[(trp)^2(t-r)(r-p)(p-t)(s-u)^4 \\
\quad -(tsp)^2(t-s)(s-p)(p-t)(r-u)^4 \\
\quad +(tsr)^2(t-s)(s-r)(r-t)(p-u)^4], & t \le u \le s \\
48[-(tsp)^2(t-s)(s-p)(p-t)(r-u)^4 \\
\quad +(tsr)^2(t-s)(s-r)(r-t)(p-u)^4], & s \le u \le r \\
48[(tsr)^2(t-s)(s-r)(r-t)(p-u)^4], & r \le u \le p,
\end{cases}
$$

$$\hat{K}_2 = \begin{cases} 96[-srp(s-r)(r-p)(p-s)(sr+rp+ps)(t-u)^4 \\ \quad +trp(t-r)(r-p)(p-t)(tr+rp+pt)(s-u)^4 \\ \quad -tsp(t-s)(s-p)(p-t)(ts+sp+pt)(r-u)^4 \\ \quad +tsr(t-s)(s-r)(r-t)(ts+sr+rt)(p-u)^4], & 0 \le u \le t \\ \\ 96[trp(t-r)(r-p)(p-t)(tr+rp+pt)(s-u)^4 \\ \quad -tsp(t-s)(s-p)(p-t)(ts+sp+pt)(r-u)^4 \\ \quad +tsr(t-s)(s-r)(r-t)(ts+sr+rt)(p-u)^4], & t \le u \le s \\ \\ 96[-tsp(t-s)(s-p)(p-t)(ts+sp+pt)(r-u)^4 \\ \quad +tsr(t-s)(s-r)(r-t)(ts+sr+rt)(p-u)^4], & s \le u \le r \\ \\ 96[tsr(t-s)(s-r)(r-t)(ts+sr+rt)(p-u)^4], & r \le u \le p, \end{cases}$$

$$\hat{K}_3 = \begin{cases} 288[-srp(s-r)(r-p)(p-s)(s+r+p)(t-u)^4 \\ \quad +trp(t-r)(r-p)(p-t)(t+r+p)(s-u)^4 \\ \quad -tsp(t-s)(s-p)(p-t)(t+s+p)(r-u)^4 \\ \quad +tsr(t-s)(s-r)(r-t)(t+s+r)(p-u)^4], & 0 \le u \le t \\ \\ 288[trp(t-r)(r-p)(p-t)(t+r+p)(s-u)^4 \\ \quad -tsp(t-s)(s-p)(p-t)(t+s+p)(r-u)^4 \\ \quad +tsr(t-s)(s-r)(r-t)(t+s+r)(p-u)^4], & t \le u \le s \\ \\ 288[-tsp(t-s)(s-p)(p-t)(t+s+p)(r-u)^4 \\ \quad +tsr(t-s)(s-r)(r-t)(t+s+r)(p-u)^4], & s \le u \le r \\ \\ 288[tsr(t-s)(s-r)(r-t)(t+s+r)(p-u)^4], & r \le u \le p, \end{cases}$$

$$\hat{K}_4 = \begin{cases} 1152[-srp(s-r)(r-p)(p-s)(t-u)^4 \\ \quad +trp(t-r)(r-p)(p-t)(s-u)^4 \\ \quad -tsp(t-s)(s-p)(p-t)(r-u)^4 \\ \quad +tsr(t-s)(s-r)(r-t)(p-u)^4], & 0 \le u \le t \\ 1152[trp(t-r)(r-p)(p-t)(s-u)^4 \\ \quad -tsp(t-s)(s-p)(p-t)(r-u)^4 \\ \quad +tsr(t-s)(s-r)(r-t)(p-u)^4], & t \le u \le s \\ 1152[-tsp(t-s)(s-p)(p-t)(r-u)^4 \\ \quad +tsr(t-s)(s-r)(r-t)(p-u)^4], & s \le u \le r \\ 1152[tsr(t-s)(s-r)(r-t)(p-u)^4], & r \le u \le p\,. \end{cases}$$

It is obvious that $K_i = K_i(t,s,r,p;u) \ge 0$, $i = 1,2,3,4$, for every $u \in [0,p]$ $(0 < t < s < r < p)$, and that the following equalities

$$\int_0^p K_1 du = -\frac{tsrp}{120}, \qquad \int_0^p K_2 du = -\frac{tsr+tsp+trp+rsp}{60}, \tag{13}$$

$$\int_0^p K_3 du = -\frac{ts+tr+tp+sr+sp+rp}{20}, \qquad \int_0^p K_4 du = -\frac{t+s+r+p}{5}, \tag{13'}$$

hold. Note that (13)–(13′) hold because identities

$$\begin{aligned} &(r-p)[(s-r)(p-s)t^3 - (t-r)(p-t)s^3] \\ &+(t-s)[(s-p)(p-t)r^3 - (s-r)(r-t)p^3] \\ &= -D/1152tsrp\,, \end{aligned} \tag{14}$$

$$\begin{aligned} &(r-p)[(s-r)(p-s)(sr+rp+ps)t^4 - (t-r)(p-t)(tr+rp+pt)s^4] \\ &+(t-s)[(s-p)(p-t)(ts+sp+pt)r^4 - (s-r)(r-t)(ts+sr+rt)p^4] \\ &= -D(tsr+tsp+trp+srp)/1152tsrp\,, \end{aligned} \tag{14'}$$

$$\begin{aligned} &(r-p)[(s-r)(p-s)(s+r+p)t^4 - (t-r)(p-t)(t+r+p)s^4] \\ &+(t-s)[(s-p)(p-t)(t+s+p)r^4 - (s-r)(r-t)(t+s+r)p^4] \\ &= -D(ts+tr+tp+sr+sp+rp)/1152tsrp\,, \end{aligned} \tag{14''}$$

$$\begin{aligned} &(r-p)[(s-r)(p-s)t^4 - (t-r)(p-t)s^4] \\ &+(t-s)[(s-p)(p-t)r^4 - (s-r)(r-t)p^4] \\ &= -D(t+s+r+p)/1152tsrp\,, \end{aligned} \tag{14'''}$$

hold.

Therefore from formulas (12)–(12‴), (13)–(13′), and triangle inequality, I get inequalities (7)–(7‴). This completes the proof of Theorem 1.

Proof of Theorem 2. Setting

$$s = m_1 t, \quad r = m_2 t, \quad p = m_3 t, \qquad m_3 > m_2 > m_1 > 1, \quad t > 0 \tag{15}$$

in (7)–(7‴), I obtain the following inequalities

$$\|Ax\| \le a_1 \frac{1}{t} + b_1 t^4, \qquad \|A^2 x\| \le a_2 \frac{1}{t^2} + b_2 t^3 \,, \tag{16}$$

$$\|A^3 x\| \le a_3 \frac{1}{t^3} + b_3 t^2, \qquad \|A^4 x\| \le a_4 \frac{1}{t^4} + b_4 t \,, \tag{16'}$$

where

$$\begin{aligned} a_1 = \Big\{ & M[(m_1 m_2 m_3)^2 (m_1 - m_2)(m_2 - m_3)(m_3 - m_1) \\ & + (m_2 m_3)^2 (1 - m_2)(m_2 - m_3)(m_3 - 1) \\ & + (m_1 m_3)^2 (1 - m_1)(m_1 - m_3)(m_3 - 1) \\ & + (m_1 m_2)^2 (1 - m_1)(m_1 - m_2)(m_2 - 1)] / \\ & m_1 m_2 m_3 (1 - m_1)(1 - m_2)(1 - m_3)(m_1 - m_2)(m_1 - m_3)(m_2 - m_3) \\ & + \frac{m_1 m_2 + m_1 m_3 + m_2 m_3 + m_1 m_2 m_3}{m_1 m_2 m_3} \Big\} \|x\| \,, \end{aligned}$$

$$b_1 = M \frac{m_1 m_2 m_3}{120} \|A^5 x\| \,,$$

$$\begin{aligned} a_2 = 2 \Big\{ & M[(m_1 m_2 m_3)(m_1 - m_2)(m_2 - m_3)(m_3 - m_1) \\ & \times (m_1 m_2 + m_2 m_3 + m_3 m_1) \\ & + m_2 m_3 (1 - m_2)(m_2 - m_3)(m_3 - 1)(m_2 + m_2 m_3 + m_3) \\ & + m_1 m_3 (1 - m_1)(m_1 - m_3)(m_3 - 1)(m_1 + m_1 m_3 + m_3) \\ & + m_1 m_2 (1 - m_1)(m_1 - m_2)(m_2 - 1)(m_1 + m_1 m_2 + m_2)] / \\ & m_1 m_2 m_3 (1 - m_1)(1 - m_2)(1 - m_3)(m_1 - m_2)(m_1 - m_3)(m_2 - m_3) \\ & + \frac{m_1 + m_2 + m_3 + m_1 m_2 + m_1 m_3 + m_2 m_3}{m_1 m_2 m_3} \Big\} \|x\| \,, \end{aligned}$$

$$b_2 = M \frac{m_1 m_2 + m_1 m_3 + m_2 m_3 + m_1 m_2 m_3}{60} \|A^5 x\| \,,$$

$$a_3 = 6\Big\{ M[(m_1m_2m_3)(m_1-m_2)(m_2-m_3)(m_3-m_1)(m_1+m_2+m_3)$$
$$+ m_2m_3(1-m_2)(m_2-m_3)(m_3-1)(1+m_2+m_3)$$
$$+ m_1m_3(1-m_1)(m_1-m_3)(m_3-1)(1+m_1+m_3)$$
$$+ m_1m_2(1-m_1)(m_1-m_2)(m_2-1)(1+m_1+m_2)]/$$
$$m_1m_2m_3(1-m_1)(1-m_2)(1-m_3)(m_1-m_2)(m_1-m_3)(m_2-m_3)$$
$$+ \frac{1+m_1+m_2+m_3}{m_1m_2m_3}\Big\}\|x\|\,,$$

$$b_3 = M\frac{m_1+m_2+m_3+m_1m_2+m_1m_3+m_2m_3}{20}\|A^5x\|\,,$$

$$a_4 = 24\Big\{ M[m_1m_2m_3(m_1-m_2)(m_2-m_3)(m_3-m_1)$$
$$+ m_2m_3(1-m_2)(m_2-m_3)(m_3-1)$$
$$+ m_1m_3(1-m_1)(m_1-m_3)(m_3-1)$$
$$+ m_1m_2(1-m_1)(m_1-m_2)(m_2-1)]/$$
$$m_1m_2m_3(1-m_1)(1-m_2)(1-m_3)(m_1-m_2)(m_1-m_3)(m_2-m_3)$$
$$+ \frac{1}{m_1m_2m_3}\Big\}\|x\|\,,$$

$$b_4 = M\frac{1+m_1+m_2+m_3}{5}\|A^5x\|\,.$$

Minimizing the right-hand side functions of t of (16)–(16′), I get the sharper inequalities

$$\|Ax\|^5 \le \frac{3125}{256}a_1^4b_1, \qquad \|A^2x\|^5 \le \frac{3125}{108}a_2^3b_2^2\,, \tag{17}$$

$$\|A^3x\|^5 \le \frac{3125}{108}a_3^2b_3^3, \qquad \|A^4x\|^5 \le \frac{3125}{256}a_4b_4^4\,. \tag{17′}$$

But

$$a_1^4 b_1 = \frac{M}{120} g_1(m_1, m_2, m_3) \|x\|^4 \|A^5 x\|,$$

$$a_2^3 b_2^2 = \frac{M^2}{450} g_2(m_1, m_2, m_3) \|x\|^3 \|A^5 x\|^2,$$

$$a_3^2 b_3^3 = \frac{9M^3}{2000} g_3(m_1, m_2, m_3) \|x\|^2 \|A^5 x\|^3,$$

$$a_4 b_4^4 = \frac{24M^4}{625} g_4(m_1, m_2, m_3) \|x\| \; \|A^5 x\|^4 .$$

Therefore from (15), (16)–(16′), and (17)–(17′), I obtain inequalities (8)–(8‴). This completes the proof of Theorem 2.

Proof of Theorem 3. Taking $M = 1$, I find that the functions $g_i = g_i(m_1, m_2, m_3)$, $i = 1, 2, 3, 4$, $m_3 > m_2 > m_1 > 1$, attain their minimum at

$$m_1 = \frac{5+\sqrt{5}}{2} \qquad (= 3 + \text{ "golden section number"}), \tag{18}$$

$$m_2 = \frac{7+3\sqrt{5}}{2} \qquad (= \text{ positive root of equation: } 7 - \frac{1}{x} = x), \tag{18'}$$

$$m_3 = 5 + 2\sqrt{5} \,, \tag{18''}$$

so that

$$\min g_1(m_1, m_2, m_3) = 2^4 \, 5^5, \qquad \min g_2(m_1, m_2, m_3) = 2^4 \, 3^3 \, 5^5, \tag{19}$$

$$\min g_3(m_1, m_2, m_3) = 3^3 \, 7^5, \qquad \min g_4(m_1, m_2, m_3) = 2^{12} . \tag{19'}$$

Therefore, inequalities (8)–(8‴) with $M = 1$, and minima (19)–(19′) yield (9)–(9‴). This completes the proof of Theorem 3.

3. Groups

Let $t \to T(t)$ be a uniformly bounded ($\|T(t)\| \leq M < \infty, t \in \mathbb{R} = (-\infty, \infty)$) strongly continuous group of linear operators on X with infinitesimal generator A. It is clear that analogous inequalities (to those in the above-mentioned Theorem 1–3) hold for every $t, s, r, p \in \mathbb{R}^- = (-\infty, 0)$, $t < s < r < p < 0$.

Consider: $s < p < 0 < r < t$, such that

$$s = m_1 t, \quad r = m_2 t, \quad p = m_3 t, \quad m_1 = -1, \quad m_2 = \frac{1}{2}, \quad m_3 = -\frac{1}{2}, \quad (15')$$

$$t > 0,$$

and

$$x_1 = 24tAx, \quad x_2 = 12t^2A^2x, \quad x_3 = 4t^3A^3x, \quad x_4 = t^4A^4x\,, \qquad (20)$$

as well as

$$a = 24T(t)x - 24x - \int_0^t (t-u)^4\, T(u)A^5 x du\,,$$

$$b = 24T(-t)x - 24x + \int_0^t (t-u)^4\, T(-u)A^5 x du\,,$$

$$c = 24T\left(\frac{1}{2}x\right) - 24x - \int_0^t \left(\frac{1}{2}\right)^5 (t-u)^4\, T\left(\frac{1}{2}u\right) A^5 x du\,,$$

$$d = 24T\left(-\frac{1}{2}x\right) - 24x + \int_0^t \left(\frac{1}{2}\right)^5 (t-u)^4\, T\left(-\frac{1}{2}u\right) A^5 x du\,,$$

because

$$\int_0^{nt} (nt-u)^4 T(u) A^5 x du = n^5 \int_0^t (t-u)^4 T(nu) A^5 x du\,, \qquad (20')$$

$n = \pm\frac{1}{2}$, and $n = -1$.

Then system (10) takes the following form:

$$x_1 + x_2 + x_3 + x_4 = a, \qquad -x_1 + x_2 - x_3 + x_4 = b, \qquad (10')$$

$$\frac{1}{2}x_1 + \frac{1}{4}x_2 + \frac{1}{8}x_3 + \frac{1}{16}x_4 = c, \qquad -\frac{1}{2}x_1 + \frac{1}{4}x_2 - \frac{1}{8}x_3 + \frac{1}{16}x_4 = d. \qquad (10'')$$

Solving system (10′)–(10″), I find the unique solution

$$x_1 = \frac{-a + b + 8c - 8d}{6}, \qquad x_2 = \frac{-a - b + 16c + 16d}{6},$$

$$x_3 = \frac{2a - 2b - 4c + 4d}{3}, \qquad x_4 = \frac{2a + 2b - 8c - 8d}{3}.$$

Theorem 4. *Let $t \to T(t)$ be a strongly continuous contraction ($\|T(t)\| \leq 1, t \in \mathbb{R}$) group of linear operators on a complex Banach space X with infinitesimal generator A, such that $A^5 x \neq 0$. Then the following inequalities*

$$\|Ax\|^5 \leq \frac{(15)^4}{2^{11}} \|x\|^4 \|A^5 x\| \, , \tag{21}$$

$$\|A^2 x\|^5 \leq \frac{5^3 \, (22)^2}{3^8} \|x\|^3 \|A^5 x\|^2 \, , \tag{21'}$$

$$\|A^3 x\|^5 \leq \frac{5^2 \, (17)^3}{2^8 \, 3^4} \|x\|^2 \|A^5 x\|^3 \, , \tag{21''}$$

$$\|A^4 x\|^5 \leq 5 \left(\frac{3}{2}\right)^4 \|x\| \, \|A^5 x\|^4 \, , \tag{21'''}$$

hold for every $x \in D(A^5)$.

Proof. From (15′), (20)–(20′), and the solution of system (10′)–(10″), I find the following formulas

$$\begin{aligned} Ax = {} & \left(\frac{T(t)x - T(-t)x - 8T(\frac{t}{2})x + 8T(-\frac{t}{2})x}{6} \right) \frac{1}{t} \\ & - \left(\int_0^t (t-u)^4 \frac{T(u) - T(-u) - 8T(\frac{u}{2}) + 8T(-\frac{u}{2})}{144} A^5 x du \right) \frac{1}{t} \, , \end{aligned} \tag{22}$$

$$\begin{aligned} A^2 x = {} & \left(\frac{-T(t)x - T(-t)x + 16T(\frac{t}{2})x + 16T(-\frac{t}{2})x}{3} - 30x \right) \frac{1}{t^2} \\ & - \left(\int_0^t (t-u)^4 \frac{-32T(u) + 32T(-u) - T(\frac{u}{2}) + T(-\frac{u}{2})}{2304} A^5 x du \right) \frac{1}{t^2} \, , \end{aligned} \tag{22'}$$

$$\begin{aligned} A^3 x = {} & 4 \left(T(t)x - T(-t)x - 2T\left(\frac{t}{2}\right) x + 2T\left(-\frac{t}{2}\right) x \right) \frac{1}{t^3} \\ & - \left(\int_0^t (t-u)^4 \frac{16T(u) + 16T(-u) - T(\frac{u}{2}) - T(-\frac{u}{2})}{96} A^5 x du \right) \frac{1}{t^3} \, , \end{aligned} \tag{22''}$$

$$\begin{aligned} A^4 x = {} & 16 \left(T(t)x + T(-t)x - 4T\left(\frac{t}{2}\right) x - 4T\left(-\frac{t}{2}\right) x + 6x \right) \frac{1}{t^4} \\ & - \left(\int_0^t (t-u)^4 \frac{8T(u) - 8T(-u) - T(\frac{u}{2}) + T(-\frac{u}{2})}{12} A^5 x du \right) \frac{1}{t^4} \, . \end{aligned} \tag{22'''}$$

Therefore from (22)–(22‴), the fact that $||T(t)|| \leq 1, t \in \mathbb{R}$, and the triangle inequality, I find the estimates

$$||Ax|| \leq 3||x||\frac{1}{t} + \frac{1}{40}||A^5x||t^4, \quad ||A^2x|| \leq \frac{64}{3}||x||\frac{1}{t^2} + \frac{11}{1920}||A^5x||t^3, \tag{23}$$

$$||A^3x|| \leq 24||x||\frac{1}{t^3} + \frac{17}{240}||A^5x||t^2, ||A^4x|| \leq 256||x||\frac{1}{t^4} + \frac{3}{10}||A^5x||t \; . \tag{23'}$$

From (23)–(23′) and minimization techniques, I get inequalities (21)–(21‴). This completes the proof of Theorem 4.

4. Cosine functions

Let $t \to T(t)$ $(t \geq 0)$ be a uniformly bounded $(||T(t)|| \leq M < \infty, t \geq 0)$ strongly continuous cosine function with infinitesimal operator A, such that $T(0) = I$ (:= Identity) in $B(X)$, $\lim_{t \downarrow 0} T(t)x = x$, for all x, and A is defined as the strong second derivative of T at zero:

$$Ax = T''(0)x \tag{24}$$

for every x in a linear subspace $D(A)$, which is dense in X, [5].

For every $x \in D(A)$, I have the formula

$$T(t)x = x + \int_0^t (t-u)T(u)Axdu \; . \tag{25}$$

Using integration by parts, I get from (25) the formula

$$\int_0^t (t-u)\left(\int_0^u (u-v)f(v)dv\right)du = \frac{1}{6}\int_0^t (t-u)^3 f(v)dv \; , \tag{26}$$

where $f(v) = TvA^2x$. Note the *Leibniz's formula*:

$$\frac{d}{du}\left(\int_0^u (u-v)^n f(v)dv\right) = n\left(\int_0^u (u-v)^{n-1} f(v)dv\right) \; . \tag{27}$$

Employing (26)–(27) and iterating (25), I find for every $x \in D(A^2)$ that

$$T(t)x = x + \frac{t^2}{2!}Ax + \frac{1}{3!}\int_0^t (t-u)^3 T(u)A^2xdx \; . \tag{25'}$$

Similarly iterating (25′), I obtain for every $x \in D(A^5)$ that

$$T(t)x = x + \frac{t^2}{2!}Ax + \frac{t^4}{4!}A^2x + \frac{t^6}{6!}A^3x + \frac{t^8}{8!}A^4x + \frac{1}{9!}\int_0^t (t-u)^9 T(u)A^5xdu \; . \tag{25''}$$

Theorem 5. *Let $t \to T(t)$ be a uniformly bounded ($\|T(t)\| \le M < \infty, t \ge 0$) strongly continuous cosine function on a complex Banach space X with infinitesimal generator A, such that $A^5 x \ne 0$. Then the following inequalities*

$$\begin{aligned}\|Ax\| \le (2!)\Big\{ & M[(srp)^2(s-r)(r-p)(p-s) + (trp)^2(t-r)(r-p)(p-t) \\ & + (tsp)^2(t-s)(s-p)(p-t) + (tsr)^2(t-s)(s-r)(r-t)]/ \\ & tsrp(t-s)(t-r)(t-p)(s-r)(s-p)(r-p) \\ & + \frac{tsr+tsp+trp+srp}{tsrp}\Big\}\|x\| + M\frac{2!}{10!}tsrp\|A^5x\|\,, \end{aligned} \tag{28}$$

$$\begin{aligned}\|A^2x\| \le (4!)\Big\{ & M[srp(s-r)(r-p)(p-s)(sr+rp+ps) \\ & + trp(t-r)(r-p)(p-t)(tr+rp+pt) \\ & + tsp(t-s)(s-p)(p-t)(ts+sp+pt) \\ & + tsr(t-s)(s-r)(r-t)(ts+sr+rt)]/ \\ & tsrp(t-s)(t-r)(t-p)(s-r)(s-p)(r-p) \\ & + \frac{ts+tr+tp+sr+sp+rp}{tsrp}\Big\}\|x\| \\ & + M\frac{4!}{10!}(tsr+tsp+trp+srp)\|A^5x\|\,, \end{aligned} \tag{28'}$$

$$\begin{aligned}\|A^3x\| \le (6!)\Big\{ & M[srp(s-r)(r-p)(p-s)(s+r+p) \\ & + trp(t-r)(r-p)(p-t)(t+r+p) \\ & + tsp(t-s)(s-p)(p-t)(t+s+p) \\ & + tsr(t-s)(s-r)(r-t)(t+s+r)]/ \\ & tsrp(t-s)(t-r)(t-p)(s-r)(s-p)(r-p) + \frac{t+s+r+p}{tsrp}\Big\}\|x\| \\ & + M\frac{6!}{10!}(ts+tr+tp+sr+sp+rp)\|A^5x\|\,, \end{aligned} \tag{28''}$$

$$\begin{aligned}\|A^4x\| \le (8!)\Big\{ & M[srp(s-r)(r-p)(p-s) + trp(t-r)(r-p)(p-t) \\ & + tsp(t-s)(s-p)(p-t) + tsr(t-s)(s-r)(r-t)]/ \\ & tsrp(t-s)(t-r)(t-p)(s-r)(s-p)(r-p) + \frac{1}{tsrp}\Big\}\|x\| \\ & + M\frac{8!}{10!}(t+s+r+p)\|A^5x\|\,, \end{aligned} \tag{28'''}$$

hold for every $x \in D(A^5)$, *and for every* $t, s, r, p \in \mathbb{R}^+$, $0 < t < s < r < p$.

Theorem 6. *Let* $t \to T(t)$ *be a uniformly bounded* ($\|T(t)\| \leq M < \infty, t \geq 0$) *strongly continuous cosine function on a complex Banach space* X *with infinitesimal generator* A, *such that* $A^5 x \neq 0$. *Then the following inequalities hold*

$$\|A^i x\|^5 \leq \frac{((2i)!)^5}{(10!)^i} \begin{bmatrix} 5 \\ i \end{bmatrix} M^i g_i(m_1, m_2, m_3) \|x\|^{5-i} \|A^5 x\|^i, \tag{29}$$

with $i = 1, 2, 3, 4$, $\begin{bmatrix} 5 \\ i \end{bmatrix} = \frac{5^5}{i^i (5-i)^{5-i}}$, $g_i = g_i(m_1, m_2, m_3)$ *as in Theorem 2, for every* $x \in D(A^5)$, *and for some* $m_1, m_2, m_3 \in \mathbb{R}^+$, $m_3 > m_2 > m_1 > 1$.

Theorem 7. *Let* $t \to T(t)$ *be a strongly continuous contraction* ($\|T(t)\| \leq 1, t \geq 0$) *cosine function on a complex Banach space* X *with infinitesimal generator* A, *such that* $A^5 x \neq 0$. *Then the following inequalities hold for every* $x \in D(A^5)$:

$$\|A^i x\|^5 \leq R_i \|x\|^{5-i} \|A^5 x\|^i \ , \tag{30}$$

$i = 1, 2, 3, 4$, *and*

$$R_1 = \frac{5^9}{9!}, \quad R_2 = (4!)^5 \frac{5^8}{(9!)^2}, \quad R_3 = \frac{(35(6!))^5}{4(10!)^3}, \quad R_4 = \frac{(5(8!))^5}{(10!/2)^4}.$$

Proof of Theorem 5. In fact, setting $t(> 0)$ instead of t^2 in (25″), I get

$$T(\sqrt{t})x = x + \frac{t}{2!}Ax + \frac{t^2}{4!}A^2 x + \frac{t^3}{6!}A^3 x + \frac{t^4}{8!}A^4 x + \frac{1}{9!}\int_0^{\sqrt{t}} (\sqrt{t} - u)^9 T(u) A^5 x du \ . \tag{25‴}$$

Formula (25‴) yields

$$\left.\begin{aligned}
&\frac{9!}{2!}tAx + \frac{9!}{4!}t^2A^2x + \frac{9!}{6!}t^3A^3x + \frac{9!}{8!}t^4A^4x \\
&\quad = 9!T(\sqrt{t})x - 9!x - \int_0^{\sqrt{t}} (\sqrt{t}-u)^9 T(u)A^5 x du \\
&\frac{9!}{2!}sAx + \frac{9!}{4!}s^2A^2x + \frac{9!}{6!}s^3A^3x + \frac{9!}{8!}s^4A^4x \\
&\quad = 9!T(\sqrt{s})x - 9!x - \int_0^{\sqrt{s}} (\sqrt{s}-u)^9 T(u)A^5 x du \\
&\frac{9!}{2!}rAx + \frac{9!}{4!}r^2A^2x + \frac{9!}{6!}r^3A^3x + \frac{9!}{8!}r^4A^4x \\
&\quad = 9!T(\sqrt{r})x - 9!x - \int_0^{\sqrt{r}} (\sqrt{r}-u)^9 T(u)A^5 x du \\
&\frac{9!}{2!}pAx + \frac{9!}{4!}p^2A^2x + \frac{9!}{6!}p^3A^3x + \frac{9!}{8!}p^4A^4x \\
&\quad = 9!T(\sqrt{p})x - 9!x - \int_0^{\sqrt{p}} (\sqrt{p}-u)^9 T(u)A^5 x du
\end{aligned}\right\} . \tag{10'}$$

The coefficient determinant D^+ of system (10′) is

$$D^+ = \frac{(9!)^4}{2!4!6!8!} tsrp(t-s)(t-r)(t-p)(s-r)(s-p)(r-p) \,. \tag{11'}$$

It is clear that $D^+ > 0$ because $0 < t < s < r < p$. Therefore there is a unique solution of system (10′) of the form

$$
\begin{aligned}
Ax = (2!)[&(srp)^2(s-r)(r-p)(p-s)T(\sqrt{t})x \\
&- (trp)^2(t-r)(r-p)(p-t)T(\sqrt{s})x \\
&+ (tsp)^2(t-s)(s-p)(p-t)T(\sqrt{r})x \\
&- (tsr)^2(t-s)(s-r)(r-t)T(\sqrt{p})x]/ \\
&tsrp(t-s)(t-r)(t-p)(s-r)(s-p)(r-p) \\
&- (2!)\frac{tsr+tsp+trp+srp}{tsrp}x - \int_0^{\sqrt{p}} K_1^+(t,s,r,p;u)T(u)A^5x\,du,
\end{aligned}
\tag{31}
$$

$$
\begin{aligned}
A^2x = -(4!)[&srp(s-r)(r-p)(p-s)(sr+rp+ps)T(\sqrt{t})x \\
&- trp(t-r)(r-p)(p-t)(tr+rp+pt)T(\sqrt{s})x \\
&+ tsp(t-s)(s-p)(p-t)(ts+sp+pt)T(\sqrt{r})x \\
&- tsr(t-s)(s-r)(r-t)(ts+sr+rt)T(\sqrt{p})x]/ \\
&tsrp(t-s)(t-r)(t-p)(s-r)(s-p)(r-p) \\
&+ (4!)\frac{ts+tr+tp+sr+sp+rp}{tsrp}x + \int_0^{\sqrt{p}} K_2^+(t,s,r,p;u)T(u)A^5x\,du,
\end{aligned}
\tag{31$'$}
$$

$$
\begin{aligned}
A^3x = (6!)[&srp(s-r)(r-p)(p-s)(s+r+p)T(\sqrt{t})x \\
&- trp(t-r)(r-p)(p-t)(t+r+p)T(\sqrt{s})x \\
&+ tsp(t-s)(s-p)(p-t)(t+s+p)T(\sqrt{r})x \\
&- tsr(t-s)(s-r)(r-t)(t+s+r)T(\sqrt{p})x]/ \\
&tsrp(t-s)(t-r)(t-p)(s-r)(s-p)(r-p) \\
&- (6!)\frac{t+s+r+p}{tsrp}x - \int_0^{\sqrt{p}} K_3^+(t,s,r,p;u)T(u)A^5x\,du,
\end{aligned}
\tag{31$''$}
$$

$$
\begin{aligned}
A^4x = -(8!)[&srp(s-r)(r-p)(p-s)T(\sqrt{t})x \\
&- trp(t-r)(r-p)(p-t)T(\sqrt{s})x \\
&+ tsp(t-s)(s-p)(p-t)T(\sqrt{r})x \\
&- tsr(t-s)(s-r)(r-t)T(\sqrt{p})x]/ \\
&tsrp(t-s)(t-r)(t-p)(s-r)(s-p)(r-p) \\
&+ (8!)\frac{1}{tsrp}x + \int_0^{\sqrt{p}} K_4^+(t,s,r,p;u)T(u)A^5x\,du,
\end{aligned}
\tag{31$'''$}
$$

where

$$K_1^+ = \begin{cases} \frac{(9!)^3}{4!6!8!}[-(srp)^2(s-r)(r-p)(p-s)(\sqrt{t}-u)^9 \\ \quad +(trp)^2(t-r)(r-p)(p-t)(\sqrt{s}-u)^9 \\ \quad -(tsp)^2(t-s)(s-p)(p-t)(\sqrt{r}-u)^9 \\ \quad +(tsr)^2(t-s)(s-r)(r-t)(\sqrt{p}-u)^9]/ \\ \quad (-D^+), & 0 \le u \le \sqrt{t} \\ \frac{(9!)^3}{4!6!8!}[(trp)^2(t-r)(r-p)(p-t)(\sqrt{s}-u)^9 \\ \quad -(tsp)^2(t-s)(s-p)(p-t)(\sqrt{r}-u)^9 \\ \quad +(tsr)^2(t-s)(s-r)(r-t)(\sqrt{p}-u)^9]/ \\ \quad (-D^+), & \sqrt{t} \le u \le \sqrt{s} \\ \frac{(9!)^3}{4!6!8!}[-(tsp)^2(t-s)(s-p)(p-t)(\sqrt{r}-u)^9 \\ \quad +(tsr)^2(t-s)(s-r)(r-t)(\sqrt{p}-u)^9]/ \\ \quad (-D^+), & \sqrt{s} \le u \le \sqrt{r} \\ \frac{(9!)^3}{4!6!8!}[(tsr)^2(t-s)(s-r)(r-t)(\sqrt{p}-u)^9]/ \\ \quad (-D^+), & \sqrt{r} \le u \le \sqrt{p}, \end{cases}$$

$$K_2^+ = \begin{cases} \frac{(9!)^3}{2!6!8!}[-srp(s-r)(r-p)(p-s)(sr+rp+ps)(\sqrt{t}-u)^9 \\ \quad +trp(t-r)(r-p)(p-t)(tr+rp+pt)(\sqrt{s}-u)^9 \\ \quad -tsp(t-s)(s-p)(p-t)(ts+sp+pt)(\sqrt{r}-u)^9 \\ \quad +tsr(t-s)(s-r)(r-t)(ts+sr+rt)(\sqrt{p}-u)^9]/ \\ \quad (-D^+), & 0 \le u \le \sqrt{t} \\ \frac{(9!)^3}{2!6!8!}[trp(t-r)(r-p)(p-t)(tr+rp+pt)(\sqrt{s}-u)^9 \\ \quad -tsp(t-s)(s-p)(p-t)(ts+sp+pt)(\sqrt{r}-u)^9 \\ \quad +tsr(t-s)(s-r)(r-t)(ts+sr+rt)(\sqrt{p}-u)^9]/ \\ \quad (-D^+), & \sqrt{t} \le u \le \sqrt{s} \\ \frac{(9!)^3}{2!6!8!}[-tsp(t-s)(s-p)(p-t)(ts+sp+pt)(\sqrt{r}-u)^9 \\ \quad +tsr(t-s)(s-r)(r-t)(ts+sr+rt)(\sqrt{p}-u)^9]/ \\ \quad (-D^+), & \sqrt{s} \le u \le \sqrt{r} \\ \frac{(9!)^3}{2!6!8!}[tsr(t-s)(s-r)(r-t)(ts+sr+rt)(\sqrt{p}-u)^9]/ \\ \quad (-D^+), & \sqrt{r} \le u \le \sqrt{p}, \end{cases}$$

$$K_3^+ = \begin{cases} \frac{(9!)^3}{2!4!8!}[-srp(s-r)(r-p)(p-s)(s+r+p)(\sqrt{t}-u)^9 \\ \quad +trp(t-r)(r-p)(p-t)(t+r+p)(\sqrt{s}-u)^9 \\ \quad -tsp(t-s)(s-p)(p-t)(t+s+p)(\sqrt{r}-u)^9 \\ \quad +tsr(t-s)(s-r)(r-t)(t+s+r)(\sqrt{p}-u)^9]/ \\ \quad (-D^+), & 0 \le u \le \sqrt{t} \\ \frac{(9!)^3}{2!4!8!}[trp(t-r)(r-p)(p-t)(t+r+p)(\sqrt{s}-u)^9 \\ \quad -tsp(t-s)(s-p)(p-t)(t+s+p)(\sqrt{r}-u)^9 \\ \quad +tsr(t-s)(s-r)(r-t)(t+s+r)(\sqrt{p}-u)^9]/ \\ \quad (-D^+), & \sqrt{t} \le u \le \sqrt{s} \\ \frac{(9!)^3}{2!4!8!}[-tsp(t-s)(s-p)(p-t)(t+s+p)(\sqrt{r}-u)^9 \\ \quad +tsr(t-s)(s-r)(r-t)(t+s+r)(\sqrt{p}-u)^9]/ \\ \quad (-D^+), & \sqrt{s} \le u \le \sqrt{r} \\ \frac{(9!)^3}{2!4!8!}[tsr(t-s)(s-r)(r-t)(t+s+r)(\sqrt{p}-u)^9]/ \\ \quad (-D^+), & \sqrt{r} \le u \le \sqrt{p}, \end{cases}$$

$$K_4^+ = \begin{cases} \frac{(9!)^3}{2!4!6!}[-srp(s-r)(r-p)(p-s)(\sqrt{t}-u)^9 \\ \quad +trp(t-r)(r-p)(p-t)(\sqrt{s}-u)^9 \\ \quad -tsp(t-s)(s-p)(p-t)(\sqrt{r}-u)^9 \\ \quad +tsr(t-s)(s-r)(r-t)(\sqrt{p}-u)^9]/ \\ \quad (-D^+), & 0 \le u \le \sqrt{t} \\ \frac{(9!)^3}{2!4!6!}[trp(t-r)(r-p)(p-t)(\sqrt{s}-u)^9 \\ \quad -tsp(t-s)(s-p)(p-t)(\sqrt{r}-u)^9 \\ \quad +tsr(t-s)(s-r)(r-t)(\sqrt{p}-u)^9]/ \\ \quad (-D^+), & \sqrt{t} \le u \le \sqrt{s} \\ \frac{(9!)^3}{2!4!6!}[-tsp(t-s)(s-p)(p-t)(\sqrt{r}-u)^9 \\ \quad +tsr(t-s)(s-r)(r-t)(\sqrt{p}-u)^9]/ \\ \quad (-D^+), & \sqrt{s} \le u \le \sqrt{r} \\ \frac{(9!)^3}{2!4!6!}[tsr(t-s)(s-r)(r-t)(\sqrt{p}-u)^9]/ \\ \quad (-D^+), & \sqrt{r} \le u \le \sqrt{p}. \end{cases}$$

Note that $K_i^+ \geq 0$, $i = 1, 2, 3, 4$, for every $u \in [0, \sqrt{p}]$ $(0 < t < s < r < p)$. Besides

$$\left.\begin{aligned} \int_0^{\sqrt{p}} K_1^+ du &= -\frac{2!}{10!} tsrp \\ \int_0^{\sqrt{p}} K_2^+ du &= -\frac{4!}{10!}(tsr + tsp + trp + srp) \end{aligned}\right\}, \tag{32}$$

$$\left.\begin{aligned} \int_0^{\sqrt{p}} K_3^+ du &= -\frac{6!}{10!}(ts + tr + tp + sr + sp + rp) \\ \int_0^{\sqrt{p}} K_4^+ du &= -\frac{8!}{10!}(t + s + r + p) \end{aligned}\right\}. \tag{32'}$$

From (31)–(31‴), triangle inequalities, (32)–(32′), and similarly as "on semigroups", I get (28)–(28‴). This completes the proof of Theorem 5.

Proof of Theorem 6. From (15) and (28)–(28‴) and similar calculations as in the previous section "on semigroups", I find inequalities (29). This completes the proof of Theorem 6.

Proof of Theorem 7. Setting $M = 1$, using (29), and minimizing $g_i = g_i(m_1, m_2, m_3)$, $i = 1, 2, 3, 4$, as in section "on semigroups", I obtain inequalities (30). This completes the proof of Theorem 7.

References

1. Z. Ditzian, *Some remarks on inequalities of Landau and Kolmogorov*, Aequat. Math. 12 (1975), 145–151.
2. E. Hille and R. S. Phillips, *Functional Analysis and Semigroups*, Amer. Math. Soc. Colloq. Publ., Vol. 31, AMS, Providence, RI, 1957.
3. R. R. Kallman and G. C. Rota *On the inequality* $\|f'\|^2 \leq 4\|f\|\|f''\|$. In *Inequalities II*, ed. O. Shisha, Acad. Press, New York, London, 1970, 187–192.
4. H. Kraljević and S. Kurepa, *Semigroups on Banach Spaces*, Glas. Mat. **5** (1970), 109–117.
5. H. Kraljević and J. Pečarić, *Some Landau's type inequalities for infinitesimal generators*, Aequat. Math, **40** (1990), 147–153.
6. E. Landau, *Einige Ungleichungen für zweimal differentzierbare Funktionen*, Proc. London Math. Soc., (2) **13** (1913), 43–49.
7. J. M. Rassias, *Landau's type inequalities*, World Scientific, Singapore, 1994, *Functional Analysis, Approximation Theory, Numerical Analysis*, (to appear).